Block Copolymer Nanocomposites

Block Copolymer Nanocomposites

edited by

Galder Kortaberria
Agnieszka Tercjak

Pan Stanford Publishing

Published by

Pan Stanford Publishing Pte. Ltd.
Penthouse Level, Suntec Tower 3
8 Temasek Boulevard
Singapore 038988

Email: editorial@panstanford.com
Web: www.panstanford.com

British Library Cataloguing-in-Publication Data
A catalogue record for this book is available from the British Library.

Block Copolymer Nanocomposites

ISBN 978-981-4669-54-2 (Hardcover)
ISBN 978-981-4669-55-9 (eBook)

Printed in the USA

Contents

Preface

The self-assembly of block copolymers (BCPs) into different nanoscale structures makes them interesting polymeric macromolecules from both academic and industrial points of view. This class of macromolecules consists of two or more covalently linked polymers which are thermodynamically incompatible, giving rise to a variety of microstructures. As it is well known, BCPs can self-assemble to form nanoscale structures with domain spacing that depends strongly on the molecular weight, segment size, and interaction between the blocks among others. Consequently, the microphase separation of BCPs is determined by the degree of polymerization, N, the volume fraction of each blocks, f, and the Flory–Huggins interaction parameter, χ, which depends on the temperature. The typical size of the microphase-separated BCP domains is in the range of 10–200 nm.

As confirmed by many research groups, BCPs can be microphase-separated to form stable structures such as lamellar, hexagonal-packed cylinder, body-centred cubic, close-packed spherical, or bicontinuous cubic gyroid structures. The ability to control both the length scale and the spatial organization of BCP morphologies makes these polymeric materials attractive candidates for use as templates for the fabrication of novel multifunctional materials with the application of many fields of nanotechnology and advanced materials. On the other hand, BCPs can also act as nanostructured agents for different homopolymers and thermosetting systems.

One of the most important classes of polymeric materials based on BCPs with a wide range of application are nanocomposites based on both block copolymers and nanostructured thermosetting materials. These materials consist of a polymeric matrix used as a template for dispersion and localization of different nanoobjects such as nanoparticles, carbon nanotubes, microfibrillated cellulose, and azobenzene.

As visualized in the graph from the ISI Web of Knowledge, the research activity in the field of nanocomposites based on block copolymers has increased progressively from 1995 and received high interest as confirmed by the number of citations.

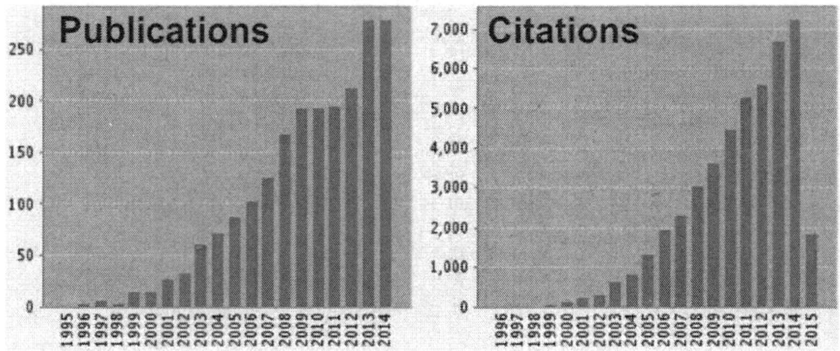

Number of publications and corresponding citation per year related to nanocomposites based on BCPs (ISI Web of Knowledge).

The Materials+Technologies Group of the University of the Basque Country, Spain, founded in 1988 by Prof. Iñaki Mondragon, has significantly contributed to the development and characterization of novel multifunctional materials based on block copolymers and nanostructured thermosetting systems. Prof. Mondragon encouraged his students to go further in their research, increasing the group's knowledge about block copolymers. As a tribute to his memory, some of his students, senior researchers in different countries now, have presented in this book most of the work done on nanocomposites based on block copolymers.

The first four chapters present the authors' work on nanocomposites based on block copolymers and various nanofillers such as nanoparticles and carbon nanotubes. Chapter 1 discusses the preparation of nanocomposites based on block copolymers and nanoparticles with the aid of surfactants. Chapter 2 focuses on nanocomposites based on block copolymers and nanoparticles with polymeric brushes, and Chapter 3 discusses block copolymer/carbon nanotube nanocomposites. Chapter 4 shows the block copolymer–assisted sol gel templating, a way to prepare block copolymer–based nanocomposites in a single step.

The next five chapters are dedicated to nanocomposites obtained by nanostructuring thermosetting matrices with block copolymers. Chapter 5 discusses nanostructured epoxy–based thermosetting materials modified with amphiphilic block copolymers. Chapters 6 and 7 present chemically functionalized block copolymers as reactive modifiers for nanostructuring epoxy matrices and nanostructuring of polyester matrices with block copolymers, respectively. Chapter 8 is devoted to ternary systems based on epoxy matrices nanostructured with block copolymers and inorganic fillers. The book ends with a chapter on epoxy systems nanostructured with block copolymers containing azobenzene groups for optical applications.

We hope that readers will enjoy this book and it will prove to be useful addition to the knowledge in the field of block copolymers and nanostructured thermosetting systems based nanocomposites.

Chapter 1

Surfactant-Treated Nanoparticles Confinement into Block Copolymer Domains

Laura Peponi

Institute of Polymer Science and Technology,
ICTP-CSIC, Juan de la Cierva, 3, 28006 Madrid, Spain

lpeponi@ictp.csic.es

This chapter presents a review on the confinement of surfactant-treated nanoparticles into one block of the block copolymer matrix. While most of surfactant-treated nano-elements confined into the block copolymer domains are isotropic in shape (i.e., 3D nanoparticles), the selective incorporation of 2D and 1D surfactant-treated nanoparticles into the nanostructured block copolymer domains is not so easy due to steric problems between the geometry of the nanoparticles and the nanostructure domains of the block copolymer host.

1.1 Introduction

Block copolymers can be defined as the more important self-assembling materials due to their ability to allow controlling both

Block Copolymer Nanocomposites
Edited by Galder Kortaberria and Agnieszka Tercjak

ISBN 978-981-4669-54-2 (Hardcover), 978-981-4669-55-9 (eBook)
www.panstanford.com

"scalar" behaviors over self-assembling and advanced "vectorial" aspects regarding the application of the resulting nanostructured morphologies [1]. In particular, the attention has been focused on the nanostructuration behavior of block copolymer melts, thin films and block copolymer solutions trying to understand their phase separation theories.

In the recent years, due to the progress in the understanding self-assembly processes, and to the possibility to tailor block copolymer morphologies as well as their functionality, they are used also as matrices to develop hybrid nanocomposites with different nanoentities [2].

As it is well known, one of the most usual problems of nanoentities is their aggregation, which may affect negatively to the final properties of the hybrid materials [3]. To overcome the problem of the tendency of nanoobjects to aggregate due to their high surface area and surface energy and to facilitate their confinement in a selected block of a block copolymer different routes have been used [4, 5].

This book chapter focuses on the nanocomposites based on block copolymer matrix and reinforced with nanoparticles. In particular, attention will be paid to the confinement of nanoparticles into the block copolymer domains by means of surfactants. In this regard, a short summary on nanoparticles geometry and on the use of surfactants is given. Then, a review on the confinement of nanoparticles is presented, as well as their treatment with surfactants, into the block copolymers domains.

1.2 Nanoparticle Geometry

It is widely known that the nanoparticles (NP) can considerably improve the nanocomposite properties thanks to the strong and large polymer-nanofiller interactions as well as good particles dispersion into the polymer matrix. Polymer nanocomposites, including BC nanocomposites, refer to multiphase materials where at least one of the constituent phases has one dimension in the nanoscale range (below 100 nm). As a consequence of their nanoscale dimensions, nanoparticles possess a large surface area for a given volume [6]. However, when working with block copolymer, it should be taken into account that the NP should be confined into just one block of the BC leading the geometrical relationship between BC nanophases and nanoparticles a key factor in the preparation of nanostructured nanocomposites. Before studying the nanoparticles' confinement into the block copolymer domains, a brief summary on the different nanoparticle geometries is presented, in order to understand which kind of nanoparticles can be confined in the different block copolymer domains.

It is possible to classify the nanoparticles into three main categories depending on their aspect ratio and geometry: (a) isodimensional nanoparticles (3D) (Fig. 1.1a), (b) acicular or fibrous ones (2D) (Fig. 1.1b), and (c) layered particles (1D) (Fig. 1.1c) [7].

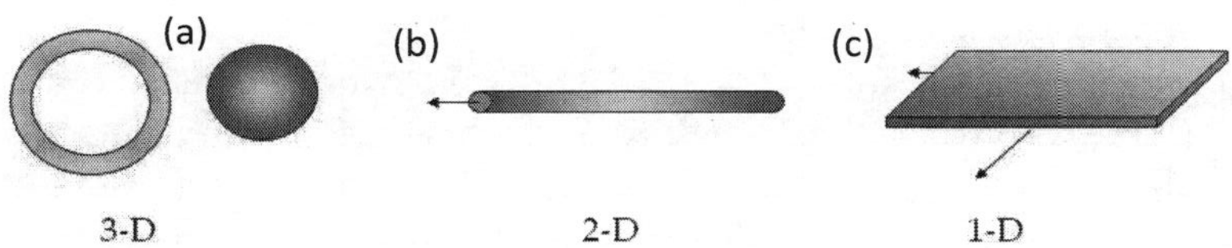

Figure 1.1 Nanoparticle geometries: (a) isodimensional nanoparticles (3D), (b) acicular or fibrous ones (2D) and (c) layered particles (1D).

Isodimensional nanoparticles (Fig. 1.2a) are structures without any "infinite" dimensions (with three nano dimensions), such as vesicles, spherical micelles, fullerenes, nanoparticles (metal and ceramic particles, metal oxides) [6]. 3D geometry NP are very versatile because they can be incorporated into all the domains of the BC presenting the most tightly confining geometry [8].

For instance, metal nanoparticles have been extensively incorporated in polymer matrices to produce functional nanocomposites for numerous applications, such as the use of iron nanoparticles to provide electromagnetic properties [6, 9], silver nanoparticles for antibacterial activity [10] or gold nanoparticles for electrical conductivity [11]. Metals undergo the most considerable property change by size reduction. For this reason, nano-sized metals have special characteristics that can be tuned by changing the nanoparticle dimension. Although metallic NP were traditionally produced by the top-down approach, the bottom-up strategy proves to be better for producing this kind of size-dependent particles via chemical processes [12]. Metal NP require a narrow range of particle sizes distribution and dispersion. Moreover, since nano-sized metals are very instable, they trend to aggregate, and thus, the use of surfactant is considerable interesting.

In the case of 2D nanoparticles (Fig. 1.1b), one dimension is "infinite" compared with the others two, which are nanodimensions. Some examples of 2D nanoparticles are the cylindrical micelles, DNA, whiskers, some proteins and carbon nanotubes [12]. In this sense, carbon nanotubes (CNT) are the most investigated 2D nanofillers due to its exceptional mechanical resistance as well as electrical and thermal conductivity, which can be potentially used in many applications ranging from nanodevices to macroscopic materials [6].

In 1D layered particles (Fig. 1.1c), two dimensions are "infinite" compared with the third one, which is finite and constant typically with a thickness on the order of 1 nm [6]. An example is corresponded to the lamellar solid. The most popular 1D particles are graphene sheets and layered silicates (smectic clays and layered double hydroxides) [6, 7].

Among all layered silicates, montmorillonites (MMT) are the most investigated for their potentially high aspect ratio and the unique intercalation/exfoliation characteristics [12]. In this sense, clay particles should be exfoliated to increase the surface area. Nanocomposites with MMT are able to achieve remarkable thermal, barrier, thermomechanical and fire-resistance properties at low nanofiller content [6]. To improve the clay/matrix affinity the use of hydrophilic surfactants, such as carbohydrate surfactants, has been proposed to modify the nanofiller polarity and reach exfoliation [7].

Graphene exhibits exceptional properties due to its unique structural and electrical characteristics [6]. However, as a consequence of its hydrophobic nature, producing single layer graphene by solution-based processes is a great challenge [12]. Instead of pristine graphene results easier to obtain graphene oxide which is a water dispersible intermediary obtained by chemical modification of graphite [12]. Graphene oxide reaches better dispersion within polymeric materials, but it exhibits poor electrical conductivity [6].

1.3 Surfactants

Nanoparticles are thermodynamically unstable system and tend to aggregate to reduce free energy. In order to avoid the aggregation of NP both functionalization technique and surfactant can be used. In this chapter, we focus on the use of surfactant. In particular, it is important to note that several types of surfactants are used to stabilize NP dispersions obtaining homogenous dispersions [13].

As we see before, in order to achieve a good dispersion of nanoparticles into a polymeric matrix it is necessary to compatibilize the NP with the matrix, by using two different techniques: functionalizing by chemical modification either the nanoparticle surface or the host polymeric chains or by using a surfactant. However, due to their block architecture, block copolymers in many cases are used as surfactants, as reported recently by Bodratti et al., in which they studied the influence of low molecular weight poly(ethylene oxide) (PEO) homopolymer on the adsorption of a representative poly(ethylene oxide)-poly(propylene oxide)-poly(ethylene oxide) PEO-PPO-PEO block copolymer (at the surface of protonated silica nanoparticles) [14].

Wang et al. studied the self-assembly of binary mixtures of architecturally and compositionally diverse surfactant-like polystyrene-poly(2-vinylpyridine) (PSPVP) block copolymers in solution and in thin films [15]. Recently, Wang et al used the triblock copolymer poly(ethylene glycol)-block-poly(propylene glycol)-block-poly(ethylene glycol) as surfactant in the synthesis of ZnO-Au hybrid nanoparticles [16]. Meanwhile, Kos et al. used poly(methyl methacrylate)-b-poly(methacrylic acid) PMMA-b-PMAA as reactive surfactant for the functionalization of ZnO nanoparticles [17].

Surfactants, amphiphiles or popularly known as soaps are compounds that consist of a hydrophobic group, usually a long hydrocarbon chain, and a hydrophilic ionic or polar group [18]. Examples of anionic and cationic surfactants are sodium dodecylsulfate (SDS) (Fig. 1.2a) and cetyl trimethyl ammonium bromide (CTAB) (Fig. 1.2b), respectively. As well there are neutral surfactants such as the well-known dodecanethiol (DT) (Fig. 1.2c) [19].

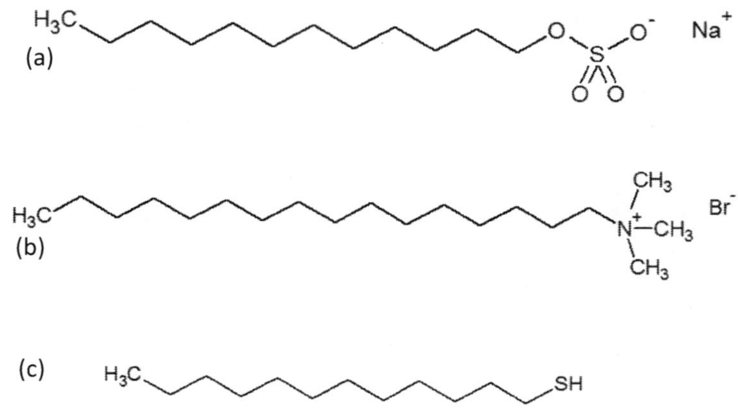

Figure 1.2 Anionic, cationic and neutral surfactants used to confine NP into BC domains (a) sodium dodecylsulfate (SDS), (b) cetyl trimethyl ammonium bromide (CTAB) and (c) dodecanethiol (DT).

When a BC matrix is used to obtain nanocomposites with confined nanoparticles, an adequate surfactant can be used to selectively disperse nanoparticles in just one block of the BC matrix instead of using directly the block copolymer as dispersant [19, 20].

However, in this book chapter, the focus will be on the use of block copolymers as polymeric matrices and on the surfactants to confine the nanoparticles into one of the block copolymer nanostructured domains. It is very important to underline that when working with block copolymer, the surfactant used has to act as a selective bridge for the nanoparticles confinement. This means that the surfactant not only act as a bridge between the hydrophilic and the hydrophobic parts but also that it is chosen in order to selectively confine the nanoparticles into just one block of the block copolymer matrix. The surfactant is selective with only one block of the block copolymer matrix. In this regard, an important role is playing by the affinity between the solubility parameters of the different block of the block copolymer matrix and the surfactant used [21].

In order to confine the nanoparticles into the self-assembled phase separated domains of the block copolymers, it is important to take into account the geometry of both the nanoparticles and the self-organized domains of the block copolymers.

Table 1.1 summarizes the geometry relationship between BC nanophases and the nanoparticles.

- NP with 3D geometry can be incorporated into all the domains of the BC;
- 2D nanoparticles theoretically can be introduced into cylindrical and lamellar domains;
- 1D NP only can be introduced into lamellar domains [12].

Table 1.1 Relationship between BC morphologies and nanoparticle geometry [12]

Relationship between BC morphologies and nanoparticle geometry		**BC morphologies**		

Nanoparticle geometry	3D 	✓	✓	✓
	2D 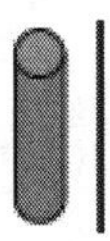	✓	•	X
	1D 	•	X	X

1.4 Nanostructuring in Nanocomposites Based on Block Copolymer Matrices and Surfactant

Block-copolymer self-assembling provides a unique tool for realizing large-area ordered metamaterials, with desired optical, electrical, magnetic, etc. properties. The benefits of using BC as templates for metamaterials come from two main aspects: first, BC show a rich range of available nano-morphologies, whose domains can be conveniently tuned in size, shape and periodicity, by changing molecular parameters; second, the chemical properties of the block polymers can lead to the selective inclusion of functionalized nanoparticles of different materials in specific nanodomains, generating periodic arrays of NP according to the geometry of the BC acting as template [22, 23].

Before reviewing in details the nanocomposites obtained by using block copolymer as host for the confinement of the NP with different geometries, we can summarize the advantages on working with self-assembled nanostructured block copolymer nanocomposites. In particular, block copolymers lead to

- precise control over length scale of the nanostructured domain dimensions;
- control over morphology thus considering their phase diagrams;
- prediction on the final nanocomposites equilibrium structures, taking into account both the morphologies of the self-assembled nanodomains and the nanoparticles morphologies;
- control over domain functionality and properties of the resulting nanocomposites, thanks to the functionality of the confined nanoparticles;
- maintenance of the traditional advantages of polymeric materials in terms of cost, effectiveness and flexibility.

In order to study the morphological changes in the self-assembled domains of the block copolymers induced to the nanoparticles confinements, TEM, AFM, scanning electron microscopy (SEM) analysis are required, not neglecting others techniques as SAXS analysis. Moreover, in order to study the functional properties of the final nanocomposites, such as conductive and magnetic properties among others, electrostatic force microscopy (EFM) is used.

In conclusion, due to the progress in the understanding self-assembly processes, and to the possibility to tailor BC morphologies as well as their functionality, they can be used as template for nanocomposite materials [12].

1.4.1 Surfactant-Treated 3D Nanoparticles

3D nanoparticles, intuitively, when driven by a proper surfactant, can be confined in all the self-assembled nanostructures of a block copolymer. Many are the examples reported in the scientific literature on the confinement of metallic spherical nanoparticles into the block copolymer domains by using surfactant. It is also important to underline that in order to achieve nanocomposite with good final behavior, in term of mechanical, thermal, rheological, etc. properties, an adequate amount of surfactant has to be used. In this regard, Adhikari et al. studied the effects of surfactant-treated-boehmite NPs on the thermal and mechanical properties of block copolymer nanocomposites [24]. Whilst, Peponi et al. reported a study on the minimum concentration of surfactant necessary to confine silver nanoparticles into the polystyrene PS domains of a poly(styrene-b-isoprene-b-styrene) SIS block copolymer matrix without affecting the mechanical as well as the rheological properties of the block copolymer nanocomposites [19].

Moreover Peponi et al. studied the influence of the addition of different amount of silver nanoparticles on the self-assembled nanostructuration of commercial SIS block copolymer matrices by using dodecanethiol (DT) as surfactant. In particular, DT-Ag nanoparticles, from 0.5 up 10 wt% in content, have been confined into the polystyrene domains of the SIS block copolymer [12, 19, 25–29]. Figure 1.3 reports the different morphologies obtained maintaining constant the ratio between surfactant and nanoparticles and varying the amount of nanoparticles. In this regard, in order to understand the change in the self-assembled nanostructured morphology of the SIS block copolymer matrix, the phase diagram of block copolymer has to be taken into account. Owing to the fact that the DT is selective for the PS block, the DT-Ag nanoparticles are confined in the PS block, creating a gradient of the PS phase. Therefore, the DT-Ag NP is able to increase the volume fraction of PS phase respect to the volume fraction of this block in the neat block copolymer. In particular, in this case, enlarging the amount of the DT-Ag NP, the PS gradient is moving horizontally along the phase diagram yielding to design different nanostructured morphologies as follows:

- The neat SIS block copolymer presents PS cylindrical domains parallel to the free surface with a diameter of about 15–20 nm.
- When 0.5 and 1 wt% of DT-Ag nanoparticles have been added the self-assembled morphology changes from ordered cylinders to a lamellar structure with an interlamellar periodicity of about 24–28 nm confirming that the gradient of PS phase increased at the expense of the PI phase, moving

throughout the lamellar zone of the characteristic phase diagram.

- When 2 and 3 wt% of DT-Ag nanoparticles have been added to the block copolymer matrix, well-ordered hexagonal-packed cylindrical domains perpendicular to the free surface were obtained. This means that the "phase inversion" phenomenon occurred [30]. The formed ordered hexagonal-packed cylindrical domains are of the PI-block with diameter of about 30 nm.
- When 5 wt% of DT-Ag NPs was added to the SIS block copolymer, PI cylindrical domains parallel to the free surface have been obtained.
- Finally, when 10 wt% of DT-Ag nanoparticles have been added to the SIS matrix, a good dispersion of the nanoparticles is achieved but the ordered self-assembled the nanostructuration of the polymeric matrix is not obtained, yielding to the disordered state.

Figure 1.3 shows what happens when 10 wt% of DT-Ag nanoparticles have been added to the SIS matrix. It is possible to see a large area (10 × 10 um) image compared with the others (2 × 2 um) in order to analyze the absence of nanoparticle agglomeration.

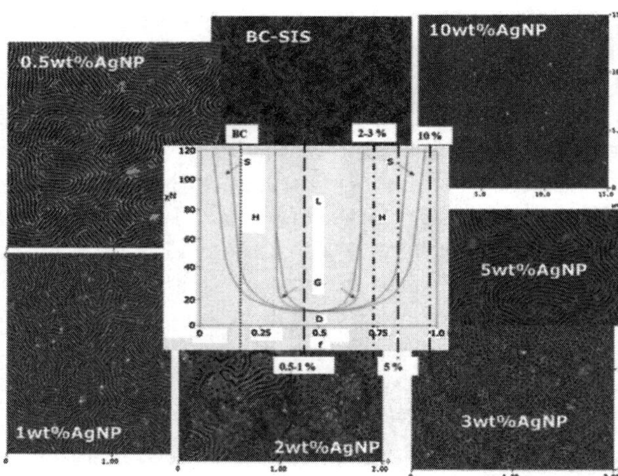

Figure 1.3 TM-AFM images of different nanostructured morphologies obtained adding different amount (0.5, 1, 2, 3, 5, 10 wt%) of DT-Ag NPs to the SIS block copolymer matrix. Reproduced with permission of [12] Copyright 2014, Elsevier.

So, with this example, it is easy to understand that it is possible to change the self-assembled nanostructuration on the block copolymer matrix by adding the proper amount of confined nanoparticles [26, 27, 31].

However, numerous examples have been reported in the literature about the confinement of nanoparticles into the block copolymers domains by using thermoplastic elastomers, such as Peponi et al., who studied the confinement of different amount of DT-Ag nanoparticles into two different poly(styrene-b-butadiene-b-styrene) SBS block copolymers. Meanwhile, Zhao et al. studied the order-disorder transition in nanocomposites based on palladium nanoparticles dispersed in poly(styrene-b-isoprene) microdomain [32].

Zalakain et al. studied also the confinement of DT-NP into the SIS domains [33], and Suntivich et al. studied the growth of gold NP in poly(styrene/(2-vinyl pyridine)) star-shaped block copolymer monolayers [34]. Adhikari et al. studied the effects of the incorporation of surfactant-treated ceramic nanoparticles into block copolymer matrices [35]. Ocando et al. studied the effect of the addition of surfactant-treated alumina nanoparticles on the domain orientation of SBS triblock copolymers [36].

However, block copolymers show capability to self-assemble into ordered nanostructured not only as thin films but also in solutions. This means that block copolymers in solution with a selective solvent, for only one block of the block copolymer, lead to the formation of block copolymer micelles. These micelles influence the structure as well as the flow characteristic of the BC solution leading to a material with different behavior than the corresponding homopolymers [37].

To summarize, the BC micelle formation occurs because, in selective solvents, some block copolymers form micelle-like aggregates consisting of an insoluble polymeric core surrounded by a solvent swollen corona. The thermodynamically favored morphology and aggregate dimensions are determined by a force balance relates to the free energy contributions of the core, the corona, and the interface as deeply studied by Eisenberg et al. [38–40].

From the discussion above, it is clear that there are numerous applications of the self-assembling nanostructured BC, but the common factor is to understand their phase separation behavior.

For this reason, many scientists focused on the study of the morphological changes produced for the addition of surfactant-embedded nanoparticles in block copolymer solution. Peponi et al. studied the morphological change in the micelles of nanocomposite gels based on poly(styrene-b-butadiene-b-styrene) SBS BC when 1 wt% Ag NP treated with dodecanethiol as surfactant were added to the physical gel of SBS [25]. Haupt et al. created nanohole-patterned gold films by using the self-assembly capability of metal-loaded poly(styrene-b-(2-vinylpyridine)) micelles [41]. Banerjee et al. studied the vesicles formation driven by the self-assembly of cationic block copolymer and anionic surfactant reinforced with gold nanoparticles [42]. Also Ahn et al. found to be a strong correlation between the self-assembled template structure and the formed gold nanoparticles [43].

As it was previously commented, the possibility to confine nanoparticles with different functionality into both, domains and micelles of block copolymer matrix, confers to the new nanocomposites unique properties, depending on the functionality of the nanoparticles themselves. For instance, conducting or magnetic nanoparticles lead to conductive or magnetic nanostructured nanocomposites, respectively.

Hu et al. reported a study on the fabrication of organic/inorganic hybrid micelles of amphiphilic block copolymer and superparamagnetic iron oxide [44].

Recently, Di Mauro et al. reported a study on the conductive properties of nanostructured nanocomposites obtained throughout the selective confinement of surfactant-coated Au NP in the polystyrene block of amphiphilic polystyrene-block-polyethylene oxide (PS-b-PEO) adding up to 17 wt% of NPs [34].

Peponi et al. studied the conductive response of nanocomposite based on poly(styrene-b-butadiene-b-styrene) SBS BC and DT-Ag nanoparticles by EFM [28]. In particular, they analyzed the conductive abilities of SBS-Ag nanocomposites to act as template for conductive materials. In this regard, the surface potential of the samples were measured by EFM and the localization of nanoparticles confined into the polystyrene cylinders was detected by applying a minimum voltage of 5 V (Fig. 1.4).

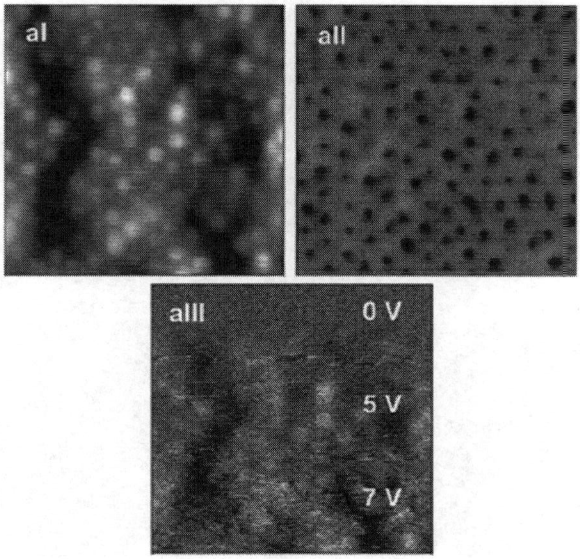

Figure 1.4 500 × 500 nm topographical, height and phase I and II, respectively, and surface potential images, III carried out on DT-Ag/SBS nanocomposite containing 7 wt% Ag nanoparticles. Reproduced with permission of John Wiley and Sons [44].

Gutierrez et al. also studied by EFM the conductive properties of nanocomposites based on TiO_2 nanoparticles and poly(styrene-b-(ethylene oxide)) BC matrices [45]. Askari et al. studied the optical properties of nanocomposites based on gold spherical nanoparticles dispersed in a cylindrical di-block copolymer [46]. Garcia et al. reported a deep study on the nanostructuration of BC nanocomposites reinforced with modified magnetic nanoparticles [47].

The study reported by Davidi et al. [48] is also interesting. In fact, they studied the effects on the weak surfactant–polymer interactions on the morphological behavior of hierarchical structures of polystyrene-block-poly(2-vinylpyridine)/palladium-pincer surfactants. In particular, they reported that the result of this behavior is initially the persistence of the lamellar phase with increasing amounts of the surfactant and then the dominance of a perforated lamellar phase, as indicated in Fig. 1.5.

So, from the pioneer studies [49–51] on the confinement of surfactant-treated 3D nanoparticles into block copolymer matrix up to now, enormous progress has been made in order to use a simple strategy to tailor the surface of the nanoparticles to be localize into the block copolymer domains, maintaining this topic very important and current in the scientific panorama.

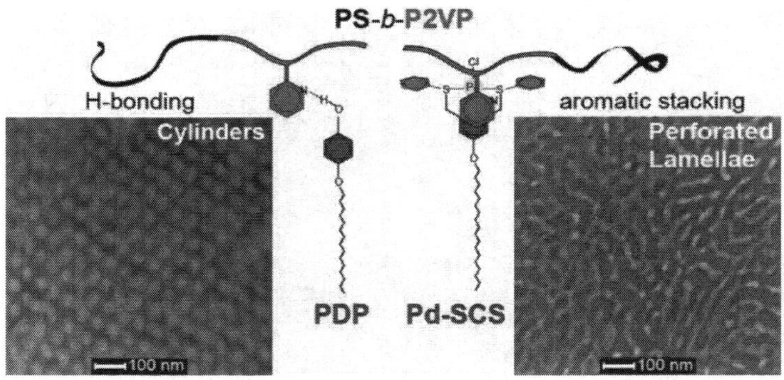

Figure 1.5 TEM images of the morphological change on the self-assembled nanostructuration of the block copolymer domains. Reproduced with permission of [48]. Copyright 2014, American Chemical Society.

1.4.2 Surfactant Treated 2D Nanoparticles

In most cases reported on the scientific literature, when 2D nanoparticles are used with block copolymers, these ones act as surfactant, and not as polymeric matrices, as reported recently by Ansos-Casaos et al. [52]. They, in fact, reported a study on the electrical conductivity and the mechanical properties of composite materials based on block copolymer-wrapped single walled carbon nanotube (SWCNT) and poly(methyl methacrylate) matrix.

In this regard, also Esposito et al. used PEO-PPO-PEO block copolymer to disperse carbon nanotubes in unsaturated polyester obtaining microphase separation [53]. Granite et al. evidenced that the amphiphilic copolymers of the Pluronic family are excellent dispersants for single-walled carbon nanotubes (SWCNT) in water [54].

Another method used to incorporate the carbon nanotubes into the block copolymer matrix is by functionalizing them in order to increase the compatibility between the nanoparticle and the polymeric matrix [55–58]. However, just a few scientific works have been reported in the international panorama on the use of surfactant material in order to drive the sequestering of the CNT into one domain of the self-assembled nanostructured block copolymer matrices, main goal of this book chapter. With respect to this approach, Park et al. reported for the first time the selective sequestering of multi-walled carbon nanotubes (MWCNT) in self-assembled block copolymer, even they used non-ionic surfactant in order to functionalize the carbon nanotubes and producing PS-functionalized MWCNT to be sequestered into the PS domains of a lamellar polystyrene-b-isoprene (PS-b-PI) [59].

Peponi et al. [60, 61] studied the selective confinement of functionalized carbon nanotubes in poly(styrene-b-isoprene-b-styrene) SIS block copolymer matrix by using dodecanethiol as surfactant underlining that the surfactant-treated functionalized CNT are able to well adapted to the lamellar morphology of the SIS block copolymer even when the lamellae is not straight, as reported in Fig. 1.6.

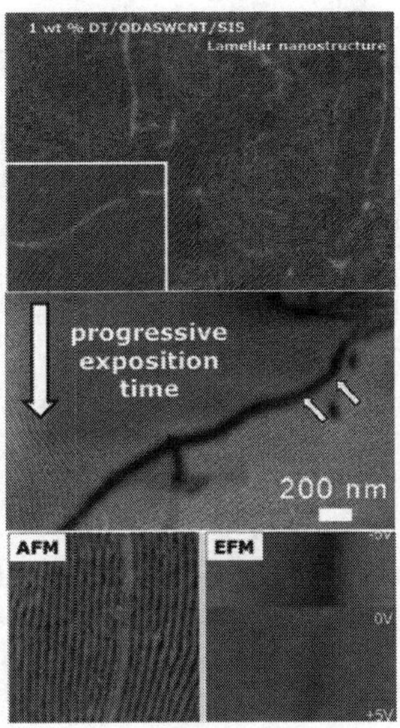

Figure 1.6 FE-SEM images at different magnifications (upper images) and AFM and EFM images (bottom images) of the confinement of functionalized CNT into the PS block of the SIS BC. Reproduced with permission of [76] and [77]. Copyright 2009–2010, Elsevier.

Moreover, in order to localize the surfactant-treated functionalized carbon nanotubes into the PS domains of the SIS block copolymer matrix, they used both field-emission SEM (FE-SEM) and EFM techniques.

In the first case, the CNT is localized by a continuing exposure of the nanocomposite to the electron beam effect. Inthe second one, the contrast between the conducting CNT and the insulating polymer is evidenced by EFM.

Also, Garate et al. selectively confined functionalized carbon nanotubes into the SIS block copolymer domains by using dodecanethiol as surfactant [62].

Nie et al. recently demonstrated the efficient incorporation of PS-tethered cadmium sulfide nanorods into polystyrene-block-poly(ethylene oxide) (PS-b-PEO) ultralong cylindrical micelles by solvent evaporation-driven self-assembly by using hexadecyl trimethyl ammonium bromide (CTAB) as surfactant in chloroform-in-water emulsions [63].

1.4.3 Surfactant-Treated 1D Nanoparticles

Regarding the confinement of 1D nanoparticles (layered silicates and graphene) into the block copolymer domains, very few articles have been published in the scientific literature, due to steric problems between the nanoparticles geometry and the nanostructure domains of the block copolymer host [12].

Regarding the dispersion of layered silicate into the block copolymer domains we can evidence the works of Choi et al. in which they studied the generation of micelle-templated organosilicate nanostructures resulting from the self-assembly of a block copolymer/organosilicate mixture [64]. Moreover, Adhikari et al. studied the mechanical response of nanocomposites based on a SBS tri-block copolymer and optimum dispersed organo-modified layered silicates [65]. Ganguly et al. studied the morphological changes of thermoplastic elastomeric tri-block copolymers when clays have added founding well-ordered phase separated morphology [66]. However, they did not use surfactant. Only Vazquez et al. studied the self-assembled lamellar SBS tri-block copolymer containing layered silicate treated with surfactant [67].

Regarding the use of graphene with block copolymers, most of the times the block copolymers have been used to functionalize the graphene sheets as reported by Cao et al. [68] and by Chen et al. [69] among others. Additionally, when working with graphene, block copolymers can be used to obtain a direct exfoliation of graphite flakes as reported by Liu et al. [70] and Skaltsas et al. [71]. In other cases, block copolymers nanotemplates have been obtained by using graphene, as reported by Kim et al. [72]

However, only in the article of Peponi et al., a nanostructuration of a self-assembled SIS block copolymer has been obtained by using surfactant-modified graphene sheets. They used also EFM analysis in order to localize the graphene sheets into the nanostructured block copolymer domains, as reported in Fig. 1.7.

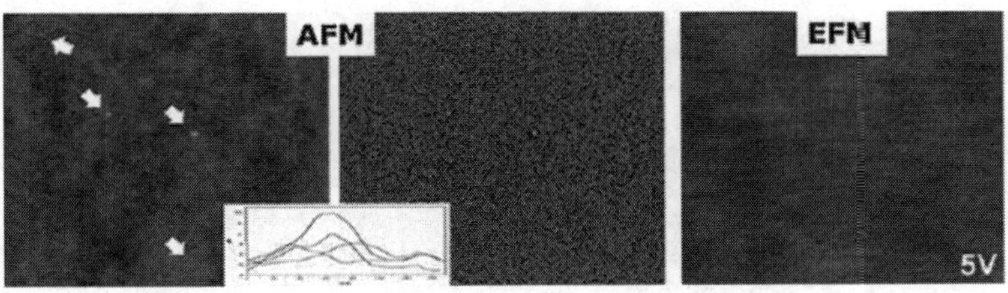

Figure 1.7 Nanocomposite based on SIS and graphene nanosheets. AFM and EFM analysis. Reproduced with permission of [73]. Copyright 2009, American Chemical Society.

References

1. Dobrynin, A. V., and Erukhimovich, I. Y. (1993). Computer-aided comparative investigation of architecture influence on block copolymer phase diagrams, *Macromolecules*, **26**, pp. 276–281.

2. Fahmi, A., Pietsch, T., Mendoza, C., and Cheval, N. (2009). Functional hybrid materials, *Mater. Today*, **12**, pp. 44–50.

3. Lin, Y., Boker, A., He, J., Sill, K., Xiang, H., Abetz, C., Li, X., Wang, J., Balazs, A., and Russel, T. P. (2005). Self-directed self-assembly of nanoparticle/copolymer mixtures, *Nature*, **434**, pp. 55–59

4. Tercjak, A., Peponi, L., Garcia, I., Gutierrez, J., and Mondragon, I. (2008). Methodologies for confining nanoparticles in block copolymers, in *White Book of Network of Excellence NANOFUN-POLY* (Kenny, J. M. ed.), Chapter 3, Morphema Editrice, Rome (Italy), ISBN: 978-88-96051-00-9).

5. Darling, S. B. (2007). Directing the self-assembly of block copolymers. *Prog. Polym. Sci.*, **32**, pp 1152–1204.
6. Raquez, J. M., Habibi, Y., Murariu, M., and Dubois, P. (2013). Polylactide (PLA)-based nanocomposites, *Prog. Polym. Sci.*, **38**, pp. 1504–1542.
7. Chivrac, F., Pollet, E., and Avérous, L. (2009). Progress in nano-biocomposites based on polysaccharides and nanoclays, *Mater. Sci. Eng. R Rep.*, **67**, pp. 1–17.
8. Jin, Z., and Fan, H. (2014). Self-assembly of nanostructured block copolymer nanoparticles, *Soft Matter.*, **10**, pp. 9212–9219.
9. Wilson, J. L., Poddar, P., Frey, N. A., Srikanth, H., Mohomed, K., Harmon, J. P., Kotha, S., and Wachsmuth, J. (2004). Synthesis and magnetic properties of polymer nanocomposites with embedded iron nanoparticles, *J. Appl. Phys.*, **95**, pp. 1439–1443.
10. Dallas, P., Sharma, V. K., and Zboril, R. (2011). Silver polymeric nanocomposites as advanced antimicrobial agents: Classification, synthetic paths, applications, and perspectives, *Adv. Colloid Interface*, **166**, pp. 119–135.
11. Sih, B. C., and Wolf, M. O. (2005). Metal nanoparticle-conjugated polymer nanocomposites, *Chem. Commun.*, **27**, pp. 3375–3384.
12. Peponi, L., Puglia, D., Torre, L., Valentini, L., and Kenny, J. M. (2014). Processing of nanostructured polymers and advanced polymeric based nanocomposites, *Mater. Sci. Eng. R Rep.*, **85**, pp. 1–46.
13. Singh, R. P., Jain, S., and Ramarao, P. (2013). Surfactant-assisted dispersion of carbon nanotubes: Mechanism of stabilization and biocompatibility of the surfactant, *J. Nanopart. Res.*, **15**, p. 1985.
14. Bodratti, A. M., Sarkar, B., Song, D., Tsianou, M., and Alexandridis, P. (2015). Competitive adsorption between PEO-containing block copolymers and homopolymers at silica, *J. Disper Sci. Technol.*, **36**, pp. 1–9.
15. Wang, X., Davis, J. L., Hinestrosa, J. P., Mays, J. W., and Kilbey, S. M. (2014). Control of self-assembled structure through architecturally and compositionally complex block copolymer surfactant mixtures, *Macromolecules*, **47**, pp. 7138–7150.
16. Wang, X. H., Zhang, X. Y., Cheng, W. Z., Shao, H. Q., Liu, X., Li, X. M., Liu, H. L., and Wu, J. H. (2014). Facile synthesis and optical properties of polymer-laced ZnO-Au hybrid nanoparticles, *Nanoscale Res. Lett.*, **9**, p. 109.

17. Kos, T., Anzlovar, A., Zagar, E., Orel, Z. C., and Zigon, M. (2014). PMMA-b-PMAA diblock copolymer as a reactive polymeric surfactant for the functionalization of ZnO nanoparticles, *Acta. Chim. Slov.*, **61**, pp. 497–505.
18. Reynhout, I. C., Cornelissen, J. J. L. M., and Nolte, R. J. M. (2009). Synthesis of polymer-biohybrids: From small to giant surfactants, *Acc. Chem. Res.*, **42**, pp. 681–692.
19. Peponi, L., Tercjak, A., Torre, L., Kenny J. M., and Mondragon, I. (2009). Surfactant effects on morphology-properties relationships of silver-poly(styrene-b-isoprene-b-styrene) block copolymer nanocomposites, *J. Nanosci. Nanotechnol.*, **9**, pp. 2128–2139.
20. Peponi, L., Valentini, L., Torre, L., Mondragon, I., and Kenny, J. M. (2009). Surfactant assisted selective confinement of carbon nanotubes functionalized with octadecylamine in a poly(styrene-b-isoprene-b-styrene) block copolymer matrix, *Carbon*, **47**, pp. 2474–2480.
21. Van K. (1990). *Properties of Polymers* (Elsevier), p. 875.
22. Bockstaller, M. R. Mickiewicz, R. A., and Thomas, E. L. (2005). Block copolymer nanocomposites: Perspectives for tailored functional materials, *Adv. Mater.*, **17**, pp. 1331–1349.
23. Capretti, A., Auriemma, F., De Rosa, C., Di Girolamo, R., Forestiere, C., Miano, G., and Pepe, G. P. (2013). Block-copolymer-based plasmonic metamaterials, *Proc. SPIE* 8771, Metamaterials VIII, 87710V.
24. Adhikari, R., Brostow, W., Datashvili, T. Henning, S., Menard, B., Menard K. P., and Michler, G. H. (2012). Effect of surfactant treated boehmite nanoparticles on properties of block copolymers, *Mater. Res. Innov.*, **16**, pp. 19–24.
25. Peponi, L., Tercjak, A., Torre, L., Mondragon, I., Kenny, J. M. (2009). Nanostructured physical gel of SBS block copolymer and Ag/DT/SBS nanocomposites, *J. Mater. Sci.*, **44**, pp. 1287–1293.
26. Peponi, L., Tercjak, A., Torre, L., Kenny, J. M., and Mondragon, I. (2008). Influence of the amount of silver nanoparticles in the properties of nanocomposites based on poly(styrene-b-isoprene-b-styrene) and poly(styrene-b-butadiene-b-styrene) block copolymer matrix, *J. Nanostruct. Polym. Nanocompos.*, **4**, pp. 76–83.
27. Peponi, L., Tercjak, A., Torre, L., Kenny, J. M., and Mondragon, I. (2008). Morphological analysis of self-assembled SIS block copolymer matrices containing silver nanoparticles, *Compos. Sci. Technol.*, **68**, pp. 1631–1636.

28. Peponi, L., Tercjak, A., Gutierrez, J., Stadler, H., Torre, L., Kenny, J. M., and Mondragon, I. (2008). Self-assembling of SBS block copolymers as templates for conductive silver nanocomposites, *Macromol. Mater. Eng.*, **293**, pp. 568–573.
29. Torre, L., and Mondragon, I. (2008). Morphology and rheological behaviour of Ag-SBS nanocomposite gels, AIP Conference Proceedings, (1042), pp. 272–275.
30. Adedeji, A., Jamieson, A. M., and Hudson, S. D. (1995). Phase inversion in compatibilized immiscible polymer blends with exothermic interfacial mixing, *Macromolecules,* **28**, pp. 5255.
31. Peponi, L., Tercjak, A., Martin, L., Torre, L., Mondragon I., and Kenny J. M. (2011). Morphology-properties relationship on nanocomposite films based on poly(styrene-block-diene-block-styrene) copolymers and silver nanoparticles, *Express Polym. Lett.*, **5**, pp. 104–118.
32. Zhao, Y., Saijo, K., Takenaka, M., Koizumi, S., and Hashimoto T. (2009). Order-disorder transition of nanocomposites: Pd nanoparticles in polystyrene-block-polyisoprene microdomain templates, *Macromolecules*, **42**, pp. 5272–5277.
33. Zalakain, I., Politakos, N., Ramos, J. A., Fernandez, R., Etxeberria, H., Mondragon, I., Angeles Corcuera, M., and Eceiza, A. (2013). Synthesis and incorporation of dodecanethiol capped silver nanoparticles into poly(styrene-b-isoprene-b-styrene) block copolymer and their influence in the morphology, *Colloid Surf. A-Physicochem. Eng. Asp.*, **436**, pp. 170–177.
34. Di Mauro, A. E., Striccoli, M., Depalo, N., Fanizza, E., Cano, L., Ingrosso, C., Agostiano, A., Curri, M. L., and Tercjak, A. (2014). Selective confinement of oleylamine capped Au nanoparticles in self-assembled PS-b-PEO diblock copolymer templates, *Soft Matter,* **10**, pp. 1676–1684.
35. Adhikari, R., Brostow, W., Datashvili, T., Henning, S., Menard, B., Menard, K. P., and Michler, G. H. (2012). Effect of surfactant treated boehmite nanoparticles on properties of block copolymers, *Mater. Res. Innov.*, **16**, pp. 19–24.
36. Ocando, C., Tercjak, A., and Mondragon, I. (2011). Surfactant addition effects on dispersion and microdomain orientation in SBS triblock copolymer/alumina nanoparticle composites, *Eur. Polym. J.*, **47**, pp. 1240–1249.
37. Hamley, I. W. (2004). *Developments in Block Copolymer Science and Technology* (John Wiley & Sons).

38. Zhang, L. F., and Eisenberg, A. (1998). Formation of crew-cut aggregates of various morphologies from amphiphilic block copolymers in solution, *Polym. Adv. Technol.*, **9**, pp. 677–699.
39. Zhang, L. F., and Eisenberg, A. (1999). Thermodynamic vs kinetic aspects in the formation and morphological transitions of crew-cut aggregates produced by self-assembly of polystyrene-b-poly(acrylic acid) block copolymers in dilute solution, *Macromolecules*, **32**, pp. 2239–2249.
40. Rizis, G., Van de Ven, T. G. M., and Eisenberg, A. (2014). Crystallinity-driven morphological ripening processes for poly(ethylene oxide)-block-polycaprolactone micelles in water, *Soft Matter*, **10**, pp. 2825–2835.
41. Haupt, M., Miller, S., Glass, R., Arnold, M., Sauer, R., Thonke, K., Möller M., and Spatz, J. P. (2003). Nanoporous gold films created using templates formed from self-assembled structures of inorganic-block copolymer micelles, *Adv. Mater.*, **15**, pp. 829–831.
42. Banerjee, R., Dutta, S., Pal, S., and Dhara, D. (2013). Spontaneous formation of vesicles by self-assembly of cationic block copolymer in the presence of anionic surfactants and their application in formation of polymer embedded gold nanoparticles, *J. Phys. Chem. B*, **117**, pp. 3624–3633.
43. Ahn, S., Jung, S. Y., and Lee, S. J. (2011). Self-assembly change by gold nanoparticle growth, *J. Phys. Chem. C*, **115**, pp. 22301–22308.
44. Hu, J., Qian, Y., Wang, X., Liu, T., and Liu, S. (2012). Drug-loaded and superparamagnetic iron oxide nanoparticle surface-embedded amphiphilic block copolymer micelles for integrated chemotherapeutic drug delivery and MR imaging, *Langmuir*, **28**, pp. 2073–2082.
45. Gutierrez, J., Tercjak, A., Peponi, L., and Mondragon, I. (2009). Conductive properties of inorganic and organic TiO_2/polystyrene-block-poly(ethylene oxide) nanocomposites, *J. Phys. Chem. C*, **113**, pp. 8601–8605.
46. Askari, A. A., Rahimi, L., Bahrampour, A. R., and Pepe, G. P. (2013). Optical properties of a metalo-dielectric nanocomposite: Small gold spherical nanoparticles in a cylindrical di-block copolymer, *Eur. Phys. J. B*, **86**, pp. 150–158.
47. Garcia, I., Tercjak, A., Gutierrez, J., Rueda, L., and Mondragon, I. (2008). Nanostructuration via solvent vapor exposure of poly(2-vinyl pyridine-b-methyl methacrylate) nanocomposites using modified magnetic nanoparticles, *J. Phys. Chem. C*, **112**, pp. 14343–14347.

48. Davidi, I., Patra, D., Herrnida-Merino, D., Portale, G., Rotello, V. M., Raviv, U., and Shenhar, R. (2014). Hierarchical structures of polystyrene-block-poly(2-vinylpyridine)/palladium-pincer surfactants: Effect of weak surfactant-polymer interactions on the morphological behavior, *Macromolecules*, **47**, pp. 5774–5783.

49. Kim, B. J., Bang, J., Hawker, C. J., Chiu, J. J., Pine, D. J., Jang, S. G., Yang, S.-M., and Kramer, E. J. (2007). Creating surfactant nanoparticles for block copolymer composites through surface chemistry, *Langmuir*, **23**, pp. 12693–12703.

50. Hu, S., Brittain, W. J., Jacobson, S., and Balazs, A. C. (2006). Selective ordering of surfactant modified gold nanoparticles in a diblock copolymer, *Eur. Polym. J.*, **42**, pp. 2045–2052.

51. Weng, C. C., and Wei, K. H. (2003). Selective distribution of surface-modified TiO_2 nanoparticles in polystyrene-b-poly (methyl methacrylate) diblock copolymer, *Chem. Mater.*, **15**, 2936–2941.

52. Anson-Casaos, A., Pascual, J. F., Ruano, C., Fernandez-Huerta, N., Fernandez-Pato, I., Otero, J. C., Puertolas, A. J., and Martinez, T. M. (2015). Electrical conductivity and tensile properties of block-copolymer-wrapped single-walled carbon nanotube/poly(methyl methacrylate) composites, *J. Appl. Polym. Sci.*, **132**, pp. 41547.

53. Esposito, L. H., Ramos, J. A., Builes, D. H., Kortaberria, G., and Mondragon, I. (2013). Microphase separation in unsaturated polyester/PEO-PPO-PEO block copolymer mixtures containing carbon nanotubes, *Adv. Polym. Tech.*, **32**, pp. E572–E581.

54. Granite, M., Radulescu, A., Pyckhout-Hintzen, W., and Cohen, Y. (2011). Interactions between block copolymers and single-walled carbon nanotubes in aqueous solutions: A small-angle Neutron Scattering Study, *Langmuir*, **27**, pp. 751–759.

55. Albuerne, J., Boschetti-de-Fierro, A., Abetz, C., Fierro, D., and Abetz V. (2011). Block copolymer nanocomposites based on multiwall carbon nanotubes: Effect of the functionalization of multiwall carbon nanotubes on the morphology of the block copolymer, *Adv. Eng. Mater.*, **13**, pp. 803–810.

56. Zou, J. H., Liu, L. W., Chen, H., Khondaker, S. I., Mc Cullough, R. D., Huo, Q., and Zhai, L. (2008). Dispersion of pristine carbon nanotubes using conjugated block copolymers, *Adv. Mater.*, **20**, pp. 2055–2060.

57. Zou, J. H., Khondaker, S. I., Huo, Q., and Zhai, L. (2009). A general strategy to disperse and functionalize carbon nanotubes using conjugated block copolymers, *Adv. Funct. Mater.*, **19**, pp. 479–483.

58. Li, C. X., Hsu, J. C., Sugiyama, K., Hirao, A., Chen W. C., and Mezzenga, R. (2009). Synthesis and self-assembly behavior of poly(fluorenylstyrene)-block-poly(2-vinylpyridine) block copolymers and their blends with single wall carbon nanotubes (SWCNTs), *Macromolecules*, **42**, pp. 5793–5801.

59. Park, I., Lee, W., Kim, J., Park, M., and Lee, H. (2007). Selective sequestering of multi-walled carbon nanotubes in self-assembled block copolymer, *Sens Actuat. B-Chem.*, **126**, pp. 301–305.

60. Peponi, L., Valentini, L., Torre, L., Mondragon, I., and Kenny, J. M. (2009). Surfactant assisted selective confinement of carbon nanotubes functionalized with octadecylamine in a poly(styrene-b-isoprene-b-styrene) block copolymer matrix, *Carbon*, **47**, pp. 2474–2480.

61. Peponi, L., Tercjak, A., Gutierrez, J., Cardinali, M., Mondragon, I., Valentini, L., and Kenny, J. M. (2010). Mapping of carbon nanotubes in the polystyrene domains of a polystyrene-b-polyisoprene-b-polystyrene block copolymer matrix using electrostatic force microscopy, *Carbon*, **48**, pp. 2590–2595.

62. Garate, H., Fascio, M. L., Mondragon, I., D'Accorso, N. B., and Goyanes, S. (2012). Surfactant-aided dispersion of polystyrene-functionalized carbon nanotubes in a nanostructured poly(styrene-b-isoprene-b-styrene) block copolymer, *Polymer*, **52**, pp. 2214–2220.

63. Nie, X., Cui, J., and Jiang W. (2014). Ultralong cylindrical micelles precisely located with semiconductor nanorods by solvent evaporation-driven self-assembly, *Soft Matter.*, **10**, pp. 8051–8059.

64. Choi, J., Hermans, T. M., Lohmeijer, B. G. G., Pratt, R. C., Dubois, G., Frommer, J., Waymouth, R. M., and Hedrick, J. L. (2006). Monolayered organosilicate toroids and related structures: A phase diagram for templating from block copolymers, *Nano Lett.*, **6**, pp. 1761–1764.

65. Adhikari, R., Damm, C., Michler, G. H., Munstedt, H., and Balta-Calleja, F. J. (2008). Processing and mechanical performance of SBS block copolymer/layered silicate nanocomposites, *Compos. Interface*, **15**, pp. 453–463.

66. Ganguly, A., De Sarkar, M., and Bhowmick A. K. (2007). Morphological mapping and analysis of poly[styrene-b-(ethylene-co-butilene)-b-styrene] triblock copolymer and its clay nanocomposites by atomic force microscopy, *J. Polym. Sci. B-Polym. Phys.*, **45**, pp. 52–66.

67. Vazquez, A., Lopez, M., Serrano, E., Valea, A., Zafeiropoulos, N. E., Mondragon, I., and Zafeiropoulos, N. (2008). Structural and

Ordering Behavior of Lamellar Polystyrene-block-polybutadiene-block-polystyrene triblock copolymer containing layered silicates, *J. Appl. Polym. Sci.,* **110**, pp. 3624–3637.

68. Cao, Y. W., Lai, Z. L., Feng, J. C., and Wu, P. Y. (2011). Graphene oxide sheets covalently functionalized with block copolymers via click chemistry as reinforcing fillers, *J. Mater. Chem.*, **21**, pp. 9271–9278.
69. Chen, N., Liu, Y. T., Xie, X. M., Ye, X. Y., Feng, X., Chen, Y. F., and Wang, Y. H. (2012). High-concentration aliphatic and aromatic dispersions of single- and few-layer graphene noncovalently modified by block copolymer crystallization, *Carbon*, **50**, pp. 4760–4764.
70. Liu, Z., Liu, J. Q., Cui, L., Wang, R., Luo, X., Barrow, C. J., and Yang, W. R. (2013). Preparation of graphene/polymer composites by direct exfoliation of graphite in functionalised block copolymer matrix, *Carbon*, **51**, pp. 148–155.
71. Skaltsas, T., Karousis, N., Yan, H. J., Wang, C. R., Pispas, S., and Tagmatarchis, N. (2012). Graphene exfoliation in organic solvents and switching solubility in aqueous media with the aid of amphiphilic block copolymers, *J. Mater. Chem.,* **22**, pp. 21507–21512.
72. Kim, B. H., Kim, J. Y., Jeong, S. J., Hwang, J. O., Lee, D. H., Shin, D. O., Choi, S. Y., and Kim, S. O. (2010). Surface energy modification by spin-cast, large-area graphene film for block copolymer lithography, *ACS Nano*, **4**, pp. 5464–5470.
73. Peponi, L., Tercjak, A., Verdejo, R., Lopez-Manchado M. A., Mondragon, I., and Kenny, J. M. (2009). Confinement of functionalized graphene sheets by triblock copolymers, *J. Phys. Chem. C*, **113**, pp. 17973–17978.

Chapter 2

Nanocomposites Based on Block Copolymers and Metallic Nanoparticles Grafted with Polymer Brushes

Galder Kortaberria

Materials + Technologies Group (GMT),
Department of Chemical and Environmental Engineering,
Polytechnic School, University of the Basque Country (UPV/EHU),
Plaza Europa 1, 20018 Donostia-San Sebastián, Spain

galder.cortaberria@eh.eus

One of the most effective ways to properly disperse and selectively place metallic nanoparticles into microdomains of self-assembled block copolymer, preventing their agglomeration, is to chemically modify their surface with polymeric chains compatible with one of the blocks of the copolymer. This so-called grafting process can be done by three different ways: *grafting from*, *grafting to* and *grafting through*. Each process presents its advantages and drawbacks, but all of them could make possible to place nanoparticles on the desired domains. Depending on the molecular weight and grafting density of brushes, self-assembled morphology of the copolymer and properties of nanocomposites

Block Copolymer Nanocomposites
Edited by Galder Kortaberria and Agnieszka Tercjak

ISBN 978-981-4669-54-2 (Hardcover), 978-981-4669-55-9 (eBook)
www.panstanford.com

can be tuned in order to obtain periodic structures to form materials with enhanced mechanical strength as well as to achieve unique optical, electronic and magnetic properties at the nanometer scale, for applications in solar cells, catalysts or high density magnetic storage media.

2.1 Introduction

Engineering the self-assembly of inorganic nanoparticles within block copolymer nanodomains is useful for the design of periodic structures to form materials with enhanced mechanical strength as well as to achieve unique optical, electronic and magnetic properties at the nanometer scale, for applications in solar cells, catalysts or high density magnetic storage media. Block copolymers are a versatile platform material because they can self-assemble into various periodic structures for proper compositions and under adequate conditions, owing to the microphase separation between dissimilar blocks [1–3]. To overcome the problem of the tendency of nanoparticles to aggregate due to their high surface area and surface energy and to facilitate their dispersion in a selected block of a block copolymer different routes have been used [1, 3–5]. One of them has been the use of surfactants. In that way, Peponi et al. [1] used surfactants to disperse conductive silver nanoparticles in the desired domains of poly(styrene-b-butadiene-b-styrene) (SBS) copolymer. Emrick et al. [4] controlled the surface hydrophobicity by using different surfactants in order to disperse CdSe nanoparticles in poly(styrene-b-2-vinylpiridine) (PS-b-P2VP) copolymer, creating hierarchically ordered patterns with CdSe nanoparticles located in PS or P2VP domains depending on the surfactant. Electrophoretic deposition of nanoparticles [5] has been another method to disperse nanoparticles. Zhang et al. [5] used this method for placing CdSe nanoparticles in diblock copolymer templates. The so-called in situ approach has also been used by several authors [6–9] for incorporating inorganic nanoparticles into block copolymer nanostructures: Nanoparticles are directly synthesized within a block copolymer domain from metal precursors. Preformed micelles of block copolymers containing metal precursors are used as nanoreactors to synthesize nanoparticles selectively in block copolymers. Due to its chemical affinity, the salt selectively infiltrates the hydrophilic copolymer domain. Nanoparticles then form selectively, upon reduction within the precursor-loaded domains. In that way Chan et al. [6] prepared nanocomposites with block copolymers and Pb or Pt nanoparticles. Saito et al. [7, 8] synthesized silver nanoparticles in the lamellar and spherical domains of PS-b-P2VP copolymer. Cohen et al. [9] prepared nanocomposites based on poly(styrene-b-acrylic acid) (PS-b-PAA) copolymer and metallic nanoparticles of Pd, Cu, Au and Ag.

Among different methods used for dispersing nanoparticles into block copolymers, one of the most effective ones is the grafting of polymers either through physical adsorption [10, 11] or by covalent bonding of the polymer chain [12–14]. Non-covalent physisorption makes the processing of particles difficult and hence, covalent bonding is preferable in many cases [12]. There are essentially three techniques to chemically graft polymers on nanoparticles surface: *grafting to* [15–17], *grafting from* [18–20] and *grafting through* [21–27]. In the *grafting to* method previously synthesized end-functionalized polymers (such as thiol-terminated ones) are directly grafted onto the nanoparticle surface. This technique is relatively easy to perform but once some polymeric chains are grafted to the nanoparticle surface, it becomes difficult for the rest of chains to reach nanoparticle surface, due mainly to steric effects. Obtained grafting densities are usually low [28, 29]. In the *grafting from* method, polymers are grown from the nanoparticle surface, previously modified with polymerization initiators. Higher grafting densities are usually obtained with this technique [30, 31]. In the *grafting through* technique, molecules attached to the surface also present a group suitable for polymerization (usually a silane with terminal vinyl groups, which are subsequently used for the polymerization). Nanoparticles present in the polymerization medium are covered by the polymer. Obtained grafting densities are usually higher than those obtained by *grafting to* and the technique is easier to carry out than *grafting from* one. As nanoparticle surface is multifunctional, polymeric chains also present bonds among them, creating a sort of network [32]. Figure 2.1 shows the schematic depiction of those grafting techniques.

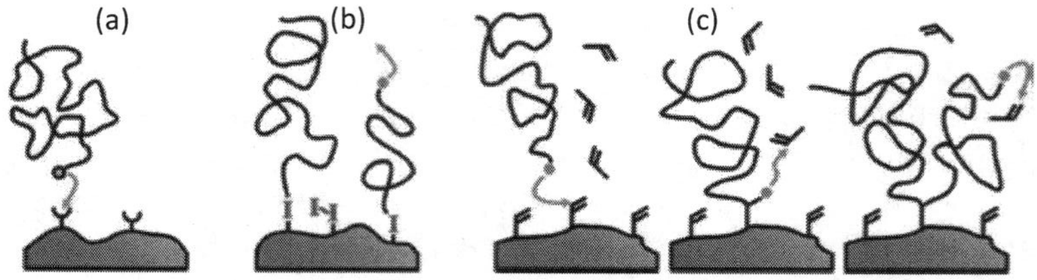

Figure 2.1 Schematic depiction of different techniques of chemical grafting of polymers to nanoparticle surfaces: (a) *grafting to*, (b) *grafting from*, and (c) *grafting through* consisting of the attachment step and further chain growth. Reproduced with permission of *Macromolecules*, 2014, 47(9), 2929. Copyright 2014, ACS.

Different polymerization techniques have been used, such as anionic [33, 34], cationic [35, 36], ring opening polymerization [37, 38] or living radical polymerizations such as atom transfer radical polymerization (ATRP) [39–41], or reversible addition fragmentation chain transfer (RAFT) [42], among others.

In the present chapter, works of several authors on preparation and characterization of nanocomposites based on different block copolymers with mainly metallic or metal oxide nanoparticles (Au, CdSe, Fe_2O_3, etc.) with polymeric chains on their surface grown by different techniques are presented. Surface modification with polymer brushes similar to one of the blocks of the host copolymer matrix has been carried out by several authors improving nanoparticle dispersion and selectively placing them in self-assembled block copolymer domains.

2.2 Nanocomposites Based on Block Copolymers and Polymer-Grafted Nanoparticles

2.2.1 Nanocomposites with Nanoparticles Modified by Grafting from Method

In this modification method, chains grow in situ from initiator molecules that have been pre-grafted onto the surface of nanoparticles. There are different radical techniques that allow the generation of an elevated grafting density of polymer brushes whose molecular weight and polydispersity can be controlled. Different living radical polymerization methods such as ATRP, reversible addition-fragmentation chain transfer and nitroxide mediated polymerization exist. Among them, ATRP has been the most used for generating brushes from nanoparticle surface. Authors that have prepared nanocomposites based on block copolymers and nanoparticles with brushes by *grafting from* have used this method [40, 41, 43–47]. In this technique, an atom transfer step is the key elementary reaction responsible for the uniform growth of polymer chains [19, 48]. Two methods can be applied. In the first one, a sacrificial free initiator is used, whereas in the second one a persistent radical should be added for setting up the equilibrium and thereby leading to the controlled polymerization.

Q. Lan et al. [43] prepared nanocomposites based on poly(styrene-b-butadiene) (SB) block copolymer and silica nanoparticles modified by ATRP with PS chains. They compared obtained dispersion state with that of nanocomposites prepared with the same nanoparticles and PS homopolymer. They used commercial silica nanoparticles that were first modified with (chloromethylphenylethyl)dimethylchlorosilane. They dispersed them in DMAC and added PMDETA and Cu(I)Br and Cu(II)Br as catalyst. Reaction was carried out at 110°C for different times in order to obtain brushes with different molecular weights. Reaction was finished by cooling in liquid nitrogen. They selectively cleaved Si-O-Si bonds that attach the PS to the surface and measured molecular weight of brushes. They obtained grafted chains with molecular weights between 8 and 48 kDa and grafting densities among 0.11 and 0.30 chain/nm^2: Si8 (8 kDa and a density of 0.11), Si16 (16 kDa and 0.11), Si32 (32 kDa and 0.21) and Si48 (48 kDa and 0.30). They synthesized SB copolymer by anionic polymerization with different block contents and molecular weights: SB34 (52 wt% PS and 34 kDa), SB54 (52 wt% of PS and 54 kDa), SB 342 (47 wt% of PS and 342 kDa) and SB465 (52 wt% of PS and 465 kDa). Nanocomposites were prepared by solution mixing using toluene as solvent.

First they analyzed neat block copolymer morphologies by TEM, as shown in Fig. 2.2. They found lamellar morphology for all the cases, whose periodicity increased with molecular weight.

In order to analyze the effect of SB copolymer, they analyzed by TEM morphologies obtained for nanocomposites with 1 wt% of Si8 with all the copolymers. An extensive particle aggregation was found with all but the highest molecular weight, in which a modest aggregation in the form of short strings was found. In order to analyze the effect of graft molecular weight, they analyzed by TEM morphologies obtained for nanocomposites with 1 wt% of all nanoparticles and the copolymer which showed the best dispersion of nanoparticles, SB465. Higher molecular weight resulted in massive aggregation, while for Si32 smaller aggregates were found along well-dispersed particles, and for Si16 and Si8 uniform distributions with a minor amount of string-like associations were observed. Even if they expected nanoparticles to be selectively located at PS domains, they found that nanoparticles were located in both domains with a little tendency to favor PB ones, as it can be seen in Fig. 2.3.

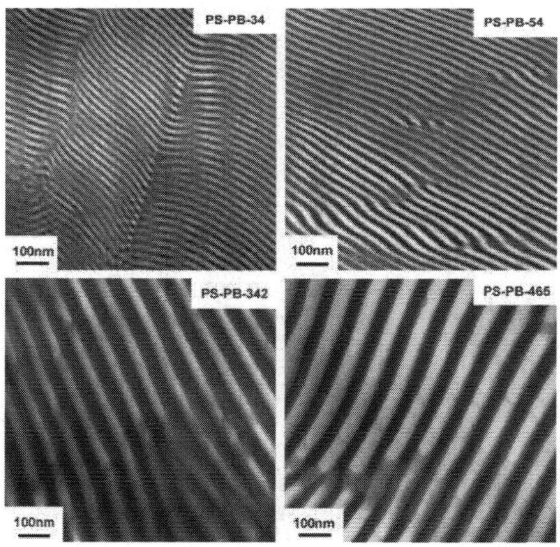

Figure 2.2 TEM images obtained from solvent cast and OsO_4 stained specimens of SB34, SB54, SB342 and SB465. Reproduced with permission of [43]. Copyright 2007, Wiley Periodicals Inc.

They compared the dispersion of nanoparticles in the block copolymer with that in a PS homopolymer, finding that dispersion was governed by different mechanisms. Uniform dispersions were obtained when the molecular weight of the homopolymer was equal or smaller than that of polymer brushes. In the case of copolymers, dispersion was governed by two factors: block copolymer molecular weight and nanoparticle diameter/microdomain size ratio. Higher-molecular-weight copolymers with lower graft molecular weights enhanced dispersion but nanoparticles were not selectively located at PS domains as it was expected.

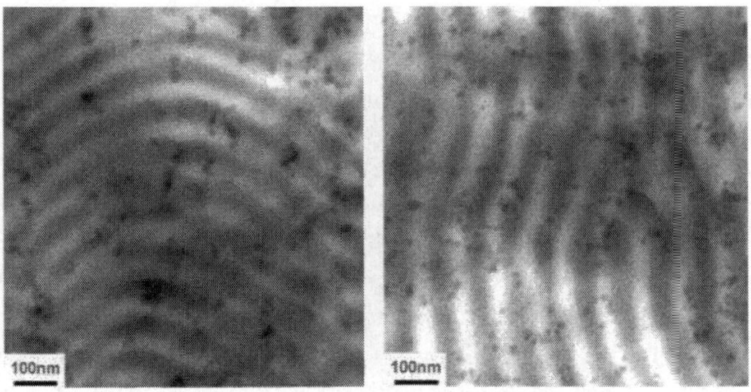

Figure 2.3 TEM images obtained from mixtures of Si8 (left) and Si16 (right) nanoparticles with SB465. Specimens were slightly stained with OsO_4. Reproduced with permission of [43]. Copyright 2007, Wiley Periodicals Inc.

The group of I. Garcia et al. [40, 41, 44, 48] also used ATRP for decorating magnetic Fe_3O_4 nanoparticles with PS and poly(methylmethacrylate) (PMMA) brushes for being selectively placed in PS or PMMA domains of various copolymers. First they prepared nanocomposites with poly(styrene-b-butadiene-b-styrene) SBS block copolymer and 1, 2 and 4 wt% of nanoparticles with PS brushes (PS-MN) grafted from their surface [40, 41]. The grafting process was carried out in two steps: first, 2-(4-chlorosulfonylphenyl) ethyltrichlorosilane (CTCS) as an initiator was anchored onto nanoparticles. Then, ATRP was carried out with CuBr/2,2′-bipyridine as catalyst system and *p*-toluenesulfonyl chloride (TsCl) as sacrificial initiator, used to obtain a satisfactory result for the control of the polymerization because its addition as a free initiator generated a sufficient concentration of persistent radicals (deactivators). They controlled the reaction conversion and obtained molecular weights, measured after cleavage of brushes. They obtained a grafting density of 0.9 chains/nm^2.

Nanocomposites with commercial SBS were prepared by solution mixing followed by solvent casting onto silicon wafers, obtaining films that were annealed for nanostructure generation. They first analyzed the effect of neat (MN) and modified nanoparticles (PS-MN) on the T_g values of both blocks. For SBS/MN samples, T_g of both blocks decreased with MN content, showing a higher mobility of polymeric chains due to the poor interfacial adhesion with fillers. For SBS/PS-MN samples, T_g of PB maintained almost constant. For PS block, two T_g values were found, one related to the PS domains without PS-MN (at around 82°C) and the other one related to PS domains with PS-MN (higher value of around 103°C). This trend was very similar for composites with 1, 2 and 4 wt% of PS-MN. The decreased mobility of PS chains seemed to indicate the location of PS-MN on them.This fact was further demonstrated by atomic force microscopy (AFM) characterization. They also observed an enhancement of mechanical properties for SBS/PS-MN composites when compared with that of SBS/MN, corroborating better interaction and interfacial adhesion. By AFM they observed neither agglomerates nor aggregates of PS-MN, indicative of good distribution and dispersion. In the case of MN, aggregates of around 100 nm more or less uniformly dispersed were found. This fact was corroborated by DSC results. Figures 2.4 and 2.5 show AFM images of nanocomposites.

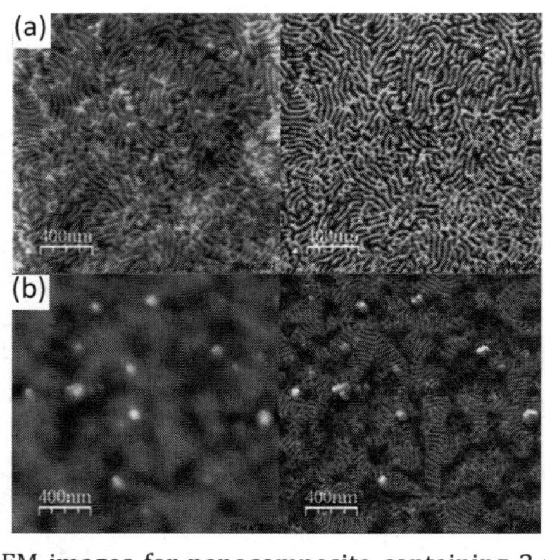

Figure 2.4 AFM images for nanocomposite containing 2 wt% of PS-MN (a) or MN (b). Reproduced with permission of [40]. Copyright 2008, American Chemical Society.

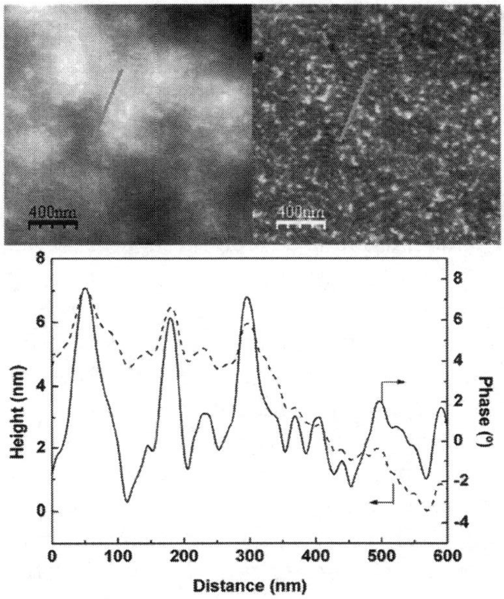

Figure 2.5 AFM images of SBS/4 wt% PS-MN: topography (left), phase (right), and height and phase profiles. Reproduced with permission of [40]. Copyright 2008, American Chemical Society.

Even for the higher PS-MN content of 4 wt%, they observed that morphology was very similar to that of neat SBS, finding two indicators that allowed them to conclude that PS-MN were assembled into PS domains: the very pronounced phase shift and the increase of thickness of lamellae that disrupted the morphology. As a conclusion, they pointed out that a very good and selective dispersion of PS-MN was obtained in SBS, maintaining the morphology of neat copolymer and increasing mechanical properties.

Secondly, they prepared nanocomposites based on poly(2-vinylpyridine-b-methylmethacrylate) (P2VP-b-PMMA) block copolymer and 2 and 4 wt% of magnetic Fe_3O_4 nanoparticles with PMMA brushes (PMMA-MN) grafted from their surface [44]. They carried out the previously showed grafting procedure by ATRP [48]. They obtained a grafting density of around 0.1 chains/nm^2 and molecular weight of around 30000 g/mol. Nanocomposites were prepared by solution mixing in toluene followed by deposition onto glass wafer by spin coating. Thin films were then exposed to saturated vapors of CCl_4 (selective for PMMA) in order to be nanostructured by solvent vapor annealing. The effect of exposure time and nanoparticles on obtained morphologies as well as on their dispersion was analyzed by AFM and magnetic force microscopy (MFM). First they analyzed the effect of annealing on the morphology of neat P2VP-b-PMMA. Before CCl_4 vapor treatment a soft microphase separation was observed with a weak hexagonal morphology due to the higher affinity of toluene with PMMA. After 3 h of exposure to vapors, a well-defined hexagonal morphology was observed. An increase of exposure time led to changes in the film structure. After 9 h, a quasi-lamellar morphology normal to the substrate was observed. The reason was that CCl_4 was a selective solvent for PMMA and provoked the migration of P2VP from the free surface to the substrate. PMMA migrated to the free surface due to its higher affinity. Increasing exposure time to 14 h, they found a new change in the morphology. PMMA migrated to the surface and appeared as the continuous phase, being P2VP the discontinuous one. The last step before reaching the thermodynamically stable state that finally would lead to PMMA completely filling the top surface of the film, as it can be seen in Fig. 2.6.

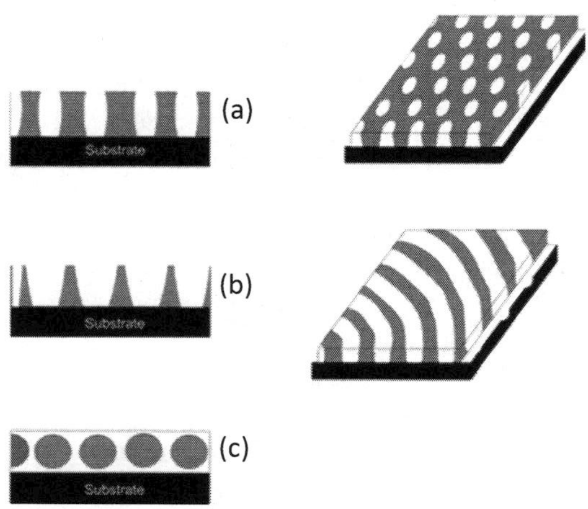

Figure 2.6 Schematic model of time evolution of nanostructure formation of thin P2VP-b-PMMA film. P2VP-black regions, bright regions-PMMA. Reproduced with permission of [44]. Copyright 2008, American Chemical Society.

Then they analyzed nanocomposites. For 2 wt% of PMMA-MN after 3 h of exposure, the same hexagonal structure than that of pure copolymer was found with a good distribution of nanoparticles in the PMMA domains. By height and phase profiles they probed that PMMA-MN were confined. For 4 wt% PMMA-MN nanocomposite, some agglomerations were observed, probably because the grafting density was not enough to prevent nanoparticle flocculation. They also used MFM in order to confirm magnetic properties of the incorporated nanofillers. Results can be seen in Fig. 2.7, where the AFM images can be seen both for neat copolymer and nanocomposites. Those AFM images were taken after performing the magnetic mode scans by applying a magnetic field with the tip.

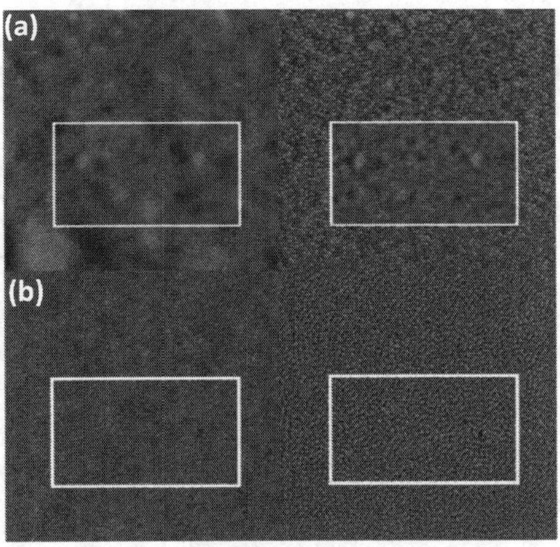

Figure 2.7 TM-AFM topography (left) and phase (right) images of thin film of (a) P2VP-b-PMMA/2 wt% PMMA-MN and (b) neat P2VP-b-PMMA after 3 h of exposure and area after magnetic scan inside the white envelope. Reproduced with permission of [44]. Copyright 2008, American Chemical Society.

For the nanocomposite (Fig. 2.7a), hexagonal morphology was vanished (area within white envelope) after applying the magnetic field, due to interactions among magnetic nanoparticles and magnetized tip, big enough to move copolymer chains as a consequence of weak interactions between nanoparticles and PMMA chains, due to the low grafting density. In the case of neat copolymer (Fig. 2.7b) the morphology was retained after the magnetic field. They concluded that the modification of nanoparticles with PMMA brushes allowed a good and selective dispersion of them into PMMA domains of nanostructured copolymer.

C. Xu et al. [45] also modified Fe_3O_4 nanoparticles with PMMA brushes by ATRP in order to better disperse them into poly(styrene-b-methyl methacrylate) (PS-b-PMMA) copolymer. Brush synthesis was carried out in a mixture of ethanol/water/ammonia with a triethoxysilane having an ATRP initiating site, (2-bromo-2-methyl)propionyloxyhexyltriethoxysilane. The surface-initiated ATRP of methyl methacrylate (MMA) mediated by a copper complex was carried out with the initiator-fixed nanoparticles in the presence of a "sacrificial" (free) initiator [49]. They synthesized brushes with 2.7, 13.3 and 35.7 kg/mol, with a grafting density of 0.73 chains/nm^2. TEM images and size distribution histograms can be seen in Fig. 2.8.

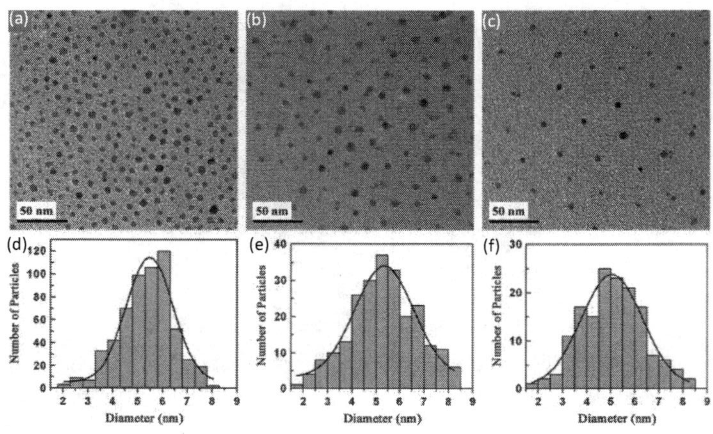

Figure 2.8 TEM images of PMMA-grafted nanoparticles with different molecular weights of (a) 2.7, (b) 13.3 and (c) 35.7 kg/mol, and the corresponding histograms (d–f) of NP core diameter distribution. Reproduced with permission of [45]. Copyright 2008, Elsevier Ltd.

Nanocomposites with 1 to 16 wt% of nanoparticles were prepared by solution mixing in toluene followed by spin casting onto silicon wafers. Films were then annealed at 160°C for different times. They found that as brush molecular weight increased, nanoparticle dispersion in the copolymer was reduced resulting in the formation of aggregates even in the as-cast nanocomposite films. Small aggregates tended to locate at defects in the lamellae, whereas larger ones greatly perturbed the morphology and became encapsulated by onion-ring copolymer structures.

Neat copolymer formed a parallel lamellar structure after annealing. For the nanocomposite with 1 wt% of nanoparticles with smallest brushes (2.7 kg/mol), they were well dispersed and positioned into PMMA domains. Increasing amount nanoparticles tended to form aggregates that avoided the formation of lamellar morphology. For nanocomposites with brushes of 13.3 kg/mol, those with 4 wt% of nanoparticles showed some individual and small aggregates able to locate within PMMA domains together with big agglomerates unable to fit within PMMA domains, located at the boundaries. Increased molecular weight of brushes induced aggregation. For the nanocomposites with brushes of 35.7 kg/mol, for all nanoparticle amounts, almost all the nanoparticles formed aggregates, the morphology of the block copolymer being greatly perturbed because the size of aggregates was greater than lamellar period, lamellae assembling around the aggregates to form onion-ring like features. They used a Flory–Huggins theory describing the phase stability of the three-component solution to predict the critical concentrations at which aggregation was found. Those predictions were in agreement with their experimental observations.

H. Etxeberria et al. [46] grafted PS brushes from CdSe nanoparticles synthesized in aqueous media in order to place them in the PS domains of SBS block copolymer. They first synthesized mercaptoethanol-stabilized CdSe quantum dots in aqueous media [50]. Then they modified nanoparticles with 3-glycidoxypropyltrimethoxysilane (GPS) for the latter attachment of the 4,4′-azobis(4-cyanopentanoic acid) (ACP) azo initiator. Polymerization was carried out with α-picoline as catalyst at 50°C. Nanocomposites with 0.5 and 3 wt% of CdSe-PS nanoparticles were prepared by solution mixing in toluene, followed by film preparation by spin casting onto glass substrates. Films were annealed at 102°C for 24 h. They obtained grafting densities of around 0.17 chains/nm^2, with brushes of 20000 g/mol. PS-modified nanoparticles maintained their optical properties, as they showed by photoluminescence measurements.

For neat SBS they found a lamellar self-assembled structure perpendicular to the surface, after annealing treatment at 102°C. For nanocomposites with 0.5 wt% of nanoparticles, they found that morphology was not significantly altered, observing small bright nanoparticles well located at the PS phase, as it can be seen in Fig. 2.9.

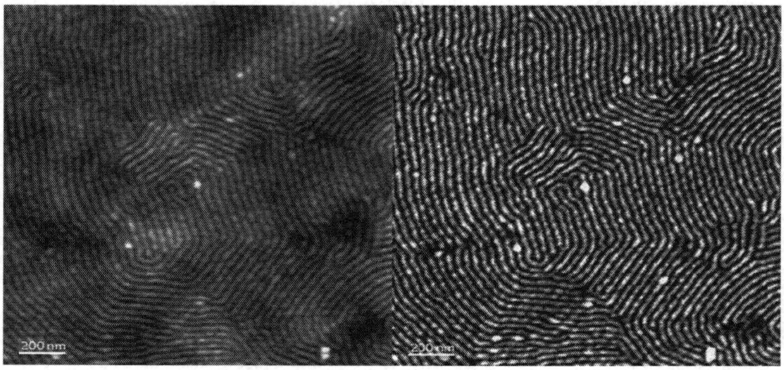

Figure 2.9 AFM images (left/right height/phase) for: 0.5 wt% (upper images) and 3 wt% (bottom images) SBS/PS-CdSe nanocomposite.

Taking into account the size of nanoparticles, they concluded that aggregates of various nanoparticles were placed in PS domains but without altering lamellar structure. In the case of 3 wt% nanocomposites, lamellar morphology was still kept but with higher wideness for lamellas due to the presence of nanoparticle agglomerates. As some agglomerates were bigger than PS lamella width, the lamella of PB had to make a curve surrounding the nanoparticles instead of following the straight lines, disrupting a little bit the nanostructure, as it can be seen in Fig. 2.9.

S. Douadi-Masrouki et al. [47] also grafted PS brushes from magnetic γ-Fe_2O_3 nanoparticles surface by ATRP in order to incorporate them into lamellar poly(styrene-b-butyl methacrylate) (PS-b-PBMA) diblock copolymer films. Nanoparticles were synthesized in aqueous medium, stabilized with oleate. Then an ATRP initiator (2-bromo-2-methylpropionic, BrMPA) was coated onto nanoparticles via ligand-exchange reaction [51]. They used CuBr as catalyst. They obtained brushes with molecular weights ranging from 15500 to 45800 g/mol. They also synthesized the diblock copolymer by ATRP. They first synthesized poly(n-butyl acrylate) (PBMA) for being used as macroinitiator for styrene polymerization. They obtained symmetrical diblock copolymers (in order to obtain lamellar morphologies) with different molecular weights ranging from 57000 to 112000 g/mol by using copper halides as catalyst. Nanocomposites were prepared by solution mixing in toluene, followed by spin coating onto silicon wafers. Obtained thin films were annealed at 150°C (well above glass transition temperature of PS and below order-disorder transition temperature) for 48 h. They first showed the lamellar ordering of thin films of diblock copolymer, with the presence of "islands" or "holes" as defects on top of the films. Regarding synthesized nanoparticles with polymer brushes, they found some clusters due to coupling reactions between growing chains from adjacent nanoparticles. As the average size of clusters remained below a threshold value of the order of lamella thickness, they could be flattened between the lamellae remaining compatible with the self-assembly process. They showed that the introduction of nanoparticles led to nanocomposites with a preserved lamellar structure, with the insertion of nanoparticles into lamellae. However, nanoparticle repartition was found to be irregular, not penetrating indistinctively all the space between PS layers. As an increase of optical refractive index was found by ellipsometry, they pointed out the possible application of those materials as optical waveguides.

2.2.2 Nanocomposites with Nanoparticles Modified with Grafting to Method

In this method, previously synthesized end-functionalized polymers (such as thiol-terminated ones), are directly grafted onto the nanoparticle surface by the functional groups present on it. Several authors have used this method to modify nanoparticles with brushes for being used as nanofillers in nanocomposites with block copolymers [17, 52–59].

E. J. Kramer et al. [56] grafted PS chains to the surface of gold nanoparticles in order to improve their dispersion and selective localization into a nanostructured poly(styrene-b-2-vinylpiridine) (PS-b-P2VP) block copolymer. They first synthesized thiol-terminated PS (PS-SH) by living anionic polymerization in benzene with a molecular weight of 1.3 kg/mol. Gold nanoparticles with a diameter of 2.5 nm and stabilized by the thiol-terminated PS were synthesized [60]. The strategy followed consisted in growing the metallic clusters with the simultaneous attachment of thiol-terminated PS chains on the growing nuclei. In order to allow the surface reaction to take place during metal nucleation and growth, nanoparticles were grown in a two-phase system. Two-phase redox reactions were carried out by an appropriate choice of redox reagents present in the adjoining phases. In the present case, $AuCl_4^-$ was transferred from aqueous solution to toluene using tetraoctylammonium bromide as the phase transfer reagent and reduced with aqueous sodium borohydride in the presence of thiol-terminated PS chains. Two symmetric PS-b-P2VP copolymers with molecular weights of 59 and 114 kg/mol were used. Nanocomposites were prepared by solution mixing in toluene followed by casting onto a epoxy substrate. They showed that below a critical amount of nanoparticles lamellar morphology of copolymers remained unchanged while for higher amounts the local concentration of nanoparticles varied along the film thickness, emerging different coexisting morphologies as a function of sample depth. They first showed the cross-sectional morphology obtained by TEM for the nanocomposite with PS-b-P2VP of 59 kg/mol and a volume fraction of nanoparticles of 0.14, below the critical concentration. The lamellae lied parallel to the film surface with gold nanoparticles aligned near the centre of PS domains. For nanoparticle amount of 0.25, higher than the critical value, they showed totally different results, as it can be seen in Fig. 2.10. Morphology changed with sample depth.

Near the air–polymer interface (Fig. 2.10a), lamellar morphology was detected. Near the substrate, a layer with spherical morphology was observed (Fig. 2.10c,d). Between these two layers (Fig. 2.10b), a layer with disordered morphology was observed. Most of the nanoparticles were located near the center of PS domain, with higher concentration in the spherical phase than in lamellar or disordered ones. They calculated from TEM images of Fig. 2.10 the volume fraction of coated nanoparticles as a function of the distance from the film surface, whose thickness was around 100 μm. They found three main regions: (1) a region with low nanoparticle concentration (around 25 μm thick) of around 0.15;

(2) a region of around 20 μm thick with an increasing volume fraction between lamellar and hexagonal morphologies; and (3) a high concentration region of around 60 μm thick with a fraction of 0.65. Thus they observed a two-phase macrostructure having a layer of low particle volume fraction with lamellar morphology (near air/film interface) and a layer of high particle volume fraction with spherical morphology near the substrate. To explain these phenomena they also prepared films of neat PS-b-P2VP and also with PS-b-P2VP and PS-SH polymers with the same volume fraction than nanoparticles. Analyzing morphologies (only lamellar at any depth for neat copolymer and spherical morphology caused by PS-SH chains swelled selectively in PS domains for samples with PS-SH polymers) they found that PS-Au behave like PS homopolymers. Moreover, morphological changes observed for the examined concentrations arose from the spatially varying concentration of nanoparticles.

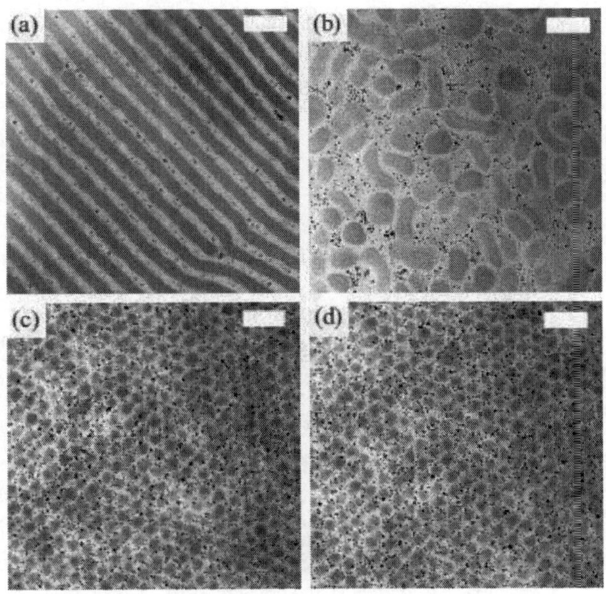

Figure 2.10 Cross-sectional TEM images of a film of gold nanoparticles /PS-b-P2VP. The overall volume fraction of nanoparticles is 0.5. The distance from the top of the film (L) is (a) 27, (b) 36, (c) 52 and (d) 95 μm. Scale bars of 100 nm. Reproduced with permission of [56]. Copyright 2005, Wiley-VCH.

They also prepared nanocomposites with the longer PS-b-P2VP copolymer. For a volume fraction of 0.30, they observed a similar two-phase macrostructure, as it can be seen in Fig. 2.11: lamellar morphology with low nanoparticle concentration near the surface and hexagonal with higher concentration near the substrate.

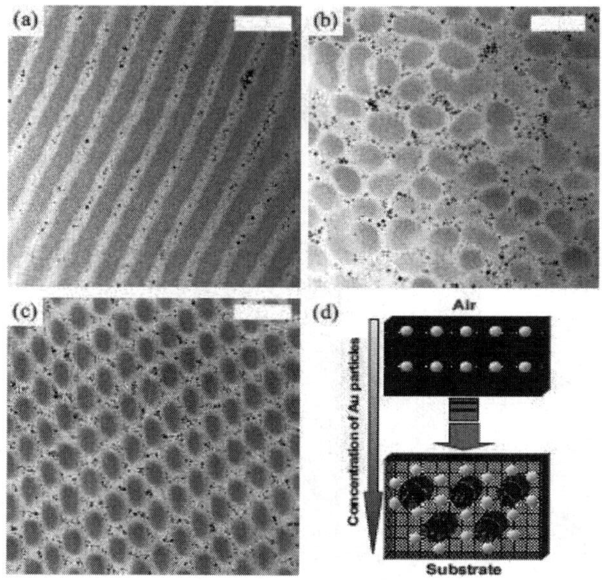

Figure 2.11 Cross-sectional TEM image of a film of PS-Au/PS-b-P2VP with a volume fraction of around 0.30. Distance from the top of film (L) is (a) 27, (b) 46, (c) 46 μm. (d) Schematic of the film structure based on TEM images. Reproduced with permission of [56]. Copyright 2005, Wiley-VCH.

Again, nanoparticles and copolymer formed layered structures with three distinctive concentration regions and morphologies depending on the depth of the film. In their opinion, several factors could cause the transport of PS-Au towards the substrate with the subsequent morphology change: (1) the diffusion towards the substrate was faster than the diffusion away from the substrate because there was more solvent near the substrate; (2) the interaction between solvent and PS-Au was more favorable than that between solvent and copolymer; and (3) lamellar layers near the sample/air interface could not accommodate a large volume fraction of PS-Au and maintained lamellar morphology. As particles concentrated towards the bottom, morphology changed to morphologies that could better accommodate having PS as the major component. This occurred when the concentration of nanoparticles exceeded a critical value. They obtained different morphologies and layered order-order transitions caused by high concentrations of PS-Au nanoparticles in one of the domains. They pointed out that the versatility of the method opened up possibilities for building a variety of self-assembled structures of inorganic nanoparticles in a single step at low cost.

In their next work, E. J. Kramer et al. [55] studied the effect of areal chain density on the location of PS and P2VP-modified gold nanoparticles in a PS-b-P2VP copolymer template. In this case they used a longer block copolymer (196 kg/mol) and gold nanoparticles in which they grafted PS-SH chains with a molecular weight of 3.4 kg/mol or P2VP-SH chains of 3 kg/mol. The areal chain density of PS-HS ligands on nanoparticles was varied by changing the mole ratio of PS-SH or thiol-terminated P2VP chains to gold atoms. They found that this density was critical to control their location in the block copolymer template. Both PS-HS and P2VP-HS were synthesized by living anionic polymerization following the same procedure previously described [56]. PS-coated Au nanoparticles were synthesized following the same procedure previously described [56, 60] using a two-phase system consisting on toluene and water by varying the initial mole feed ratio of PS ligands to (Au atoms + PS ligands) from 1/5 to 1/129. As P2VP is not soluble in toluene, one-phase method using tetrahydrofurane (THF) was employed for the synthesis of P2VP-Au: gold precursors ($HAuCl_4$) and P2VP ligands were solved into THF. Au nanoparticles were synthesized by adding a reducing agent, $(C_2H_5)_3BH$ superhydride, in THF [61]. Several different P2VP-Au were synthesized with various initial molar ratios of P2VP ligands to (Au atoms + P2VP ligands) from 2 to 0.015. Nanocomposites were prepared by solvent mixing with dichloromethane followed by casting onto epoxy substrate. They started analyzing PS-Au/PS-b-P2VP composites with a 0.15 weight fraction of nanoparticles with different grafting densities ranging from 0.83 to 1.64 chains/nm^2. Cross-sectional TEM images can be seen in Fig. 2.12.

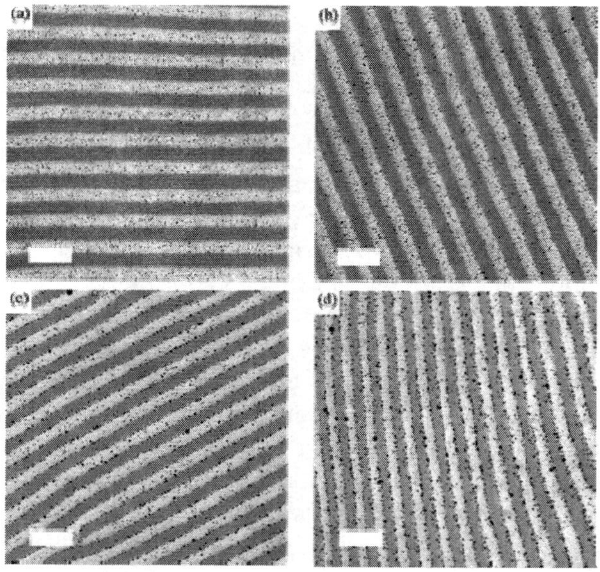

Figure 2.12 Cross-sectional TEM images of PS-b-P2VP block copolymer containing PS-coated Au nanoparticles with various chain areal densities: (a) 1.64, (b) 1.45, (c) 1.22, and (d) 0.83 chains/nm^2. Scale bar of 100 nm. Reproduced with permission of [55]. Copyright 2006, American Chemical Society.

Darker domains correspond to P2VP, as it was stained with iodine vapor. It is clear that particles were located near the center of PS domains for the highest density (Fig. 2.12a). As the grafting density decreased to 1.45, particles were distributed equally between the center of PS domains and interfaces (Fig. 2.12b). For a density of 1.22 (Fig. 2.12c), most particles were segregated along the interfaces between PS and P2VP, while for the lowest density (Fig. 2.12d) all the nanoparticles were at the interfaces. Nanoparticles with PS chain densities lower than 1.3 chains/nm^2 did not fully shield the Au nanoparticle surface for interacting with the P2VP block. There was a favorable interaction between gold and P2VP while PS/gold interaction was weaker [62]. They demonstrated such a favorable interaction by secondary ion mass spectrometry (DSIMS).

They also prepared nanocomposites with P2VP-Au nanoparticles and PS-b-P2VP block copolymer. They used nanoparticles with chain densities of 1.14 and 0.76 chains/nm^2. TEM images can be seen in Fig. 2.13.

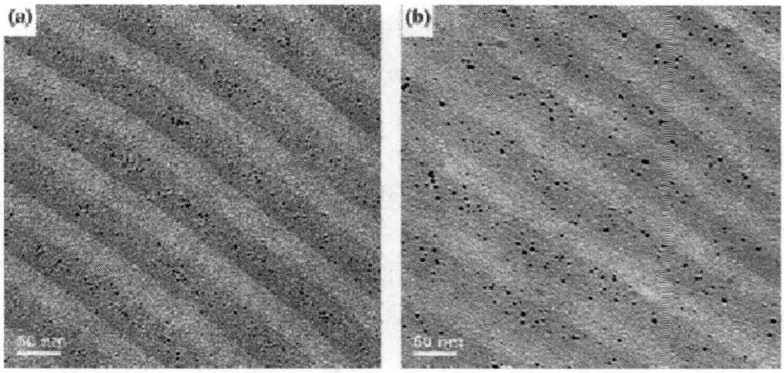

Figure 2.13 Cross-sectional TEM images of P2VP-coated gold nanoparticles with grafting densities of (a) 1.14 and (b) 0.76 chains/nm^2 in PS-b-P2VP. Reproduced with permission of [55]. Copyright 2006, American Chemical Society.

As it can be seen, P2VP-Au nanoparticles were dispersed in P2VP domains rather than segregated at interfaces, even with low chain densities, in clear contrast to the low areal chain density of PS-Au nanoparticles. This PS-Au nanoparticles with low areal chain density segregated to the interface due to the inability of the low-density PS brush to screen the favorable interaction between the P2VP block chains and the bare Au surface. This favorable interaction made possible the dispersion of P2VP-Au nanoparticles into P2VP domains even for low densities. They concluded underlying the possibility to selectively place Au nanoparticles at desirable domains of copolymer template just covering them with brushes corresponding to one of the blocks and controlling the chain density at the surface. This strategy is a simple and versatile method that can be extended to other inorganic-polymer hybrid materials.

They completed the research with another work [54] in which they also prepared Au nanoparticles coated with both PS and P2VP and dispersed into the same copolymer used in the previous work. They modified the nanoparticles in the same way previously shown, but adding a 1:1 molar mixture of PS-SH and P2VP-SH.

They found that so modified nanoparticles were located at the interfaces between PS and P2VP domains. They explained this interfacial absorption in terms of the interfacial energies of PS-b-P2VP, nanoparticle-PS and the nanoparticle-P2VP interfaces.

They also presented another work [53] in which they analyzed the effect of block copolymer molecular weight on obtained morphologies and nanoparticle distribution. In this case they used four copolymers with 60000, 114500, 196500 and 380000 g/mol, respectively, to prepare nanocomposites with a volume fraction of 0.15 of nanoparticles with a very high areal chain density of 9.3 chains/nm^2. This density was sufficiently high to prevent the contact of gold surface of partially covered gold nanoparticles with the P2VP block chains, contact that, as it was previously shown [55], caused gold nanoparticle to bind to the PS-P2VP interface. Nanoparticle synthesis and nanocomposite preparation were carried out in the same way than in previous works. PS-Au nanoparticles presented an average diameter d (core + shell) of 9.2 ± 2.2 nm. They calculated PS domain sizes (L_{PS}) for each copolymer by small angle X-ray spectroscopy (SAXS) and DSIMS. So they prepared samples with four different d/L ratios of 0.44, 0.31, 0.18 and 0.13 at a fixed nanoparticle volume fraction of 0.15. Nanoparticles were located at or near the center of PS domains, appearing P2VP domains free of particles. They pointed out that this location near the center of the particle-compatible block copolymer lamellar domain was consistent with numerical calculations carried out by Balazs et al. [63, 64]. Irrespectively of d/L value they observed the *center-filled* morphology. However, they observed a tendency for the particles to become more delocalized within PS domain as d/L decreased. Results were interpreted in terms of the relative entropic contributions of the particles and the copolymer and compared with theoretical calculations. For low molecular weights, PS-Au were located at or near the center of PS domains, while for high molecular weights particles were less confined to the center, evidenced by a broader distribution. According to numeric predictions, as this ratio decreased, the chain stretching penalty was lessened and particles were less likely to be confined in the center of domains, consistent with their observations. They observed similar results for different nanoparticle volume fractions from 0.07 to 0.44. Nanoparticle segregation towards the center of PS domain was observed for all fractions with a decrease in the distribution width as the fraction increased. TEM images of samples can be seen in Fig. 2.14.

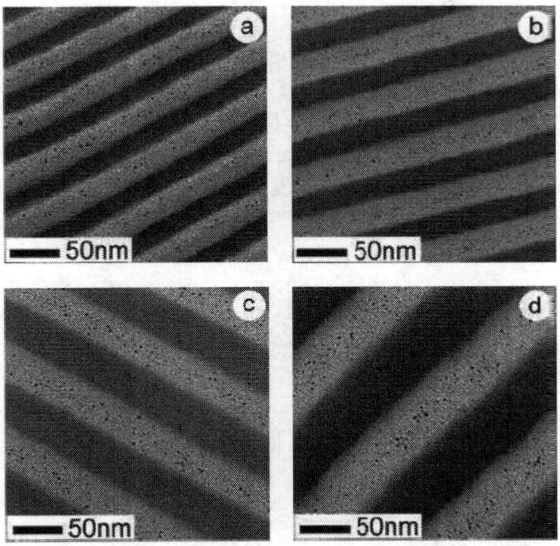

Figure 2.14 Cross-sectional TEM images of gold/block copolymer composite films of PPS-b-P2VP with various molecular weights: (a) 60, (b) 114.5, (c) 196.5, and (d) 380 kg/mol. Reproduced with permission of [53]. Copyright 2007, American Chemical Society.

In another work, E. J. Kramer et al. [52] prepared nanocomposite thin films based on poly(styrene-b-methyl methacrylate) (PS-b-PMMA) diblock copolymer and poly(ethylene oxide) (PEO)-coated gold nanoparticles. They grafted PEO chains to the surface of nanoparticles with the following procedure: thiol end-functional PEO and gold precursor $HAuCl_4$ were mixed and then a superhydride reducing agent was added to synthesize nanoparticles. Block copolymer showing a cylindrical structure with a molecular weight of 140 kg/mol was synthesized by living anionic polymerization. Nanocomposites were prepared by solution mixing in benzene followed by spin coating onto silicon substrates. Samples with 0.6, 1.9, 3.2, 4.5 and 6.4 vol% of nanoparticles were prepared, and films were annealed under saturated benzene vapor, followed by rapid solvent evaporation under controlled humidity conditions.

Their intention was to check the possibility of inducing lateral ordering and vertical orientation in hydrophobic PS-b-PMMA system by simple introduction of "external" hydrophilic moieties such as PEO-Au nanoparticles, in order to use the block copolymer as template for lithographic applications.

They first analyzed the morphology of thin films prepared without Au and with 4.5 wt% of PEO-Au, solvent annealed under 90% humidity. Neat copolymer film showed cylindrical domains aligned parallel to the substrate while films with 4.5 wt% Au-PEO showed cylindrical domains with perpendicular orientation. Black dots observed within PMMA domains were related to Au-PEO nanoparticles, due to the affinity of modified nanoparticles with this block. Next they analyzed the effect of humidity on morphologies. When films containing Au-PEO were annealed under lower humidity conditions (around 50%), cylindrical domains were very poorly ordered and in many regions oriented parallel to the substrate. They were able to see that in SEM micrographs after degradation of PMMA block by UV radiation, which can be seen in Fig. 2.15.

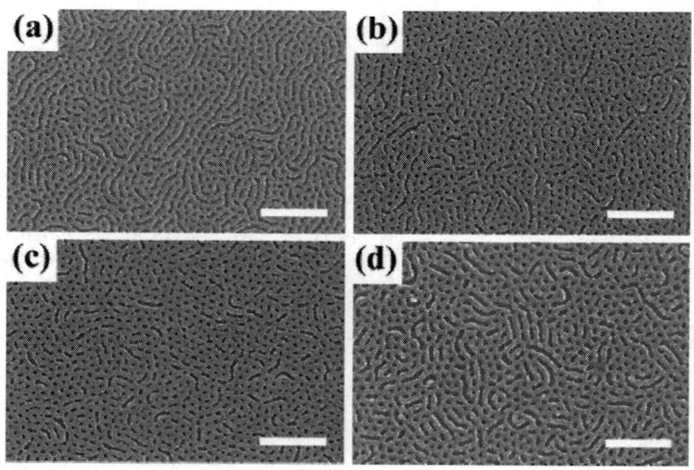

Figure 2.15 SEM images for PS-b-PMMA films containing: (a) 0, (b) 1.9, (c) 3.2, and (d) 4.5 vol% Au-PEO particles after solvent annealing and rapid solvent evaporation under low humidity conditions (50%), followed by UV radiation and then rinsing with acetic acid and water to remove PMMA domains. Reproduced with permission of [52]. Copyright 2007, American Chemical Society.

They underlyed that Au-PEO played the same role as PEO blocks in PS-b-PEO or PS-b-PEO-PS copolymers because humidity played an important role in lateral ordering presumably by a strong interaction between water vapor and PEO during annealing process [65].

Au-PEO nanoparticles did not segregate to the surface under low humidity conditions, as expected since the surface energy of PEO is higher than that of PMMA or PS [66]. The ordering would be induced by the interaction between water vapor and hydrophilic Au-PEO nanoparticles, being the ordering mechanism the same as in the case of PEO-based block copolymers: during solvent annealing, films were swollen with benzene under saturated atmosphere. Once films were removed from the chamber, solvent evaporated to the film surface, which became cooled and reached the dew point. Under high humidity conditions, water vapor condensed on the cold surface and interacted with Au-PEO present in PMMA domains, leading to hexagonal arrays on the film surface. As solvent evaporated further, PMMA domains containing Au-PEO grew vertically inducing perpendicular orientation of cylinders. Under low humidity conditions, the dew point decreased and water did not condense, obtaining the parallel orientation of cylinders observed in Fig. 2.15.

To investigate morphologies of films prepared under high humidity conditions (90%), they removed PMMA domains by UV irradiation followed by rinsing with acetic acid and water, in the same way that for films prepared under low humidity conditions. Figure 2.16 shows SEM images of so prepared films. In the absence of nanoparticles, cylinders parallel to the substrate were obtained (Fig. 2.16a). This orientation persisted when the amount was 0.6 vol% but as volume fraction increased to 1.9% film morphology changed dramatically. At 1.9 and 3.2 vol%, a perpendicular orientation of the cylinders was evident (Fig. 2.16c,d). Moreover, lateral ordering was improved with respect of that for PS-b-PMMA copolymer, so it can be attributed to the effect of Au-PEO which induced lateral ordering as well as perpendicular orientation. Increasing amount to 4.5 and 5 vol%, lateral ordering was disrupted. For 4.5 vol%, hexagonal packing of cylinders was distorted while cylinders were still oriented perpendicularly. For 6.5 vol%, a larger pore size was seen, seeming that macrophase separation induced by large amounts of Au-PEO nanoparticles took place.

Finally they studied the distribution of Au-PEO in PS-b-PMMA films by TEM, as it can be seen in Fig. 2.17 for the nanocomposite with 4.5 vol% Au-PEO.

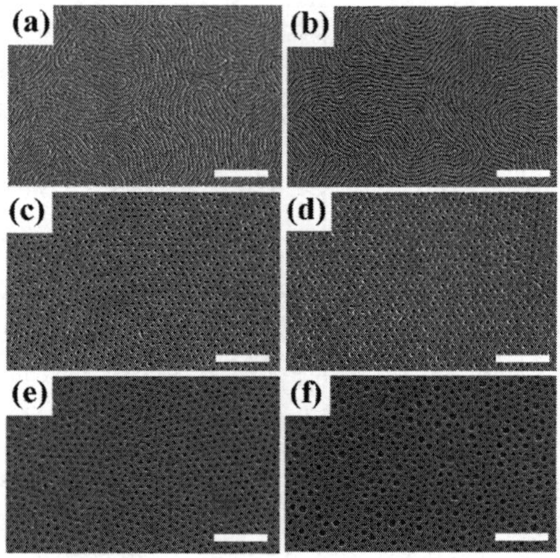

Figure 2.16 SEM images of PS-b-PMMA films containing: (a) 0, (b) 0.6, (c) 1.9, (d) 3.2, (e) 4.5 and, (f) 6.5 vol% Au-PEO after solvent annealing and rapid solvent evaporation under high humidity conditions (90%) followed by UV radiation and rinsing with acetic acid and water to remove PMMA domains. Reproduced with permission of [52]. Copyright 2007, American Chemical Society.

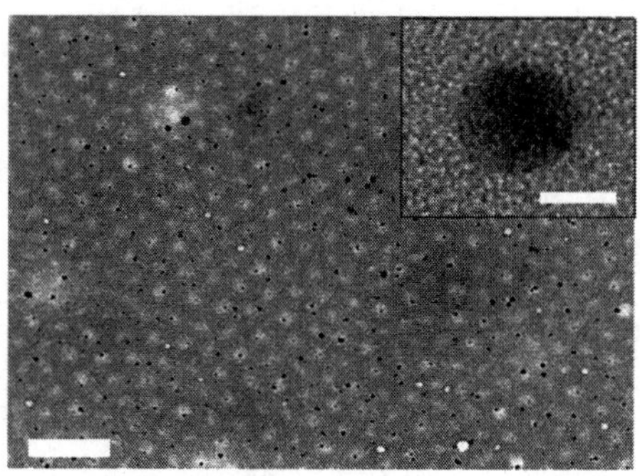

Figure 2.17 TEM images (top view) for PS-b-PMMA film containing 4.5 vol% Au-PEO. The inset is a high-magnification image of the core of a gold nanoparticle. Reproduced with permission of [52]. Copyright 2007, American Chemical Society.

The brighter areas correspond to PMMA domains whereas PS matrix appears as a darker background and Au-PEO as dark spots. Majority of Au-PEO nanoparticles were located at PMMA domains, which was consistent with previously cited theoretical predictions of Balazs et al. [63, 64]. In this system, the d/L ratio was around 0.4, so they expected Au-PEO nanoparticles to be at the center of PMMA domains, as it was pointed out in their previous works [54]. However, nanoparticles were randomly distributed in PMMA domains, as it can be seen in Fig. 2.17. They believed that Au-PEO nanoparticles were "kinetically trapped" during solvent annealing and did not reflect the true equilibrium structure for this system. Au-PEO did not achieve equilibrium localization in PMMA domains due to the fast evaporation of the solvent.

To probe this fact they also prepared films with 1.9 and 3.2 vol% Au-PEO annealed under benzene vapors but in this case the consequent solvent evaporation was carried out slowly under the same humidity conditions (90%). For slow evaporation the perpendicular orientation was not detected, finding instead poor ordering and mixed cylinder domain orientation. The slow benzene evaporation did not produced cooling of the film surface with no water condensation. So they concluded that the key controlling factor was the solvent annealing process where hydrophilic Au-PEO nanoparticles in PMMA domains interacted with water vapor under high humidity, solvent-swollen processing conditions. This strategy provided a simple route for the fabrication of ordered block copolymer templates from PS-b-PMMA without the need for a random copolymer neutralization layer.

Recently, E. Ploshnik et al. [67] also grafted PEO chains to Au nanoparticles in order to prepare nanocomposite ultrathin films with PS-b-PMMA block copolymer. The copolymer, with a molecular weight of 10^6 Da and 35 wt% PS was synthesized by standard anionic polymerization. They synthesized citrate-stabilized Au nanoparticles by Natan's method [68] to which they grafted thiol-terminated PEO (PEO-SH) chains of 5 and 20 kDa. For film preparation, they solved copolymer in toluene and nanoparticles in chloroform. After mixing solutions, they were spin casted onto silicon wafers. Films with thicknesses ranging from 17 to 30 nm were then annealed under saturated chloroform vapors for 2, 14 and 48 h. They showed that the combination of an A-B block copolymer (PS-b-PMMA) with B′-type nanoparticles (Au-PEO, with lower incompatibility of PEO with PMMA than with PS), where $\chi_{A,B'}$, $\chi_{A,B}$ and $\chi_{B,B'}$, led to the formation of hierarchical structures through two-phase separation process operating in different timescales. The block copolymer microphase-separated first due to the incompatibility among blocks. As the incompatibility among nanoparticles and PS block was even larger, PS block formed domains with the smallest local fraction of nanoparticles. Au-PEO followed the evolution of the less incompatible PMMA domains. The growth of PS domains pushed the nanoparticles with its boundary (PS/PMMA interface): during initial stages of phase separation, nanoparticles were located at the interface. After longer times, block copolymer phases annealed out defects and a much slower phase separation process between nanoparticles and their hosting domains took place. Thus nanoparticles departed from the interfaces to form aggregates at the centers of the hosting domains. The effective attractive force between the nanoparticles and net repulsion from the hosting domains led to the formation of hexagonally packed nanoparticle assemblies. The distance between neighbors was dictated by the length of the polymer chains around the nanoparticle. They pointed out that even if those structures were not in equilibrium, the macrophase separation of the nanoparticles from the copolymer took longer and could be avoided if the annealing process was halted in time. Controlling the relative incompatibility between the components could be a tool for obtaining hierarchically structured copolymer/nanoparticle assemblies.

M. R. Bockstaller et al. [57, 69] also grafted PS brushes to the surface of Au nanoparticles in order to selectively disperse them into a poly(styrene-b-ethylene/propylene) (PS-b-PEP) copolymer thin films for studying their optical properties for potential applications as metallodielectric reflectors. They used the same procedure than previous authors [52–56] to grow PS-coated Au nanoparticles, using thiol-terminated PS as ligand. Metallodielectric nanocomposites were prepared by solvent evaporation of a toluene solution of PS-PEP/Au-PS. After solvent evaporation, films were kept for 72 h in oversaturated solvent atmosphere at 60°C. PS-b-PEP microphase separated into lamellar morphology. Au-PS nanoparticles were sequestered homogeneously within the PS domain without a significant tendency for aggregation, as it can be seen in Fig. 2.18. Grey areas correspond to PS layers and bright ones to PEP layers. Au nanoparticles appear as dark spots homogeneously distributed within PS layer. This preferential sequestering of polymer-grafted Au nanocrystals provided a means to obtain nanocomposite materials for photonic applications that capitalize on both the optical properties of the periodic microstructured template copolymer and the optical properties of the sequestered nanoparticles. They studied nanocomposites by reflectometry and reflective properties were compared with model calculations. They pointed out that particle size caused strong alterations in the dielectric function of metallic inclusions that had relevance for the attainable dielectric contrast within periodic nanocomposites. They concluded that for engineering polymer-based metallodielectric materials with high rejection levels for incident light, particle sizes in the order of the electrons mean free path should be used. Their size is below the scattering limit for the range of wavelengths of interest.

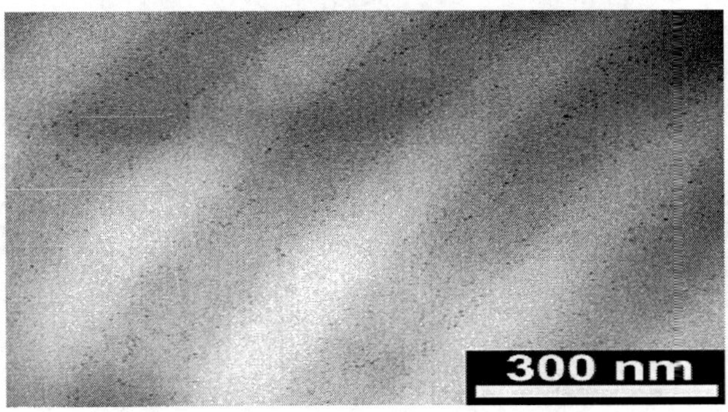

Figure 2.18 Underfocused bright field transmission electron micrograph of the Au-PS/PS-PEP film. Reproduced with permission of [57]. Copyright 2003, American Chemical Society.

Following a similar procedure for grafting polymeric brushes to Au nanoparticles, S. Kim et al. [70] prepared nanoporous bicontinuous structures by adding them to PS-b-PMMA block copolymer templates. They used the two-phase method previously described for grafting photo-cross-linkable thiol-terminated polymeric ligands containing various compositions of PS and PMMA, in order to be selective to PS or PMMA block or nonselective (neutral) to both blocks. They also prepared amphiphilic Au nanoparticles by coating the surface with equimolar mixtures of PS and PMMA selective ligands. Thiol-terminated ligands were synthesized by reversible addition-fragmentation transfer polymerization (RAFT) [71]. To compare the morphological behaviors between neutral-ligand Au and mixed-ligand Au nanoparticles, weight fractions were varied from 10 to 60 wt%. The nanocomposite samples were prepared by drop casting dichloromethane solutions of mixtures onto NaCl substrates followed by annealing at 200°C for 48 h. For the fabrication of nanoporous bicontinuous thin films, copolymer was mixed with mixed-ligand Au in toluene, followed by spin coating onto silicon substrates. Films of 200 nm thickness were annealed at 200°C for 48 h. They first observed that nanoparticles coated with ligand 1 (PMMA selective) were located at the PMMA domains of self-assembled copolymer while those coated with ligand 3 (PS selective) were located at PS ones and those coated with ligand 2 (neutral) at the interfaces, as it can be seen in Fig. 2.19. Then they focused on the morphological behaviors between neutral-ligand Au (coated with ligand 2), with neutral PS and PMMA composition, and mixed-ligand Au that were modified with equimolar mixtures of ligand 1 (PMMA selective) and ligand 3 (PS selective).

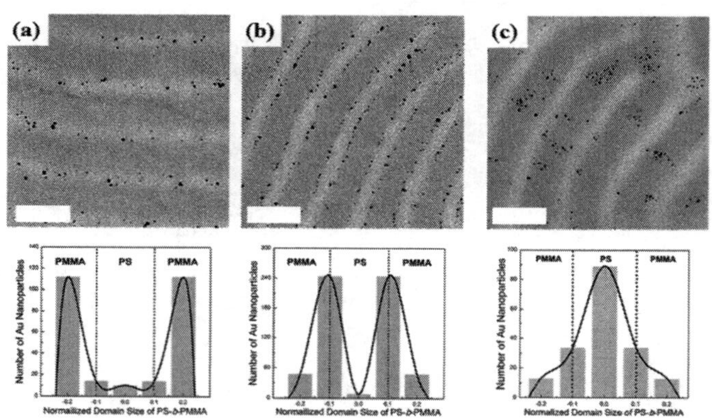

Figure 2.19 Cross-sectional TEM images and the corresponding histogram for the location of 10 wt% Au nanoparticles coated with (a) ligand 1, (b) ligand 2 and (c) ligand 3 within PS-b-PMMA copolymer. Scale bars of 100 nm. Reproduced with permission of [70]. Copyright 2013, American Chemical Society.

For nanocomposites with 10 wt% or lower, Au nanoparticles were located at the PS/PMMA interface. As the amount increased, morphological behaviors were completely different: neutral-ligand Au exhibited macrophase separation within lamellae whereas mixed-ligand Au induced the formation of bicontinuous structures without nanoparticle aggregation, as it can be seen in Fig. 2.20.

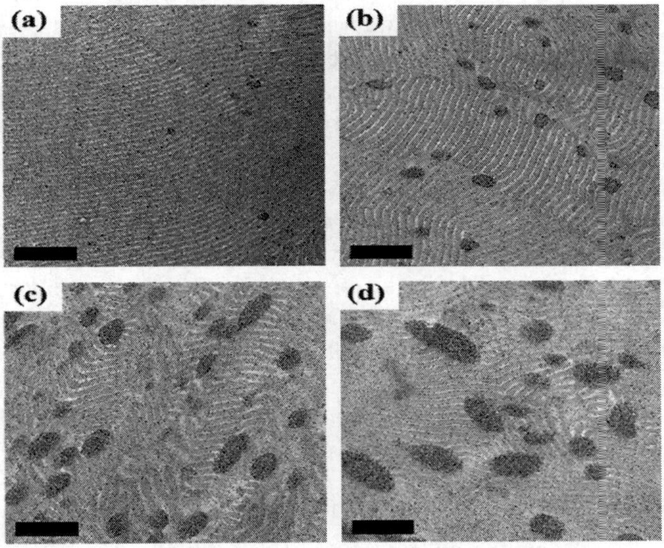

Figure 2.20 Cross-sectional TEM images of nanocomposites containing (a) 20, (b) 30, (c) 40, and (d) 50 wt% of neutral-ligand Au nanoparticles within PS-b-PMMA. Scale bars of 500 nm. Reproduced with permission of [70]. Copyright 2013, American Chemical Society.

They also demonstrated that nanoporous bicontinuous films could be fabricated after PMMA removal, in which the size of bicontinuous structures decreased with nanoparticle loading. By fitting to Teubner–Strey model they determined the transition from lamellar to bicontinuous structures. The onset of bicontinuous structure occurred at lower fraction of mixed-ligand nanoparticles.

Recently A. Ledo-Suarez et al. [72] also prepared nanocomposites with PS-b-PMMA block copolymer and Au nanoparticles after grafting thiol-functionalized PS chains of different length to their surface. After synthesizing PS homopolymer with different chain lengths, they post-treated them with a sulfur precursor for producing polymer chains bearing thiol end groups. PSSH-Au nanoparticles were synthesized by chemical reduction of hydrogen tetrachloroaurate trihydrate in the presence of thiol-functionalized PS. They synthesized PS_{26}SH-Au, PS_{75}SH-Au and PS_{167}SH-Au. As they avoided the elimination of unbounded PS ligands, the modifier employed for the mixtures with the copolymer were PS_n/PS_nSH-Au mixtures. PS-b-PMMA copolymer with 29 wt% of PMMA with a molecular weight of 67000 was used. After solvent mixing in toluene, solutions were spin coated onto silicon wafers and films (30–200 nm thick) annealed at 190°C for 1 h.

After analyzing obtained morphologies by SEM they concluded that PS_n/PS_nSH-Au mixtures were located in the PS phase during thermal annealing without macrophase separation. During copolymer nanostructuring, PS_n/PS_nSH-Au moieties experienced molecular desorption, nanocrystal core coalescence and partial molecular re-encapsulation leading to sphere-like Au nanoparticles with a larger average size without exceeding the interdomain distance. By increasing ligand length, those processes became weak, associated with major protection efficiency of the capping. Nanoparticle partitioning and arrangement on different length scales could be properly tuned by adjusting ligand length and composition. For short ligands, migration to the free surface and nanoparticle self-assembly were favored and films presented gold clusters randomly distributed in PS domains. For long ligand chains, compatibility with the host polymer was favored and migration to the surface avoided, leading to systems with a homogeneous distribution of nanoparticles throughout the entire sample.

I. Barandiaran et al. [17] used another method for grafting a poly(methyl methacrylate-b-ε-caprolactone) (PMMA-b-PCl) diblock copolymer to the surface of magnetic Fe_2O_3 nanoparticles in order to disperse them into poly(styrene-b-ε-caprolactone) (PS-b-PCl) block copolymer thin films. A PS-b-PCl copolymer with 70 wt% of PS and a molecular weight of 37000 g/mol was used. They first synthesized PMMA-b-PCl copolymer (21000 g/mol) with terminal chlorine group by ATRP [73]. Magnetic nanoparticles were modified with 3-aminopropyltryethoxysilane (APTS), reaction that implied a nucleophilic attack of OH groups at the nanoparticle surface to the Si atoms of APTS.

The anchoring of block copolymer was done by *grafting to* method. The covalent linking among silanized nanoparticles and PMMA-b-PCl with terminal chlorine group was carried out by an alkylation reaction of amine groups present at the surface of nanoparticles. They obtained a grafting density of 0.04 chains/nm^2. They prepared nanocomposites with 2 and 5 wt% of nanoparticles by solvent mixing with toluene followed by spin coating onto glass substrates. Films were annealed at 100 or 120°C under vacuum. They first analyzed neat copolymer films by AFM, finding a worm-like morphology for films annealed at 100°C and a lamellar one for films annealed at 120°C. They did not observe PCl crystalline domains at the images. Crystallization degree of PCl block (only 30 wt% of copolymer) was measured to be below 5% by DSC. For nanocomposites crystallization degrees were even lower than 1%. Then they analyzed nanocomposites with 2 and 5 wt% of nanoparticles annealed at 100 and 120°C, as shown in Fig. 2.21.

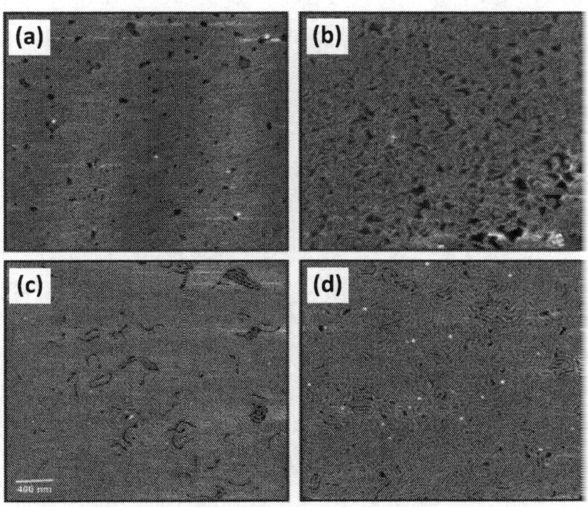

Figure 2.21 AFM phase images (3 × 3 μm) of PS-b-PCL/Fe_2O_3-g-(PMMA-b-PCL) nanocomposites annealed at different temperatures with different nanoparticle amounts: (a) 100°C and 2 wt%, (b) 100°C and 5 wt%, (c) 120°C and 2 wt%, and (d) 120°C and 5 wt%. Reproduced with permission of [17]. Copyright 2014, Elsevier Ltd.

Samples annealed at 100°C presented a worm-like morphology like neat copolymer, while those annealed at 120°C presented lamellar one like neat copolymer. Bright points observed in the images correspond to nanoparticles. Nanoparticles were well dispersed in the copolymer, much better than unmodified ones, which formed bigger aggregates disrupting copolymer morphology. The grafting process increased compatibility with matrix due to the presence of PCL blocks in both, improving dispersion. Even if some nanoparticles were located at PCL domains, most of them were located at the interface between PCL and PS domains, without altering copolymer nanostructure. Average sizes of 22 and 38 nm were obtained for nanoparticles in nanocomposites with 2 and 5 wt% annealed at 120°C, respectively, and 25 and 40 nm for those annealed at 100°C. It was clear that nanoparticles were not individual (initial size of nanoparticles was around 9 nm). They probed by TEM that during grafting process nanoparticles formed agglomerates surrounded by PMMA-b-PCL with similar size than those found in the nanocomposites. Aggregates were formed mainly during modification process and not during dispersion in the matrix. Similar to other authors, dispersion of nanoparticles was clearly improved by grafting process when compared with that of unmodified ones, maintaining nanostructured morphology of neat copolymer.

2.2.3 Nanocomposites with Nanoparticles Modified with Grafting through Method

In this technique, molecules attached to the surface also present a group suitable for polymerization (usually a silane with terminal vinyl groups, which are subsequently used for the polymerization). Nanoparticles present in the polymerization medium are covered by the polymer. Obtained grafting densities are usually higher than those obtained by *grafting to* and the technique is easier to carry out than *grafting from* one. As nanoparticle surface is multifunctional, polymeric chains also present bonds among them, creating a sort of network [32]. Regarding the preparation of nanocomposites based on block copolymers and nanoparticles grafted with brushes by this method, a work of H. Etxeberria et al. [25, 26] must be pointed out.

They prepared nanocomposites based on SBS block copolymer and CdSe quantum dots with PS brushes in their surface obtained by *grafting through* method. CdSe quantum dots stabilized with mercaptoethanol were prepared in aqueous medium optimizing their size by controlling synthesis parameters [50]. Hydroxy-coated nanoparticles were then functionalized with methacryloxypropyltrimethoxysilane (MPS) obtaining MPS-CdSe nanoparticles with vinyl groups suitable for polymerization. Then, 2,2-azobis isobutyronitrile (AIBN) and styrene monomer were added to a suspension of MPS-CdSe in dimethylformamide (DMF) to obtain PS-CdSe nanoparticles, coated with PS. They obtained a grafting density of 0.23 chains/nm^2 with PS chains of 9000 g/mol. UV-vis and photoluminescence measurements showed that optical properties of CdSe quantum dots were maintained after modification with silane and PS chains.

Then they prepared nanocomposites with PS-CdSe nanoparticles and SBS block copolymer by solvent mixing in toluene followed by spin casting onto glass substrates. Nanocomposites with 5 and 10 wt% PS-CdSe were prepared and films were annealed at 102°C for 24 h. This annealing treatment was chosen because previous rheological and morphological analysis set it as adequate in order to obtain a self-ordered structure [41]. The morphology of the copolymer was not significantly altered, obtaining a lamellar structure. Brighter dots that indicate PS-CdSe nanoparticles were selectively placed at PS domains, due to the improved affinity, without the presence of significant aggregates. Taking into account nanoparticle size values obtained by TEM, it was evident that bright points did not correspond to single nanoparticles. As polymerization was carried out with MPS-CdSe nanoparticles in the media, the functional group located in the surface could join a growing PS in which there could be more nanoparticles previously joined. As the surface of nanoparticles was multifunctional, with several double bonds, several chains could be bonded to different nanoparticles. In fact, some PS chains grew from the surface, while some others were surrounding the nanoparticles.

Due to this, CdSe quantum dots, instead of existing individually, were bonded together with several polymer chains, creating a kind of network formed by nanoparticles and PS chains, as it can be seen in the scheme shown in Fig. 2.22. When nanoparticle amount was increased to 10 wt%, the lamellar self-structure was still maintained, with nanoparticles located only in the PS phase. Due to its amount, PS-CdSe started to agglomerate in the PS lamellas. Anyway, agglomerates seemed to be small enough for being located at PS lamellas, obtaining a good dispersion. They also characterized nanocomposites by electronic force microscopy (EFM), showing that electrical semiconductive properties of CdSe nanoparticles were maintained through their functionalization and incorporation to the block copolymer [74]. Thus they presented a relatively simple method to selectively place semiconductor quantum dots with good optical properties in block copolymer templates, transferring their interesting properties to the composites.

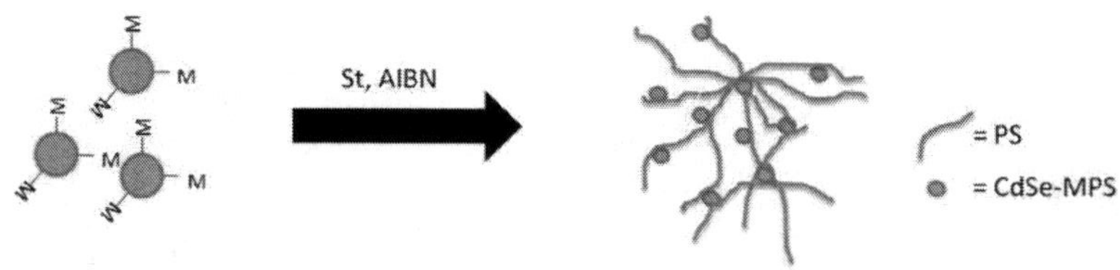

Figure 2.22 Scheme of the CdSe/PS chains network.

References

1. Peponi, L., Tercjak, A., Gutierrez, J., Stadler, H., Torre, L., Kenny, J. M., and Mondragon, I. (2008). Self-assembly of SBS block copolymers as template for conductive silver nanocomposites, *Macromol. Mater. Eng.*, **293**, pp. 568–573.
2. Lo, C. T., Lee, B., Winans, R. E., and Thiyagarajan, P. (2006). Effect of dispersion of inorganic nanoparticles on the phase behavior of block copolymer in a selective solvent, *Macromolecules*, **39**, pp. 6318–6320.
3. Lin, Y., Boker, A., He, J., Sill, K., Xiang, H., Abetz, C., Li, X., Wang, J., Balazs, A., and Russel, T. P. (2005). Self-directed self-assembly of nanoparticle/copolymer mixtures, *Nature*, **434**, pp. 55–59.
4. Zou, S., Hong, R., Emrick, T., and Walker, G. C. (2007). Ordered CdSe nanoparticles within self-assembled block copolymer domains on surfaces, *Langmuir*, **23**, pp. 1612–1614.
5. Zhang, Q., Xu, T., Butterfield, D., Misner, M. J., Ryu, D. Y., Emrick, T., and Russel, T. P. (2005). Controlled placement of CdSe nanoparticles in diblock copolymer templates by electrophoretic deposition, *Nano. Lett.*, **5**, pp. 357–361.

6. Chan, Y. C., Craig, G. S., Schrock, R. R., and Cohen, R. E. (1991). Synthesis of platinum and palladium nanoclusters within microphase-separated block copolymers, *Chem. Mater.*, **4**, pp. 885–894.
7. Saito, R., Okamura, S., and Ishizu, K. (1992). Introduction of colloidal silver into a poly(2-vinylpiridine) domain of microphase-separated poly(styrene-b-2-vinylpiridine) film, *Polymer*, **33**, pp. 1099–1101.
8. Saito, R., and Ishizu, K. (1995). Introduction of colloidal silver into a poly(2-vinylpiridine) domain of microphase-separated poly(styrene-b-2-vinylpiridine) film. 4: One-step method, *Polymer*, **36**, pp. 4119–4124.
9. Boontongkong, Y., and Cohen, R. E. (2002). Cavitated block copolymer micellar thin films: Lateral arrays of open nanoreactors, *Macromolecules*, **35**, pp. 3647–3652.
10. Hadziioannou, G., Patel, S., Granik, S., and Tirrel, M. (1986). Forces between surfaces of block copolymers adsorbed on mica, *J. Am. Chem. Soc.*, **108**, pp. 2869–2876.
11. Saleh, N., Phenrat, T., Sirk, K., Dufour, B., Ok, J., Sarbu, T., Matyjaszewski, K., Tilton, R. D., and Lowry, G. V. (2005). Adsorbed triblock copolymers deliver reactive iron nanoparticles to the oil/water interface, *Nano Lett.*, **5**, pp. 2489–2494.
12. Ladmiral, V., Morinaga, T., Ohno, K., Fukuda, T., and Tsujii, Y. (2009). Synthesis of monodisperse zinc sulfide nanoparticles grafted with concentrated polystyrene brush by surface-initiated nitroxide-mediated polymerization, *Eur. Polym. J.*, **45**, pp. 2788–2796.
13. Synytska, A., Ionov, L., Minko, S., Motornov, M., Elchorn, K., Stamm, M., and Grundle, K. (2004). Tuning wettability by controlled roughness and surface modification using core-shell particles, *Polym. Mater. Sci. Eng.*, **90**, pp. 624–625.
14. Wang, T. L., Yang, C. H., Shieh, Y. T., and Yeh, A. C. (2009). Synthesis of CdSe-poly(*N*-vinylcarbazol) composite by atom transfer radical polymerization for potential optoelectronic application *Macromol. Rapid Commun.*, **30**, pp. 1679–1683.
15. Mountrichas, G., Pispas, S., and Tagmatarchis, N. (2003). Grafting to approach for the functionalization of carbon nanotubes with polystyrene, *Mater. Sci. Eng. B*, **152**, pp. 40–43.
16. Liu, Y., Yao, Z., and Adronov, A. (2005). Functionalization of single-walled carbon nanotubes with well-defined polymers by radical coupling, *Macromolecules*, **38**, pp. 1172–1179.

17. Barandiaran, I., Cappelletti, A., Strumia, M., Eceiza, A., and Kortaberria, G. (2014). Generation of nanocomposites based on (PMMA-b-PCL)-grafted Fe_2O_3 nanoparticles and PS-b-PCL block copolymer, *Eur. Polym. J.*, **58**, pp. 226–232.
18. Park, S. J., Cho, M. S., Lim, S. T., Choi, H. J., and Jhon, M. S. (2003). Synthesis and dispersion characteristics of multi-walled carbon nanotube composites with poly(methyl methacrylate) prepared by in situ bulk polymerization, *Macromol. Rapid Commun.*, **24**, pp. 1070–1073.
19. Xu, G., Wu, W. T., Wang, Y., Pang, W., Zhu, Q., Wang, P., and You, Y. (2006). Constructing polymer brushes on multi-walled carbon nanotubes by in situ reversible addition fragmentation chain transfer polymerization, *Polymer*, **47**, pp. 5909–5918.
20. Xu, G., Wang, Y., Pang, W., Wu, W. T., Zhu, Q., and Wang, P. (2007). Fabrication of multi-walled carbon nanotubes with polymer shells through surface RAFT polymerization, *Polym. Int.*, **56**, pp. 847–852.
21. Yang, Y., Xie, X., Wu, J., and Mai, Y. W. (2006). Synthesis and self-assembly of polystyrene-grafted multi-walled carbon nanotubes with a hairy-rod nanostructure, *J. Polym. Sci. Part A: Polym. Chem.*, **44**, pp. 3869–3881.
22. Kim, M., Hong, C. K., Choe, S., and Shim, S. E. (2007). Synthesis of polystyrene brush on multi-walled carbon nanotubes treated with $KMnO_4$ in the presence of a phase-transfer catalyst, *J. Polym. Sci. Part A: Polym. Chem.*, **45**, pp. 4413–4420.
23. Du, F., Wu, K., Yang, Y., Liu, L., Gan, T., and Xie, X. (2008). Synthesis and electrochemical of water-soluble poly(sodium 4-styrenesulfonate-co-acrylic acid)-grafted multi-walled carbon nanotubes, *Nanotechnology*, **19**, 85716 (8 pp.).
24. Roghani-Mamaqani, H., Haddadi-Asl, V., Najafi, M., and Salami-Kalajahi, M. (2011). Preparation of tailor-made polystyrene nanocomposite with mixed clay-anchored and free chains via atom transfer radical polymerization, *AlChe. J.*, **57**, pp. 1873–1881.
25. Etxeberria, H., Zalakain, I., Mondragon, I., Eceiza, A., and Kortaberria, G. (2013). Generation of nanocomposites based on polystyrene-grafted CdSe nanoparticles by grafting through and block copolymer, *Colloid Polym. Sci.*, **291**, pp. 1881–1886.
26. Etxeberria, H., Zalakain, I., Tercjak, A., Eceiza, A., Kortaberria, G., and Mondragon, I. (2013). Functionalization of CdSe semiconductor nanoparticles with polystyrene brushes by radical polymerization, *J. Nanosci. Nanotechnol.*, **13**, pp. 643–648.

27. Henze, M., Madge, D., Prucker, O., and Ruhe, [illegible]. (2014). "Grafting through": Mechanistic aspects of radical polymerization reactions with surface-attached monomers, *Macromolecules*, **47**, pp. 2929–2937.

28. Mansky, P., Liu, Y., Huang, E., Russel, T. P., and Hawker, C. (1997). Controlling polymer-surface interactions with random copolymer brushes, *Science*, **275**, pp. 1458–1460.

29. Lyatskaya, Y., and Balazs, A. C. (1998). Modelling the phase behavior of clay-polymer composites, *Macromolecules*, **31**, pp. 6676–6680.

30. Edmonson, S., Osborne, V. L., and Huck, W. T. S. (2004). Polymer brushes via surface-initiated polymerization, *Chem. Soc. Rev.*, **33**, pp. 14–22.

31. Radhakrishnan, B., Ranjan, R., and Brittain, W. J. (2006). Surface inititated polymerizations from silica nanoparticles, *J. Soft Matter.*, **2**, pp. 386–396.

32. Rahimi-Razin, R., Haddadi-Asl, V., Salami-Kalajahi, M., Behboodi-Sadabad, F., and Roghani-Mamaqani, H. (2012). Matrix-grafted multiwalled carbon nanotubes/poly(methyl metharcylate) nanocomposites synthesized by in situ RAFT polymerization: A kinetic study, *Int. J. Chem. Kinet.*, **44**, pp. 555–569.

33. Jordan, R., and Ulman, A. (1999). Surface-initiated anionic polymerization of styrene by means of self-assembled monolayers, *J. Am. Chem. Soc.*, **121**, pp. 1016–1022.

34. Zhou, Q., Wang, S., Fan, X., and Advincula, R. (2002). Living anionic surface-initiated polymerization (LASIC) of a polymer on silica nanoparticles, *Langmuir*, **18**, pp. 3324–3331.

35. Zhao, B., Brittain, W. J., Zhou, W., and Cheng, S. Z. (2000). Nanopattern formation from tethered PS-b-PMMA brushes upon treatment with different solvents, *J. Am. Chem. Soc.*, **122**, pp. 2407–2408.

36. Jordan, R., and Ulman, A. (1998). Surface initiated living cationic polymerization of 2-oxazolines, *J. Am. Chem. Soc.*, **120**, pp. 243–247.

37. Joubert, M., Delaite, C., Bourgeat-Lami, E., and Dumas, P. (2005). Hairy PEO-silica nanoparticles through surface initiated polymerization of ethylene oxide, *Macromol. Rapid Commun.*, **26**, pp. 602–607.

38. Yang, Y., Wu, D., Li, C., Liu, L., Cheng, X., and Zhao, H. (2006). Poly(L-lactied) comb polymer brushes on the surface of clay layers, *Polymer*, **47**, pp. 7374–7381.

39. Li, D. X., He, Q., Cui, Y., and Li, J. B. (2007). Fabrication of pH-responsive nanocomposites of gold nanoparticles/poly(4-vinylpyridine), *Chem. Matter.*, **19**, pp. 412–417.

40. Garcia, I., Tercjak, A., Rueda, L., and Mondragon, I. (2008). Self-assembled nanomaterials using magnetic nanoparticles modified with polystyrene brushes and poly(styrene-b-butadiene-b-styrene), *Macromolecules*, **41**, pp. 9295–9298.
41. Garcia, I., Tercjak, A., Zafeiropoulos, N. E., Stamm, M., and Mondragon, I. (2007). Self-assembling nanomaterials using magnetic nanoparticles modified with polystyrene brushes, *Macromol. Rapid Commun.*, **28**, pp. 2361–2365.
42. Raula, J., Shan, J., Nuopponen, M., Niskanen, A., Jiang, H., Kauppinen, E. I., and Tenhu, H. (2003). Synthesis of gold nanoparticles grafted with thermoresponsive polymer by surface-induced reversible-addition-fragmentation chain transfer polymerization, *Langmuir*, **19**, pp. 3499–3504.
43. Lan, Q., Francis, L. F., and Bates, F. S. (2007). Silica nanoparticle dispersions in hompolymer versus block copolymer, *J. Polym. Sci. Part B: Polym. Phys.*, **45**, pp. 2284–2299.
44. Garcia, I., Tercjak, A., Gutierrez, J., Rueda, L., and Mondragon, I. (2008). Nanostructuration via solvent vapor exposure of poly(2-viny lpyridine-b-methyl methacrylate) nanocomposites using modified magnetic nanoparticles, *J. Phys. Chem. C*, **112**, pp. 14343–14347.
45. Xu, C., Ohno, K., Ladmiral, V., and Composto, R. J. (2008). Dispersion of polymer-grafted nanoparticles in homopolymers and block copolymers, *Polymer*, **49**, pp. 3568–3577.
46. Etxeberria, H., Zalakain, I., Fernandez, R., Kortaberria, G., and Mondragon, I. (2013). Controlled placement of polystyrene-grafted CdSe nanoparticles in self-assembled block copolymers, *Colloid Polym. Sci.*, **291**, pp. 633–640.
47. Douadi-Masrouki, S., Frka-Petesic, B., Save, M., Charleux, B., Cabuil, V., and Sandre, O. (2010). Incorporation of magnetic nanoparticles into lamellar poly(styrene)-b-poly(n-butyl methacrylate) diblock copolymer films: Influence of the chain end-groups on nano-structuration, *Polymer*, **51**, pp. 4673–4685.
48. Garcia, I., Tercjak, A., Zafeiropoulos, N. E., Stamm, M., and Mondragon, I. (2007). Generation of core-shell iron oxide magnetic nanoparticles with polystyrene brushes via atom transfer radical polymerization, *J. Polym. Sci. Part A: Polym. Chem.*, **45**, pp. 4744–4750.
49. Ohno, K., Morinaga, T., Koh, K., Tsujii, Y., and Fukuda, T. (2005). Synthesis of monodisperse silica particles coated with well-defined high-density polymer brushes by surface-initiated atom transfer radical polymerization, *Macromolecules*, **38**, pp. 2137–2142.

50. Etxeberria, H., Kortaberria, G., Zalakain, I., Larrañaga, A., and Mondragon, I. (2012). Effect of different aqueous synthesis parameters on the size of CdSe nanocrystals, *J. Mater. Sci.*, **47**, pp. 7167–7174.
51. Wang, Y., Teng, X., Wang, J. S., and Yang, H. (2003). Solvent-free atom transfer radical polymerization in the synthesis of Fe_2O_3@ polystyrene core-shell nanoparticles, *Nano Letters*, **3**, pp. 789–793.
52. Park, S. C., Kim, B. J., Hawker, C. J., Kramer, E. J., Bang, J., and Ha, J. S. (2007). Controlled ordering of block copolymer thin films by the addition of hydrophilic nanoparticles, *Macromolecules*, **40**, pp. 8119–8124.
53. Chiu, J. J., Kim, B. J., Bang, J., Kramer, E. J., and Pine, D. J. (2007). Distribution of nanoparticles in lamellar domains of block copolymers, *Macromolecules*, **40**, pp. 3361–3365.
54. Chiu, J. J., Kim, B. J., Bang, J., Kramer, E. J., and Pine, D. J. (2005). Control of nanoparticle location in block copolymers, *J. Am. Chem. Soc.*, **127**, pp. 5036–5037.
55. Kim, B. J., Bang, J., Hawker, C. J., and Kramer, E. J. (2006). Effect of areal chain density on the location of polymer-modified gold nanoparticles in a block copolymer template, *Macromolecules*, **39**, pp. 4108–4114.
56. Kim, B. J., Chiu, J. J., Yi, G. R., Pine, D. J., and Kramer, E. J. (2005). Nanoparticle-induced phase transitions in diblock copolymer films, *Adv. Mater.*, **17**, pp. 2618–2622.
57. Bockstaller, M. R., and Thomas, E. L. (2003). Optical properties of polymer-based photonic nanocomposite materials, *J. Phys. Chem. B*, **107**, pp. 10017–10024.
58. Bockstaller, M. R., Mickiewicz, R. A., and Thomas, E. L. (2005). Block copolymer nanocomposites: Perspectives for tailored functional materials, *Adv. Mater.*, **17**, pp. 1331–1349.
59. Oren, R., Liang, Z., Barnard, J. S., Warren, S. C., Wiesner, U., and Huck, W. T. (2009). Organization of nanoparticles in polymer brushes, *J. Am. Chem. Soc.*, **131**, pp. 1670–1671.
60. Brust, M., Walker, M., Bethell, D., Schiffrin, D. J., and Whyman, R. (1994). Synthesis of thiol-derivatised gold nanoparticles in two-phase liquid-liquid system, *J. Chem. Soc. Chem. Commun.*, **7**, pp. 801–802.
61. Yee, C. K., Jordan, R., Ulman, A., White, H., King, A., Rafailovich, M., and Sokolov, J. (1999). Novel one-phase synthesis of gold, palladium and iridium nanoparticles using superhydride, *Langmuir*, **15**, pp. 3486–3491.

62. Kunz, M., Shull, K. R., and Kellock, A. (1993). Colloidal gold dispersions in polymeric matrices, *J. Colloid Interface Sci.*, **156**, pp. 240–249.
63. Thompson, R. B., Ginzburg, V. V., Matsen, M. W., and Balazs, A. C. (2001). Predicting the mesophases of copolymer/nanoparticle composites, *Science*, **292**, pp. 2469–2472.
64. Thompson, R. B., Ginzburg, V. V., Matsen, M. W., and Balazs, A. C. (2002). Block copolymer-directed assembly of nanoparticles: Forming mesoscopically ordered hybrid materials, *Macromolecules*, **35**, pp. 1060–1071.
65. Bang, J., Kim, S. H., Drockenmuller, E., Misner, M. J., Russel, T. P., and Hawker, C. J. (2006). Defect-free nanoporous thin films from ABC triblock copolymers, *J. Am. Chem. Soc.*, **128**, pp. 7622–7629.
66. Kim, H. C., Jia, X., Stafford, C., Kim, D. H., McCarthy, T. J., Tuominen, M., Hawker, C. J., and Russel, T. P. (2001). A route to nanoscopic SiO_2 posts via block copolymer templates, *Adv. Mater.*, **13**, pp. 795–797.
67. Ploshnik, E., Langner, K. M., Halevi, A., Lulu, M. B., Muller, A. H., Fraaije, J. G., Agur-Sevink, G. J., and Shenhar, R. (2013). Hierarchical structuring in block copolymer nanocomposites through two-phase separation processes operating in different time scales, *Adv. Funct. Mater.*, **23**, pp. 4215–4226.
68. Grabar, K. C., Freeman, R. G., Hommer, M. B., and Natan, M. J. (1995). Preparation and characterization of Au colloid monolayers, *Anal. Chem.*, **67**, pp. 735–743.
69. Bockstaller, M. R., and Thomas, E. L. (2004). Proximity effects in self-organized binary particle-block copolymer blends, *Phys. Rev. Lett.*, **93**, 166106 (4 pp.).
70. Kim, S., Yoo, M., Kang, N., Moon, B., Kim, B. J., Choi, S. H., Kim, J. U., and Bang, J. (2013). Nanoporous bicontinuous structures via addition of thermally-stable amphiphilic nanoparticles within block copolymer templates, *ACS Appl. Mater. Interfaces*, **5**, pp. 5659–5666.
71. Yoo, M., Kim, S., Lim, J., Kramer, E. J., Hawker, C. J., Kim, B. J., and Bang, J. (2010). Facile synthesis of thermally-stable core-shell gold nanoparticles via photo-cross-linkable polymer ligands, *Macromolecules*, **43**, pp. 3570–3575.
72. Ledo-Suarez, A., Hoppe, C. E., Lazzari, M., Lopez Quintela, M. A., and Zucchi, I. A. (2013). Thermal annealing as an easy tool for the controlled arrangement of gold nanoparticles in block copolymer thin films, *Nanotechnology*, **24**, 255304 (8 pp.).

73. Zhang, X., and Matyjaszewski, K. (1999). Synthesis of well-defined amphiphilic block copolymers with 2-(dimethylam no) ethyl methacrylate by controlled radical polymerization, *Macromolecules*, **32**, pp. 1763–1766.

74. Etxeberria, H., Tercjak, A., Mondragon, I., Eceiza, A., and Kortaberria, G. (2014). Electrostatic force microscopy measurements of CdSe-PS nanoparticles and CdSe-PS/poly(styrene-b butadiene-b-styrene) composites, *Colloid Polym. Sci.*, **292**, pp. 229–234.

Chapter 3

Nanocomposites Based on Block Copolymers and Carbon Nanotubes

Galder Kortaberria

Materials + Technologies Group (GMT),
Department of Chemical and Environmental Engineering,
Polytechnic School, University of the Basque Country (UPV/EHU),
Plaza Europa 1, 20018 Donostia-San Sebastián, Spain

galder.cortaberria@ehu.eus

This chapter reviews the synthesis and characterization of block copolymer/carbon nanotube (BCP/CNT) nanocomposites with different potential applications. On the one hand, block copolymers can enhance the dispersability and stability of CNTs in a wide range of host polymer matrices. Moreover, as block copolymers can form a variety of ordered structures on nanometer length scale, CNTs can be organically functionalized to be confined in a specific microphase of a block copolymer. On the other hand, due to their enhanced mechanical and electrical properties, CNTs have also been used for preparing composites with block copolymers in order to obtain materials for different applications such as sensors or photoactuators. Finally, and due to their ability to self-assemble, block copolymers have also been used

Block Copolymer Nanocomposites
Edited by Galder Kortaberria and Agnieszka Tercjak

ISBN 978-981-4669-54-2 (Hardcover), 978-981-4669-55-9 (eBook)
www.panstanford.com

for preparing nanopatterned metallic nanoparticle arrays as catalysts for posterior CNT growth.

3.1 Introduction

Carbon nanotubes are allotropes of carbon with a cylindrical nanostructure. Nanotubes have been constructed with length-to-diameter ratio of up to 132,000,000:1, significantly larger than for any other material. Their name is derived from their long, hollow structure with the walls formed by one-atom-thick sheets of carbon, called graphene. These sheets are rolled at specific and discrete *chiral* angles, and the combination of the rolling angle and radius decides the nanotube properties; for example, whether the individual nanotube shell is a metal or semiconductor. Nanotubes are categorized as single-walled nanotubes (SWNTs), double-walled nanotubes (DWCNTs) and multi-walled nanotubes (MWNTs). Individual nanotubes naturally align themselves into ropes held together by van der Waals forces, more specifically, π-stacking. Iijima discovered MWCNTs in 1991, finding that they were formed by concentric cylinders of bonded carbon atoms with a diameter between 10 and 100 nm [1]. Three years later, Iijima reported the existence of SWCNTs [2]. Over recent years, CNTs have become the focus of considerable research owing to their exciting potential applications in nanocomposites, molecular devices or advanced materials with their extraordinary mechanical strength, thermal conductivity and electronic and optical properties [3–5]. However, significant effort is required in order to resolve the problem related with CNT dispersion in solvents and polymer matrices due to the poor solubility and processability [5]. Using block copolymers to wrap CNTs is a versatile and effective way for their functionalization. Compared with homopolymers, block copolymers may provide a series of attractive noncovalent wrapping and decoration methods for CNT functionalization [6]. Block copolymers can enhance the dispersability and stability of CNTs in a wide range of organic solvents and the host polymer matrices, constituting one of the major uses of block copolymers when combined with CNT-forming nanocomposites. Moreover, block copolymers can form a variety of ordered structures on the nanometer length scale, which is of particular interest to develop nanostructured materials with different properties than those provided by homopolymers. For this reason, CNTs can be organically functionalized to be confined in a specific microphase of a block copolymer.

As present, CNTs with enhanced mechanical and electrical properties have also been used for improving those properties of both homopolymers or block copolymers by preparing composites with them for many different applications such as sensors [7, 8], supercapacitors [9] or photoactuators [10]

Due to their good electric properties, CNTs have also been used to prepare polymer/CNT composites, to improve electrical behavior of matrix. However, insulating polymer/CNT composites are not the best combination for electronics because insulating polymers have a detrimental effect on the electrical properties of CNTs by acting as an interfacial resistor. Conjugated polymers (CPs) such as polypyrroles [11] or polyanilines [12] have shown efficient decoration of CNTs thereby preserving their electronic properties. However, the intrinsically rod-structured CPs are not highly soluble in most of the common (aqueous and organic) solvents. Thus the rod-coil block copolymers (RCBCs), which consist of a conjugated rod block and a soluble non-conjugated coil block, can be regarded as efficient materials to combine with CNTs for applications in electronic or electrochemical devices [13]. The π-conjugated structure in rod blocks is intimate with the CNT sidewalls, and the non-conducting coil blocks tune the solution processability. This has been another important use of block copolymer combined with CNTs.

Due to their ability to self-assemble, block copolymers also have been used for preparing nanopatterned metallic nanoparticle arrays as catalysts for posterior CNT growth [14–23]. The subnanometer-scale tunability of the catalyst size achieved by block copolymers nanostructure control enabled excellent control of wall number and diameter. The general procedure is: nanotemplate assembly, catalyst deposition and polymer removal followed by CNT growth. As they cannot be considered block copolymer/CNT nanocomposites, those works are not included in the present work, but it is hard to note the use of block copolymers for this purpose.

In the present chapter many works developed by several authors on preparation and characterization of block copolymer/CNT nanocomposites with the above-mentioned purposes, for many potential applications such as sensors, membranes, electronic devices, electrodes, etc., are presented. Firstly, nanocomposites based on thermoplastic elastomer block copolymers and CNTs are presented, followed by those based on conjugated copolymer/CNT for electronic applications. Finally, nanocomposites prepared by other block copolymers different from thermoplastic elastomers or conjugated ones with CNTs are presented.

3.2 Thermoplastic Elastomer Block Copolymer-Based Nanocomposites

Thermoplastic elastomers (TPEs) are an important class of polymers that combine the mechanical properties of rubbers with the processability and recyclability of thermoplastics [24], presenting high deformation capability and high electrical and thermal resistances [25]. These properties can be significantly modified by the addition of conductive fillers such as CNTs, with unique electrical and mechanical properties [26].

3.2.1 Poly(Styrene-b-Butadiene-b-Styrene)-Based Nanocomposites

Poly(styrene-b-butadiene-b-styrene) (SBS) triblock copolymer is a TPE known for its microphase separation into hard and soft domains, including spherical, cylindrical, gyroid and lamellar morphologies [27]. Tailoring the molecular architecture and the morphology of SBS offers the possibility to tune mechanical properties of materials from thermoplastic elastomers to tough thermoplastics [27]. This requires a detailed understanding of the interrelations between the molecular architecture, the microphase separated morphology, thermal and mechanical history and the resulting mechanical behavior [27]. The addition of CNTs into an SBS insulator matrix allows tuning electrical and mechanical properties to suit specific applications as pressure or deformation sensors [7–9, 28, 29] among others. Several authors prepared composites based on SBS and CNTs by solvent casting [7, 9, 30] or melt mixing [9, 31, 32], analyzing mechanical and electrical properties together with obtained morphologies and dispersion level of nanofiller. Moreover, the effect of block ratio and copolymer architecture on properties has also been studied, together with that of CNT type, amount and surface functionalization [7–9, 30–32].

L. Lu et al. [31] were one of the first authors to prepare SBS/CNT nanocomposites. They melt mixed commercial SBS (30 wt% of PS) with MWCNTs. Field emission scanning electron microscopy (FESEM) analysis showed that an uniform dispersion of nanotubes was achieved. Interactions among copolymer and MWCNTs were corroborated by the displacement of the α-band of UV spectra. Samples were almost transparent with improved tensile strength values (up to 66% for 7 wt% of MWCNTs). Transmission electron microscopy (TEM) and X-ray photoelectron microscope images showed that some of the nanotubes presented a coating of SBS on the surface, probably due to free radicals generated during melt mixing. On the contrary, samples prepared by solution processing with THF showed poor dispersion of nanofillers with big aggregates without any improvement of mechanical properties.

Same authors [32] studied thermal degradation of those composites by Time-resolved TGA-FTIR analysis, analyzing degradation kinetics. They found that activation energy values calculated from kinetic parameters obtained increased with MWCNT content. In the same way, samples prepared by melt mixing showed higher activation energy values than solution processed, thus indicating stronger interaction among MWCNTs and SBS for the former. Degradation of SBS occurred by a random chain scission process.

In that way, I. Pedroni et al. [9, 33] analyzed the effect of preparation method on dispersion and polymer-filler interactions in SBS/MWCNT nanocomposites for sensor or supercapacitor applications. They prepared nanocomposites by both melt mixing with a mini-extruder and by solvent casting. Nanocomposites prepared by casting method presented higher conductivity values with very low percolation threshold (0.25 wt%). Samples prepared by extrusion with 6 wt% of MWCNTs showed the same conductivity than those with 1 wt% MWCNTs prepared by casting (10^{-4}–10^{-3} S cm^{-1}). On the other side, melt-mixed samples presented better mechanical properties, Young modulus increasing with nanofiller content especially at high strain levels. Scanning electron microscopy (SEM) images showed a better filler-matrix adhesion for extruded samples. In any case, materials with higher tenacity than neat SBS were obtained.

P. Costa et al. [8] studied mechanical, electrical and electromechanical properties of SBS/MWCNT composites for being used as strain sensors. First they used pristine MWCNTs for preparing thin films by solvent casting with SBS copolymers of different block ratios (20/80, 30/70, 40/60) and architectures (radial or linear). Independently of the matrix, composites showed some degree of agglomeration of nanotubes, but with a relatively good cluster distribution. They found that both PS content and architecture affected conductivity: increased with PS content and was higher for radial architecture. Percolation threshold increased with PS content, being lower for linear architecture. They concluded that hopping between nearest fillers was the main conduction mechanism. Finally, they evaluated the electromechanical response sensibility in the following way: electrical resistance of the composites was measured during the mechanical deformation on samples with similar dimensions than those for mechanical testing. Experimental setup can be seen in Fig. 3.1. The electromechanical response sensibility was quantified by the gauge factor (GF), defined as the slope between relative charge in electrical resistance with the applied relative mechanical deformation [34]. They measured gauge values between 2 and 18 for the composites, showing its possible application as large strain sensors.

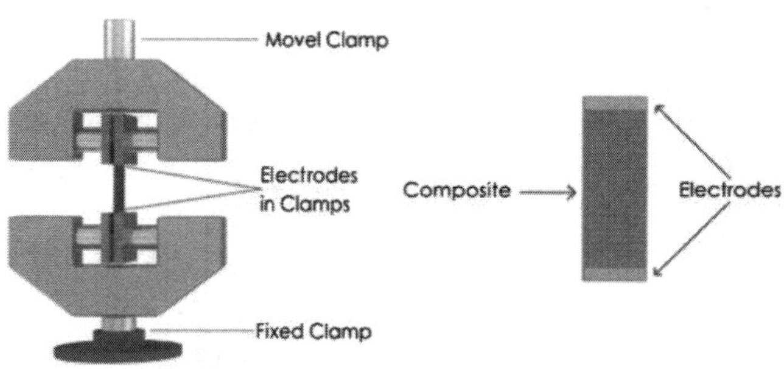

Figure 3.1 Schematic representation of the experimental configuration of the clamps for the stress-strain experiments with simultaneous electrical measurements for electromechanical response evaluation of composites. Reproduced with permission of [7]. Copyright 2014, Elsevier.

Following this way and in order to improve composite properties for potential large strain sensor applications, Costa et al. [7] prepared composites with SBS copolymers with different block ratios and architectures and functionalized nanotubes.

They used both SWCNTs and MWCNTs and modified them covalently with styrene groups (by reaction with a diazonium compound showed in Fig. 3.2) or non-covalently with $C_6H_5CH=CH_2$ for styrene and $CH_2=CH-CH=CH_2$ for butadiene. By SEM they found that modified CNTs were better dispersed than pristine ones but prevented conductivity increase for increasing filler content up to 8 wt%, therefore hindering their use as electromechanical sensors.

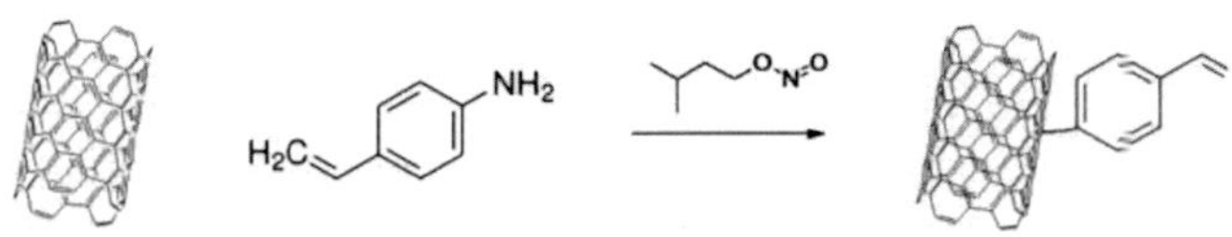

Figure 3.2 Functionalization of CNTs with styrene by diazonium salt reaction. Reproduced with permission of [7]. Copyright 2014, Elsevier.

Different pristine CNTs gave similar electrical properties, with percolation threshold near 1 wt%. Near percolation threshold they observed higher electromechanical sensibility with gauge factors up to 8, suitable for sensor applications. For composites with both pristine CNTs they found the typical stress-strain curves of TPE with maximum deformation larger than 400% and yield strain of around 3%. Electromechanical properties of composites had similar behavior before and after yielding. So they conclude highlighting the potential application of SBS/CNT composites (with different S/B ratios and architectures and pristine CNTs) as electromechanical sensors.

Following with the potential application of SBS-based composites as strain sensors, Ji et al. [35] prepared fibers of composites with SBS/thermoplastic urethane (TPU) and MWCNTs with selective localization of the nanofiller and analyzed resistivity-strain behavior of those conductive composites. They used a SBS copolymer with 40 wt% of PS, polyester-based thermoplastic polyurethane and commercial pristine MWCNTs. Composites were prepared by melt mixing via applying three different procedures: blending of premixed SBS/MWCNT or TPU/MWCNT composites with the respective unfilled polymer in a second step, direct blending of SBS/MWCNT and TPU/MWCNT in a second step, as it can be seen in Fig. 3.3. MWCNTs were selectively located depending on the procedure. With the first procedure MWCNTs were located at SBS, with the second one at TPU and with the third one MWCNTs were located in both. The fabrication of SBS/TPU/MWCNT fibers with selective nanofiller localization created different conducting network morphologies (as determined by SEM), with different interactions among elastomers and MWCNTs that led to different resistivity-strain sensitivities. By SEM and Raman spectroscopy they found that MWCNTs were oriented when they were in SBS domains. In order to better understand MWCNT dispersion and interfacial interaction between SBS/MWCNT or TPU/MWCNT, they measured wetting coefficient and adhesion work by contact angle. Work of adhesion values indicated the presence of stronger interfacial interaction between MWCNTs and SBS, probably due to the presence of double bonds and aromatic rings that enhanced interactions. The resistivity-strain sensing behavior was examined by measuring the electrical resistance as a function of uniaxial strain. Figure 3.4 shows the sketch of conductive network and phase continuity change during stretching for composites fibers with different MWCNT localization.

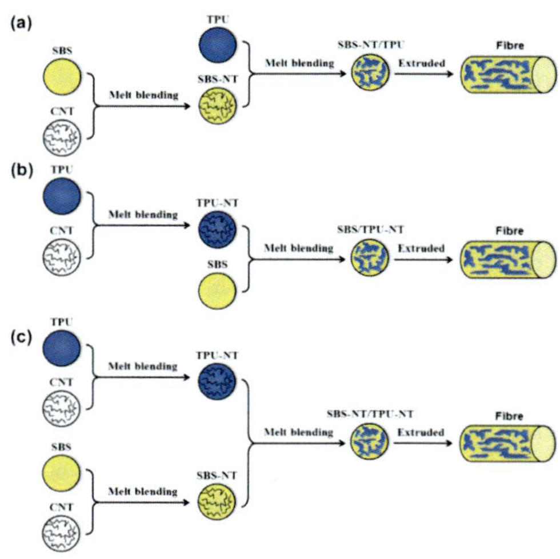

Figure 3.3 Sketch of the processing procedures: (a) SBS-NT/TPU, (b) SBS/TPU-NT and (c) SBS-NT/TPU-NT composite fibers. Reproduced with permission of [35]. Copyright 2014, Elsevier.

For SBS-MWCNT/TPU samples, nanofillers were mainly located at SBS due to interfacial interactions that realized efficient load transfer when stretched (Fig. 3.4a). As a result, conductive networks were easily deformed with increasing strain (Fig. 3.4c). For SBS/TPU-MWCNT samples (Fig. 3.4b), nanofillers were located at TPU. Interaction between MWCNTs and TPU was weaker which induced less efficient stress transfer, smaller MWCNT orientation and lower sensitivity (Fig. 3.4d). Enhancing interfacial interaction, higher amount of conductive pathways were formed. They concluded underlying that strain sensitivity can be tuned with different MWCNT localization, presenting a new and simple route for fabrication of conductive polymer composites with tunable resistivity-strain sensitivity.

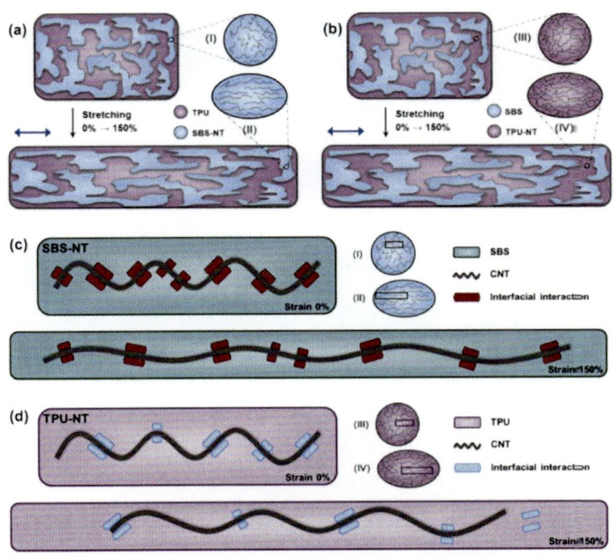

Figure 3.4 Sketch of conductive network and phase continuity change during stretching for composites fibers with different MWCNT localization. Reproduced with permission of [35]. Copyright 2014, Elsevier.

Another important aspect of nanocomposites based on block copolymers (self-assembly capability) is the possibility to confine nanofillers in specific domains. The spatial confinement of CNTs in polymer matrices provides the potential for fabricating functional hybrid materials. Self-assembly of block copolymers is an elegant and inexpensive bottom-up method for this purpose. The challenge for the selective self-assembly of CNTs in a block copolymer template is huge. To achieve quasi one-dimensional alignment of CNTs, an asymmetric (cylinder forming) block copolymer is necessary, besides a modification of the nanofiller for increasing the compatibility with one of the blocks of the matrix. In this way, Liu et al. [30] used a simple *grafting to* method to wrap DWCNTs with PS chains that were self-assembled in the cylindrical PS phase of an asymmetric SBS copolymer with 30 wt% of PS, as illustrated in Fig. 3.5. They fabricated a designed CNT array in a flexible polymer matrix, with potential applications in micro-channel reactors or integrated electronics. Samples were prepared by solvent casting using DMF and films were annealed for 120 h at 160°C in order to obtain nanoordered morphologies. Covalent modification of DWCNTs with polystyrene-*g*-(glycidyl methacrylate-*co*-styrene) (PS-*g*-(GMA-*co*-St)) was carried out by using a self-prepared method [36].

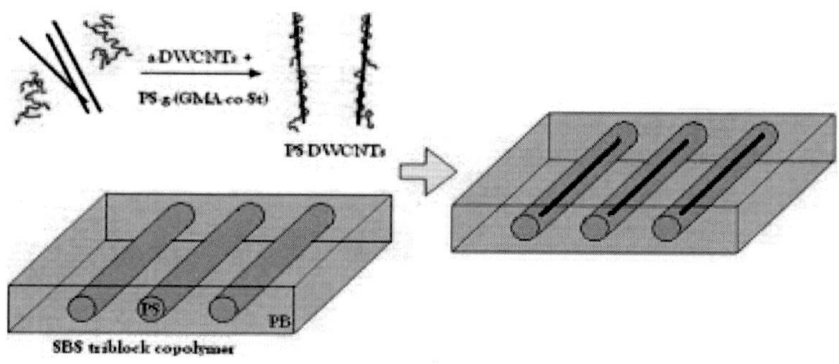

Figure 3.5 Scheme for selective self-assembly of PS-functionalized DWCNTs in the cylindrical PS phase of a microphase-separated SBS copolymer. Reproduced with permission of [30]. Copyright 2009, Elsevier.

They obtained a microphase-separated structure for neat SBS, with PS cylinders lying in the PB matrix, as revealed by TEM. For composites with PS-DWCNTs smaller than 500 nm, the majority of them were located in the PS phase, regardless of the M_w of the ligand, even when it was up to 330000, much higher than that of the PS block of SBS. This is different from the case of chemically assembling nanoparticles or nanorods with a block copolymer, in which the ligand functionalizing them should be shorter than the corresponding block. The reason for that, as was probed by high-resolution TEM, was that the chain of the ligand wrapped the DWCNTs instead of stretching out. The diameter of the PS-DWCNTs was ca. 5 nm, much smaller than the width of PS cylinders, as illustrated in Fig. 3.6. For composites with longer PS-DWCNTs, nanofillers were not confined in the PS phase. Probably, that confinement would demand a huge translational entropy penalty, which could not be compensated by the enthalpy gain facilitated by the modification ligand.

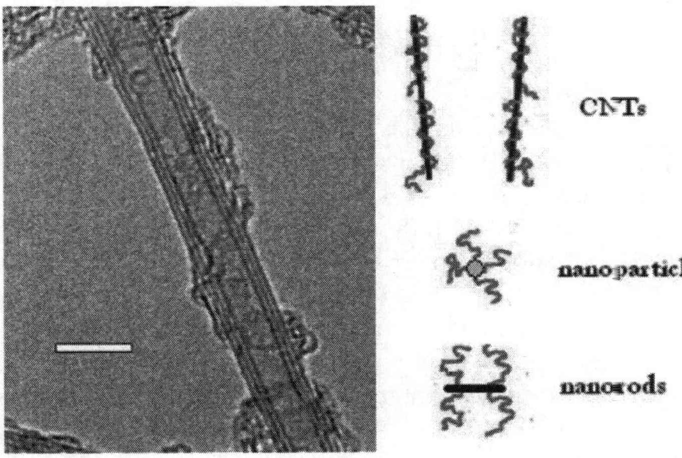

Figure 3.6 High-resolution TEM image of an individual DWCNTs wrapped by a PS-g-(GMA-co-St) chain (left) and scheme for different chain morphologies of the ligands functionalizing CNTs, nanoparticles and nanorods (right). Scale bar of 5 nm. Reproduced with permission of [30]. Copyright 2009, Elsevier.

Another attempt to achieve the confinement of MWCNTs in block copolymer domains was carried out by G. Wu et al. [37]. They prepared composites with hydroxylated SBS (HO-SBS) and acyl chloride activated MWCNTs (MWCNT-COCl) by solution processing method. By gel fraction and degree of swelling measurements they found that composites presented cross-linked structure and high solvent resistance. By SEM and TEM they found that MWCNTs were well dispersed with only some small aggregates. Dispersion was better than that of non-modified MWCNTs, but in any case they were not confined. Moreover, they disturbed microphase-separated structure of the block copolymer, decreasing order-disorder transition temperature. The average diameter of MWCNT-COCl in the nanocomposite was found to be five times larger than that of pristine ones due to the presence of polymer layers wrapped on the surface by covalent HO-COCl bond. After measuring mechanical properties they concluded that the dispersion MWCNT-COCl is an useful method to regulate mechanical properties of HO-SBS.

Due to the biostability problem of double bonds in the elastomeric block, SBS does not have any use in biomedical applications. However, after the entire saturation of butadiene segments in SBS, the obtained poly(styrene-b-(ethylene-co-butylene) (SEBS) shows a better oxidative stability [27], allowing biomedical applications such as medical gloves, catheters or transfusion bags [27]. Some authors have prepared composites with SEBS and MWCNTs.

J. G. Meier et al. [38] prepared composites with SEBS (30 wt% PS) and commercial MWCNTs by melt mixing and solution/precipitation method in order to study the effect of processing on composite conductivity. They found that mixing conditions had effect on dispersion state, effective polymer wrapping and also on the intertube distance at the tube-junctions of block copolymer/MWCNT, in which one block had stronger affinity towards adsorption onto MWCNTs. The extent of the gap determined if conduction mechanism is by electron tunneling or hopping, affecting overall conductivity and also mechanical properties. They pointed out that solution method resulted in thicker polymer layers in MWCNTs than for melt mixing below order/disorder transition of copolymer (300°C). In this conditions block copolymer was phase separated with cylindrical hexagonal closed packed structure of the PS phase. In solution, cylindrical mesophase was dissolved and the amphiphilic property respect to solvent or MWCNTs became effective. Block copolymer acted as surfactant, at the surface of nanofillers. That increased the tunnel over the gaps with an overall lower conductivity. Solution method switched off the competition between adsorption onto surface and formation of a mesophase of the copolymer. By TEM they did not observed any preferential positioning of nanofillers in one of the phases.

J. Huang et al. [39] presented the shear-flow induced hierarchical self-assembly of MWCNTs into well-ordered parallel stripes in a polymer matrix, to fabricate anisotropic conductive materials. Under the induction of shear flow, the dispersed SEBS phase and MWCNTs spontaneously assembled into well-ordered parallel stripes in a poly(propylene) (PP) thin film. PP/SEBS/MWCNT composites were prepared by melt compounding, first mixing PP/MWCNT and adding SEBS in a second step. For preparing anisotropic conductive thin films, a shear induced self-assembly of randomly distributed MWCNTs into parallel stripes was performed. Sample was held between two quartz electrodes. By changing the gap between the electrodes, electrical properties can be controlled: increasing the gap, the width and length of stripes increased. The morphological evolution under the shear flow was followed by optical microscopy. An example for a gap value of 250 μm can be seen in Fig. 3.7.

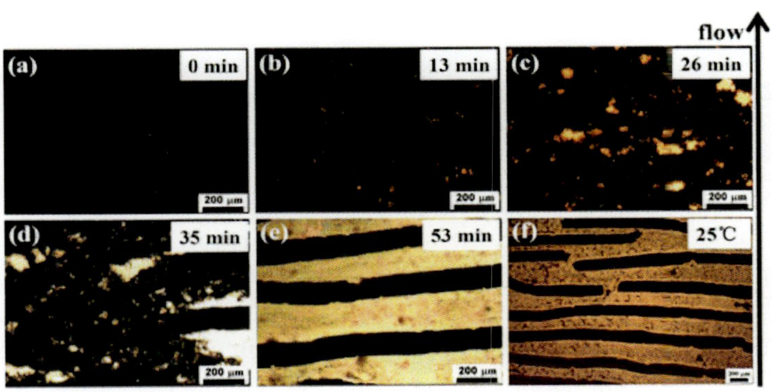

Figure 3.7 Optical micrographs of the PP/SEBS/1.5 wt% MWCNT composite at a shear rate of 0.05 s^{-1} with different shearing times and the sample after cessation of the flow and cooling to the room temperature: (a) 0 min; (b) 13 min; (c) 26 min; (d) 35 min; (e) 53 min; (f) the film at 25°C. The gap is fixed at 250 μm. Reproduced with permission of [39]. Copyright 2014, American Chemical Society.

At the initial state, before shear-induced assembly, the image was completely black because MWCNTs were randomly distributed in the sample (a). Under the shear flow, MWCNTs and SEBS formed randomly distributed loose aggregates at first (b-d) then assembled into well-ordered parallel MWCNTs stripes aligned vertical to the shear direction (e). The formation of the stripe was attributed to the cooperation of several factors: first, small SEBS droplets tended to coalesce into bigger ones. During coalescence, SEBS acted as a "glue" to adhere MWCNTs together to form aggregates. Second, the growth of clusters or stripes perpendicular to the flow direction was attributed to the existence of negative normal stress (N1). When N1 > 0, the dispersed phase was compressed in the flow direction. Moreover, the confinement effect imposed by the upper and lower disks also limited the growth in the gradient direction. As a result, MWCNT stripe could only grow and align along the vorticity direction. They also measured electrical resistance by Keithley analyzer concluding that in the direction parallel to stripes it was 6 orders of magnitude lower than that in the perpendicular one. This is by far the most striking conductive anisotropy achieved for plastic conductive materials, with electrical properties tuned by the shear cell gap.

3.2.2 Poly(Styrene-b-Isoprene) and Poly(Styrene-b-Isoprene-b-Styrene)-Based Nanocomposites

Several authors have tried to selectively place MWCNTs in self-assembled poly(styrene-b-isoprene) (SI) or poly(styrene-b-isoprene-b-styrene) (SIS) block copolymers. In that way, I. Park et al. [40] prepared composites based on ultra-high molecular weight SI synthesized by living anionic polymerization and PS-grafted (by emulsion polymerization) commercial MWCNTs. As functionalization with PS was carried out by emulsion polymerization, which generally rendered PS layers with thicknesses of tens of nanometers on MWCNTs, to accommodate large-diameter PS-MWCNTs in the PS phase, specifically synthesized ultrahigh molecular weight SI was used. Samples were prepared by casting from cumene, obtaining lamellar structure for the neat copolymer. Once PS-MWCNTs were added lamellar structure was maintained, with the alignment of PS-MWCNTs without agglomerations or macrophase separation, as it can be seen in Fig. 3.8. They pointed out that their study would promote developments in anisotropic nanosized functional materials and devices using nano-objects such as nanowires, nanorods, etc.

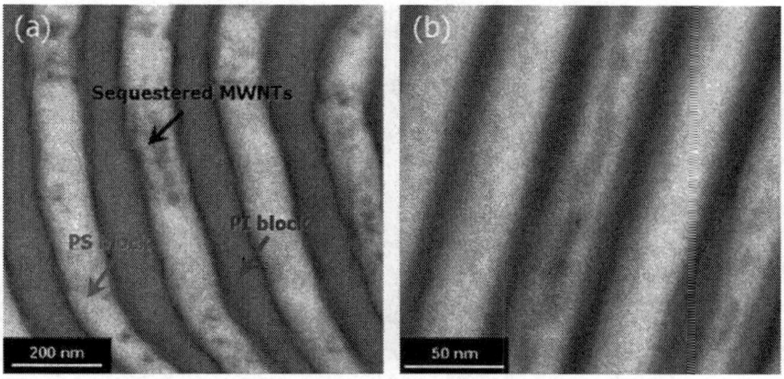

Figure 3.8 TEM images of CNT-sequestered SI block copolymer. Reproduced with permission of [40]. Copyright 2007, Elsevier.

L. Peponi et al. [41, 42] confined octadecylamine-functionalized single-walled CNTs (SWCNT-ODA) in a SIS (30 wt% PS) matrix with the use of dodecanethiol (DT) as surfactant. By FE-SEM they showed the cylindrical self-assembled nanostructure of the PS phase with the most of cylinders placed parallel to the free surface. When SIS/SWNCT-ODA composites were prepared without DT, this nanostructure was lost. If composites were prepared with DT, it switched the self-assembled structure from ordered cylinders to lamellar morphology. SWCNT-ODA were sequestered in the PS domains due to the high affinity among PS and SWCNT-ODA, as shown in Fig. 3.9.

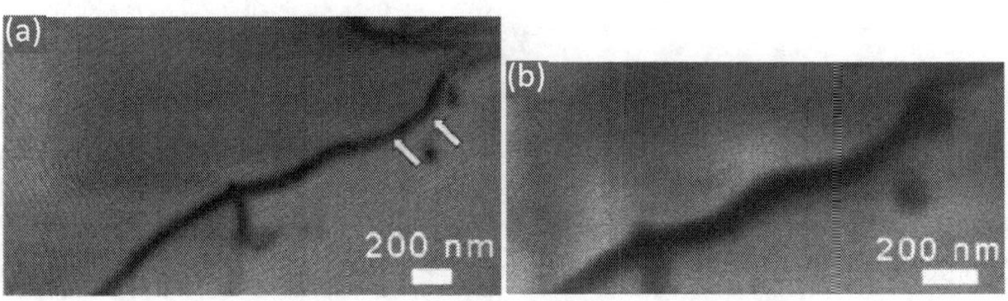

Figure 3.9 FESEM images recorded for SWCNT-ODA/SIS film after exposure to the electron beam for (a) 10 s and (b) 20 s. Reproduced with permission of [41]. Copyright 2009, Elsevier.

They also used electrostatic force microscopy (EFM) for analyzing nanocomposite surfaces because this technique brings important insights on mapping the position of conductive nanofillers inside non-conducting matrices. They probed the sequestering of SWCNT-ODA by EFM. They also made the phase diagram of SIS, pointing out that it shifted to higher concentrations of PS-rich phase. This was associated to a morphological shift in the phase diagram leading to the change from cylindrical to lamellar structure.

H. Garate et al. [43] also worked on the dispersion of MWCNTs in nanostructured SIS copolymer, by modifying MWCNTs with PS. They used SIS copolymer with 30 wt% of PS, commercial MWCNTs modified with PS chains by ATRP (scheme shown in Fig. 3.10) and dodecanethiol (DT) as surfactant. They obtained MWCNTs with grafted PS chains with an M_w of 9000, as measured after cleavage protocol.

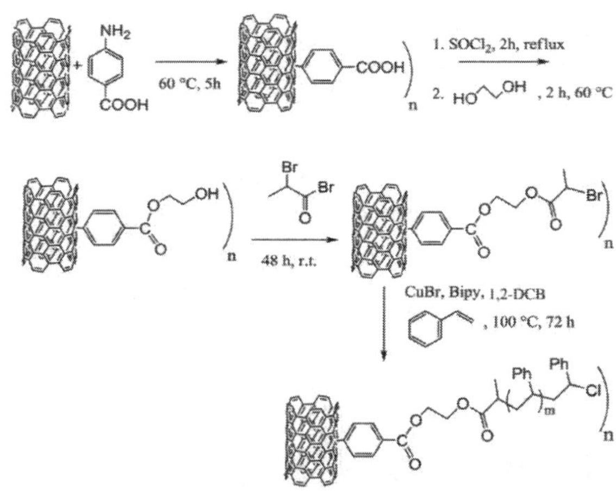

Figure 3.10 Synthetic scheme of MWCNT-PS obtention. Reproduced with permission of [43]. Copyright 2011, Elsevier.

They analyzed obtained morphologies by AFM. Neat SIS formed cylindrical morphology, with PS cylinders located parallel to the substrate. For SIS/MWCNT-PS composites without DT, MWCNTs trended to aggregate possibly due to the interaction among PS of different MWCNTs, showing that only the modification of MWCNTs was not enough. In the same way, SIS/MWCNT composites (pristine nanofillers) with DT also showed aggregates. Finally, for SIS/MWCNT-PS composites with DT, nanofillers were isolated in the matrix without affecting morphology, perpendicular to the microdomains direction extending to both phases and expanding PS cylinders due probably to an interlock of grafted PS chains and PS chains of SIS. This interlock was favored by π-stacking between phenyl groups of PS chains of MWCNT-PS and SIS, anchoring MWCNTs to the matrix. They pointed out that a good dispersion was only obtained with the synergistic effect among DT and MWCNT-PS.

M. Ilcikova et al. [10] also analyzed SIS/MWCNT-PS composites focusing on their viscoelastic and photoactuation behavior. They used a SIS copolymer with 30 wt% of PS with MWCNTs functionalized with PS by ATRP. Regarding viscoelastic properties, they found that composites with pristine MWCNTs showed deterioration of elastic properties, probably because they disturbed physical cross-linking of the matrix. For composites with MWCNT-PS, due to improved interactions with the matrix, elastic properties of SIS were retained. They also analyzed the photoactuation behavior of composites by AFM. Composite foil was thermo-mechanically shaped to obtain blister-shaped samples. This procedure resulted in an orientation of polymer chains and nanotubes on the side walls of the blister. The pre-stretching of polymer chains enabled the material to exhibit the photoactuation. To follow the photoactuation, an AFM cantilever was placed on the top of the blister sample. The blister changed height in response to LED illumination and thus moved the cantilever vertically. They measured the change in cantilever position with time for different laser power intensities. By comparing behavior of neat SIS and composites, they concluded underlying that composites presented photoactuation property due to the MWCNTs, responding quickly and reversibly to irradiation. Those composites can be considered as promising photoactuator materials.

3.2.3 Poly(Styrene-b-Butadiene-b-Methylmethacrylate)-Based Nanocomposites

Morphologies and structures of poly(styrene-b-butadiene-b-methylmethacrylate) (SBM) copolymers are various and have already been studied and established [44, 45]. T. Perie et al. [46] studied the dispersion of MWCNTs in selective solvent using SBM copolymer and used stable SBM/MWCNT suspensions to obtain masterbatches with 25 wt% of nanofiller by solvent casting to be melt blended with PVDF to obtain composites structured at the mesoscale. Thanks to its PMMA block, the use of SBM masterbatches allowed to obtain stable microstructured composites from melt mixing in a very large range of matrices, as PMMA is compatible with a wide range of relevant commercial polymers such as PVDF, polycarbonate or epoxy. They used three SBM copolymers: SBM1 (16/20/64), SBM2 (27/30/43) and SBM3 (39/11/50). After analyzing the effect of different solvents on SBM/MWCNT suspension stability, they chose acetone as the most adequate for preparing masterbatches by solvent casting, used then for preparing composites with PVDF. From SBM/MWCNT suspension in acetone, they prepared films by dip coating and analyzed morphologies by TEM, as it can be seen in Fig. 3.11.

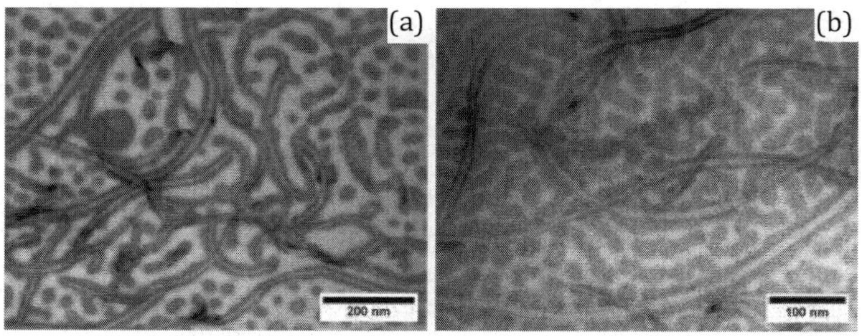

Figure 3.11 Thin films obtained by evaporation of SBM/MWCNT dispersions on TEM grids (a) SBM1/MWCNT; (b) SBM2/MWCNT. Reproduced with permission of [46]. Copyright 2011, Elsevier.

They found a continuous wrapping of MWCNTs by PS-PB blocks independently of SBM type. They found different morphologies close to nanotubes and far from their surface. Close to nanotubes, there was a repartition of PS-PB blocks onto MWCNT surface with the formation of micelles: core of PS-PB as solvophobic part and PMMA at the outer one. Far from the surface, they found characteristic morphologies for each SBM. For SBM1 they showed a stacking of micelles well separated by PMMA due to its length compared with that of PS-PB block. For SBM2, with shorter PMMA block, micelles merged allowing formation of more entangled PS-PB domains. As SBM1 presented the highest ability to disperse MWCNTs was used to prepare the masterbatch and composites with PVDF. Due to compatibility among PMMA block and PVDF, they obtained a structured composite with a mesoscopic dispersion of SBM core–shell particles and SBM copolymer coating MWCNTs. Figure 3.12 shows the micrograph and mesoscopic ordering scheme.

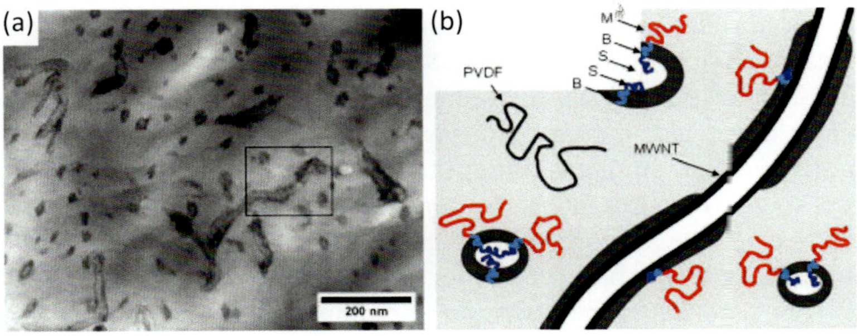

Figure 3.12 (a) Zoom in the SBM1/MWCNT/PVDF nanocomposite micrograph. An SBM coated nanotube is marked by a rectangle. (b) Scheme of mesoscopic ordering. Reproduced with permission of [46]. Copyright 2011, Elsevier.

After tensile tests, they concluded that composites exhibited improved mechanical properties with noticeable toughening compared to the direct melt-blending of MWCNTs in PVDF. The matrix nanostructuring and the increased adhesion between SBM coated MWCNTs could be the responsible for this effect. They pointed out that composites with unique strength and toughness combinations were obtained.

K. Saint-Aubin et al. [47] also prepared SBM/MWCNT composites and analyzed mechanical and electrical properties. They used commercial SBM copolymer (32/36/32), both MWCNTs and DWCNTs as nanofillers and dichloromethane (DCM), and acetone/cyclohexane (A/C) as solvents for casting. Films were annealed at different temperatures for different times. They analyzed the effect of obtained morphologies on tensile properties and electrical resistivity. First they analyzed obtained morphologies by TEM and SAXS. For neat SBM casted with DCM, lamellar microstructure was obtained while in A/C a poorly ordered hexagonal-like one was achieved. As evaporation rate of acetone is higher, system enriched in cyclohexane, which is a bad solvent for PS and PMMA and good for PB. This promoted segregation of PMMA and swelling of PB to take place, explaining the morphology with PMMA domains surrounded by PB matrix as PMMA blocks arranged in cylindrical domains with hexagonal packing. This packing could not revert towards a lamellar equilibrium structure when acetone and cyclohexane were fully evaporated. PB and PS blocks were maintained at the outer part of the micelles with frozen cores of PMMA: quenched metastable hexagonal structure. For SBM/CNT composites casted from DCM, lamellar structure was maintained (similar to that of pure SBM) while those casted from A/C gave a less ordered structure with large lamellar domains (different from that of neat SBM). Both types of CNTs promoted the formation of equilibrium lamellar microstructure. Before evaporation, PB block were anchored to CNTs. A/C was bad solvent for PB and good for PS and PMMA. While evaporating, there was an enrichment of C, solvent mixture becoming bad for PMMA and better for PB. PMMA segregation was avoided by adsorption of PB at CNT interface. For this reason morphology changed when compared with that of neat SBM. They also characterized mechanical and electrical properties. Morphology change from cylindrical to lamellar promoted by CNTs improved mechanical properties better than when CNTs were used as reinforcement agents. Morphological change also changed electrical properties: even for relatively high CNT fraction composites remained insulating. It differed from the behavior of an isotropic medium in which conductivity is controlled by filler fraction. They concluded underlying that mechanical and electrical properties can be changed by morphological change promoted by solvent and CNTs.

3.2.4 Thermoplastic Poly(Ether-Ester) Elastomer-Based Nanocomposites

Commercially important thermoplastic poly(ether-ester) elastomers (TPE-E) possess a blocky chain structure consisting of a high melting polyester block and low glass transition polyether block [48, 49]. Generally they microphase separate into high melting crystalline "hard" domains and relatively low glass transition temperature "soft" domains, on cooling from the melt or precipitation from solution. The degree to which these copolymers phase separate significantly affects many of the physical properties of these materials. Several authors studied composites based on this kind of copolymers with CNTs [49, 50–52].

S. Glenis et al. [50] studied magnetic and electronic properties of nanocomposites based on acid-treated MWCNTs dispersed in elastomeric poly(trimethylene terephtalate-b-tetramethylene oxide) (PTT-b-PTMO) segmented block copolymer with 40 wt% of PTT as rigid segment. Magnetic and electronic properties were measured in terms of electron spin resonance (ESR) and dc magnetization measurements. Nanocomposites were prepared by in situ polycondensation reaction. They found a rather uniform dispersion of MWCNTs for 0.1 and 0.3 wt%, with no large-scale agglomerates and the presence of individual and not highly entangled nanofillers. They found a drastic variation in the nanotubes magnetic response inside the polymer matrix. The diamagnetic susceptibility of MWCNT fillers and its temperature dependence was markedly reduced, suggesting the presence of effective hole-doping and substantial shift in the Fermi level, attributed to the oxygen functional groups on the nanofiller surface. The spin susceptibility of the composite was enhanced mainly due to the formation of a high density of paramagnetic defects, exceeding largely that of pristine MWCNTs.

Same authors [49] prepared and characterized structure and properties of PTT-b-PTMO nanocomposites with acid-functionalized SWCNTs by in situ polymerization, preparing samples for characterization by injection molding. First they presented the morphology of a thin film of the copolymer solvent-casted, as it can be seen in Fig. 3.13.

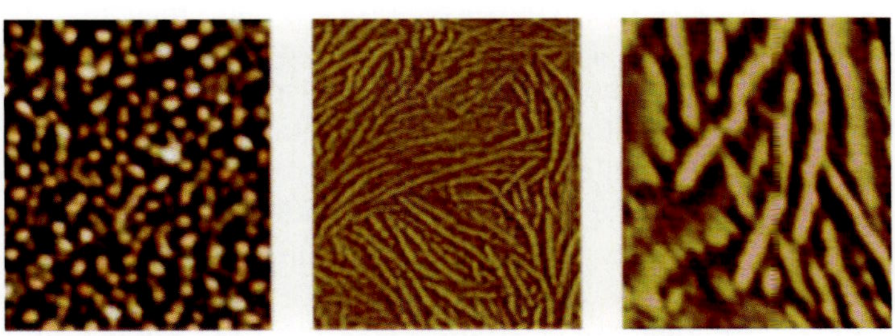

Figure 3.13 AFM phase images of (a) PBT-b-PTMO (microtome cross section) and (b) PTT-b-PTMO (casted from solution). Scan boxes are 200 × 200 nm (a and c) and 500 × 500 nm (b). Reproduced with permission of [49]. Copyright 2012, Wiley.

Morphology of PTT-b-PTMO consisted of semicrystalline PTT domains (ribbon-like crystallites) dispersed in the soft matrix of amorphous PTMO. They pointed out that nanometric structure is responsible for the mechanical properties. The addition of CNTs can change macroscopic properties of semicrystalline TPE-E, when a synergy between the crystalline lamellae of polymer and CNTs is created. They explained that this was not enough for mechanical properties improvement, as a good dispersion of the nanofiller is needed.

Previously [53] they studied the effect of SWCNTs and MWCNTs on the crystallization process of PTT, finding that SWCNTs increased crystallization rate while MWCNTs were not effective nucleation agents. For this reason they used SWCNTs for preparing composites with PTT-b-PTMO by injection molding. Fracture surface of specimens was analyzed by SEM, showing that SWCNTs were randomly dispersed with some entanglements or bundles. A strong adhesion among SWCNTs and matrix was found, due to reaction of –COOH group of SWCNTs and the ester. They found that both ends of nanotubes were tightly embedded in the matrix, resulting in a strong interfacial adhesion, with the enhancement of the tensile strength. They also concluded that thermo-oxidative stability of copolymer increased with SWCNTs. Nanofillers acted as nucleating agents, increasing crystallinity degree of PTT block.

G. Broza et al. [51] also prepared nanocomposites with thermoplastic poly(ether-ester) elastomer and MWCNTs. In this case they use poly(butylene terephtalate-b-tetramethylene oxide) (PBT-b-PTMO) copolymer (45/55) for disperse acid-functionalized MWCNTs. They used two procedures for preparing the composites. In the first one, they dispersed nanofillers in dimethyl terephatalatec (DMT) and butanediol (BD) before transesterification, while in the second one MWCNTs were dispersed in poly(tetramethylene glycol) (PTMG) after transesterification reaction. By SEM they showed a strong interfacial adhesion for both cases, being the dispersion better for the former. For the first procedure, they found a total coverage of the tubes by PBT/PTMO, while higher amount of aggregation was found for the second one.

Same authors [52] analyzed mechanical and tribological properties of composites based on PBT-b-PTMO and both SWCNTs and MWCNTs, functionalized with –COOH groups. For both types of nanofillers, tensile modulus and strain at break as function of CNT concentration showed a maximum. Elongation at break was enhanced by both CNTs, showing a plasticizing effect. Brittleness diagrams showed minima. They also determined volumetric wear after abrasion. They also found minima on the volumetric wear versus CNT concentration diagrams, similar to those on brittleness ones. They concluded underlying that mechanical properties could be tuned by CNT amount.

3.2.5 Olefin Block Copolymer-Based Nanocomposites

Olefin block copolymers (OBC) are presented as potential substitutes of thermoplastic elastomers like SBS or others previously analyzed in this chapter. They were developed by Dow Chemical using a chain shuttling technology which employed two catalysts, one forming hard or crystalline ethylene blocks with low octene concentration and the other forming amorphous blocks with high incorporation of octene in the ethylene chain. As a potential substitute of thermoplastic elastomers, they still have shortcomings like a relatively low tensile strength (around 5 MPa, while for SBS is around 28 MPa). For this purpose they are combined with nanofillers like MWCNTs or graphene. Since mechanical properties of OBC strongly depend on crystallization of hard blocks, Wu et al. [53] analyzed the effect of MWCNTs on the crystallization kinetics and melting behavior of OBC. They prepared nanocomposites by mixed solvent solution method (solutions of OBC in xylene and MWCNTs in dimethyl formamide were mixed, solvent removed and samples molded into 1 mm films). By TEM they showed a good dispersion of nanofillers, with several single MWCNTs dispersed. Crystallization kinetics and obtained structures were analyzed by DSC and synchrotron X-ray characterization, together with polarized optical microscopy. They found that MWCNTs acted as effective nucleating agents, with no remarkable change in crystal form. Size of spherulites decreased when compared to neat BC. Two melting peaks were observed for nanocomposites, related to crystals with different size and perfection, not observed for BC. Hard block chains were aligned mostly along the tube axis.

3.3 Conjugated Block Copolymer-Based Nanocomposites

Insulating polymer/CNT composites are not the best combination for electronics because insulating polymers have a detrimental effect on the electrical properties of CNTs by acting as an interfacial resistor [54]. Conjugated polymers such as polypyrroles [11], polyanilines [12], poly(3-alkylthiophene) [55], poly(phenylene vinylenes) [56] and poly(arylene ethylene) [57] have shown efficient decoration of CNTs thereby preserving their electronic properties. The strong van der Waals interactions between conjugated π-bonds in both CPs and CNTs enhance the miscibility between the respective components [11, 12, 55–57]. However, the intrinsically rod-structured CPs are not highly soluble in most of the common (aqueous and organic) solvents. From this standpoint, the RCBCs, which consist of a conjugated rod block and a soluble non-conjugated coil block, can be regarded as efficient materials to combine with CNTs for applications in electronic or electrochemical devices [58]. The π-conjugated structure in rod blocks is intimate with the CNT sidewalls, and the nonconducting coil blocks tune the solution processability [59]. A stable anchoring of a solvophilic block can be achieved via strong binding of the conjugated block onto CNT surface which sterically shields a large portion of CNT surface from attractive forces to other nanotubes.

J. Zou et al. [58] prepared superhydrophobic and conductive nanocomposite coatings based on poly(3-hexylthiophene-b-styrene) (P3HT-b-PS) conjugated block copolymer and MWCNTs. Films were prepared by solvent casting with chloroform onto glass substrate. They measured a water contact angle of around 160°. To elucidate the nature of this superhydrophobicity, they examined the morphology of the film by TEM, finding honeycomb like morphology. Different morphology and compositions were found in the ridge and the base of the honeycomb structure. The ridge was mainly formed by P3HT-b-PS polymers decorated with a small amount of MWCNTs, while the main components in the base were entangled tubes with nanoscale porosity. This fact was confirmed by fluorescent microscopy. For comparison they prepared P3HT/MWCNT films (water contact angle of 132°), observing by SEM and AFM mainly entangled nanotubes, similar to the honeycomb base morphology. This suggested that the honeycomb base with nanoscale porosity made a major contribution to the superhydrophobicity. The water contact angle of P3HT-b-PS film (that also gave honeycomb morphology) was measured as 105°, meaning that MWCNTs were crucial for the superhydrophobicity. The effect of MWCNT amount on morphology and wettability was also investigated. For lower block copolymer mass ratio (0.5:1) casted film showed a cracked honeycomb structure. By increasing copolymer mass ratio (1:1 and 1.5:1) the honeycomb structure became more continuous, pores becoming smaller and the ridge thicker and rougher. Increasing mass ratio of copolymer versus MWCNTs, contact angle increased from 155° to 157° and 158°, respectively. This confirmed that microscale roughness arising from the honeycomb morphology played an important role in the superhydrophobicity of the film. They also probed that this superhydrophobicity was stable over a long period of time at ambient conditions. Films also exhibited high electrical conductivity and excellent properties for gas sensing. They concluded that all this, together with the easy fabrication and the large surface area provided by porous morphology, make P3HT-b-PS/MWCNT conducting superhydrophobic coatings to be promising materials for many potential applications.

J. Sung et al. [60] prepared ultrathin composite films consisting of mixtures of metallic and semiconducting SWCNTs with a conjugated poly(styrene-b-para-phenylene) (PS-b-PPP) copolymer. To investigate the electrical characteristics of 2D composite films, SWCNTs were dispersed in THF solutions of PS-b-PPP under sonication. To increase the portion of the individually dispersed nanotubes, sonicated solutions were further centrifugated at different rotor speeds and the upper portions were decanted. They found that increasing centrifugation speed the amount of nanotube bundles decreased. They pointed out that the number of nanotubes can be controlled by centrifugation. The bare sonicated solutions and the decanted portions were spin-coated on various substrates as glass, PET or Si with native oxide and samples annealed at 190° to obtain uniform networked composite films. These films were integrated with a source-drain electrode into the SWCNT-networked field-effect transistors (FETs), as shown in Fig. 3.13. 2D networks of SWCNTs can act as a semiconducting channel in FETs used in large area organic electronics [61]. They found that electrical conductivity of films depended on tube density. In that way, films were modulated from metallic to semiconducting to insulating characteristics. They were able to prepare transparent SWCNT-networked TFT with mechanical flexibility by a facile and cost effective processing, with a great freedom of choice of the gate dielectric layers and source/drain electrodes for transistors containing the films. Concluding, they presented an efficient route for fabricating a semiconducting active channel polymer composite layer with a high on/off ratio and reasonable field effect mobility by a solution process easily applicable.

S. K. Hwang et al. [62] also prepared nanocomposite films of PS-b-PPP conjugated block copolymer and SWCNTs for printable non-volatile polymer memories. Films were prepared by simple solution blending. They found that PPP was strongly adhered to the tubes by π-π interactions due to the honeycomb structure present on CNT surface. PS film offered easy film formation with enough flexibility. Fabricated devices showed stable data retention at high temperatures over 100°C without performance degradation due to strong interactions among PPP and SWCNTs.

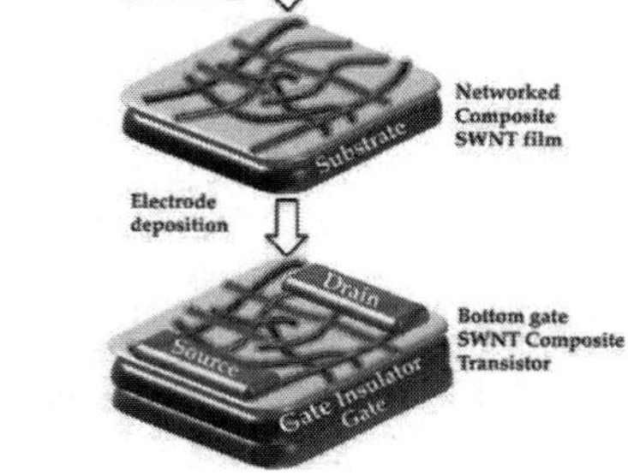

Figure 3.13 Schematics of the procedures to disperse SWCNTs in PS-b-PPP to form composite films with networked SWCNTs by spin-coating, and to fabricate TFTs. Reproduced with permission of [60]. Copyright 2010, Wiley.

H. S. Park et al. [54] also used a conjugated block copolymer for preparing flexible transparent conductive films with SWCNTs. The copolymer they used was PEDOT-b-PEO. Hybrids were prepared by solution mixing by sonication followed by centrifugation. From the supernatant they prepared the films using a membrane and then transferred to a PET substrate. In this way, hybrids were prepared by supramolecular assembly, that is, wrapping SWCNTs using functional macromolecules. Film morphology was characterized by AFM, SEM and TEM. SWCNTs were more debundled or exfoliated than pristine due to the assembly of the copolymer that coated them. Owing to this efficient coating and exfoliation of SWCNTs by the copolymer due to electronic π-π interactions among PEDOT and SWCNTs, hybrids created 3D interpenetrated networks used for percolated electron transport and to provide the mechanical robustness and integrity for the flexible transparent conductive film. By UV-vis and Raman they probed that charge transfer was by means of electronic interactions. Moreover, nanohybrids presented good solubility in organic solvents and for some compositions in water. They concluded that their work offered promising possibilities for applications in flexible electronics, biological sensors and optical devices.

3.4 Nanocomposites Based on Other Block Copolymers

Several works can be found in the literature about nanocomposites based on other block copolymers not included in previous sections and CNTs, for being used at different applications such as membranes, coatings, bio-applications, etc. [63–72].

F. Soriano-Corral et al. [63] prepared nanostructured films of poly(methyl methacrylate-b-butyl acrylate) (PMMA-b-PBA) with MWCNTs for coating applications. They synthesized different copolymers by reverse iodine transfer polymerization (RITP) with 7, 20 and 30 wt% of PBA. Films were prepared by casting copolymer/CNT mixture solutions in chloroform into glass substrates. They first analyzed the effect of PAB content on film evaporation rate, finding that for the copolymer with 30 wt% of PAB the absolute drying rate decreased in a 50%. They obtained nanostructured films in which MWCNT dispersion decreased with PAB content without affecting electrical properties, due to the induction of bridging effect and the nanofillers preference to locate into PMMA phase. By measuring the resistance of films they found that all nanostructured films presented a percolation threshold of 0.8 wt%. Dispersion of MWCNTs in nanocomposites was analyzed by SEM. They found quite good and homogeneous MWCNT dispersion but the tendency to agglomerate increased with PBA content, probably due to a poorer interaction among copolymer and nanofillers. They concluded that MWCNTs had a preference to locate at PMMA phase and/or in the interface, which caused a good distribution but poor dispersion in the copolymers with higher PBA content. It resulted in the formation of tridimensional network of MWCNTs by the bridging effect which favored the drastic decrease in the resistance at 8 wt% of MWCNTs.

J. Wang et al. [64] prepared poly(styrene-b-ethylene oxide) (PS-b-PEO) thin films as templates for MWCNT dispersion. MWCNTs were functionalized by grafting PEO into their surface. TEM images probed the formation of core–shell structures, with MWCNTs coated by PEO layer. They proposed a method for the selective self-assembly of PEO-g-MWCNTs in PS-b-PEO. In the method they developed, PEO-g-MWCNTs were prepared to form a PEO-g-MWCNT/water solution that was applied as complex substrate for the PS-b-PEO template. PEO-g-MWCNTs were absorbed by PS-b-PEO and the nanohybrid film gradually formed under toluene vapor. This method facilitated the selective assembly of nanofillers onto the PEO domains. Nanohybrids with 5 wt% of PEO-b-MWCNTs were prepared. The same method was applied for preparing PS-b-PEO films without MWCNTs with water surface as substrate under toluene vapors. Pure PS-b-PEO formed large lamellar structure on pure water substrate. For nanohybrid films after 2 h of exposure to toluene vapors, nanofillers were uniformly dispersed in the film without aggregation but PS-b-PEO did not form any microstructure. After 24 h of exposure, segregated cylindrical PS structures were found and most interestingly, MWCNT-g-PEO were found to be mainly located in the PEO microphase, due to their increased compatibility after functionalization: solubility parameters of PEO in the copolymer and PEO grafted onto MWCNTs were very close. After 48 h of exposure, more PEO-g-MWCNTs accumulated in the PEO domains.

Their developed method allowed the selective localization of the majority of nanofillers. They presented a versatile approach for the controlled assembly of anisotropic-nanostructured films with desirable patterns on soft substrate.

Z. Chen et al. [65] prepared poly(trimethylene terephtalate-b-ethylene oxide terephtalate)/MWCNT composites by in situ polymerization. They modified MWCNTs with poly(ethylene glycol) (PGE-g-MWCNTs) by *grafting to* method. By TEM they found a homogeneous dispersion of PEG-g-MWCNTs, even as single tubes. For this reason, low percolation threshold values were found (0.5 wt% by rheology and 1 wt% by conductivity). They analyzed crystallization behavior by DSC and optical microscopy, finding that nanofiller had effect on crystallization and melting, more pronounced for poly(ethylene oxide terephtalate) block. PEG-g-MWCNTs acted as nucleation agents, enhancing crystallization. Analyzing the effect of nanofiller amount, they found that for 0.5 wt%, a local network structure was formed. For 1 wt%, the network structure was found throughout the matrix (acid-modified MWCNTs gave agglomerates, better compatibility with PEG-modification). For 2 wt% of PEG-g-MWCNTs and higher amounts, agglomerates appeared. They concluded underlying the good dispersion of nanofillers due to modification, especially at low amounts, that made possible very low percolation threshold values.

R. Wang et al. [66] used cross-linkable poly(ethylene glycol-b-acrylic acid-b-styrene) (PEG-b-PAA-b-PS) amphiphilic copolymer to non-covalently capture pristine SWCNTs. Hydrophobicity of SWCNTs requires chemical functionalization with water-solubilizing moieties in order to produce stable aqueous suspensions for bioapplications. Covalent functionalization with ionic moieties confers aqueous stability but offers limited stability in vivo due to charge screening, which results in flocculation [73]. Moreover, covalent functionalization perturbs the electronic structure of the nanotube. For this reason, Wang et al. carried out the self-assembly of uncharged amphiphilic biocompatible polymers for non-covalently encapsulate SWCNTs, preserving their electric properties and obtaining stable aqueous suspensions. By non-covalently encapsulating SWCNTs within a shell-cross-linked PEGylated micelle (named as PEG-egg) they preserved electronic character, being resistant to protein displacement and retaining biostabilizing properties of PEG. Encapsulation procedure is illustrated in Fig. 3.14.

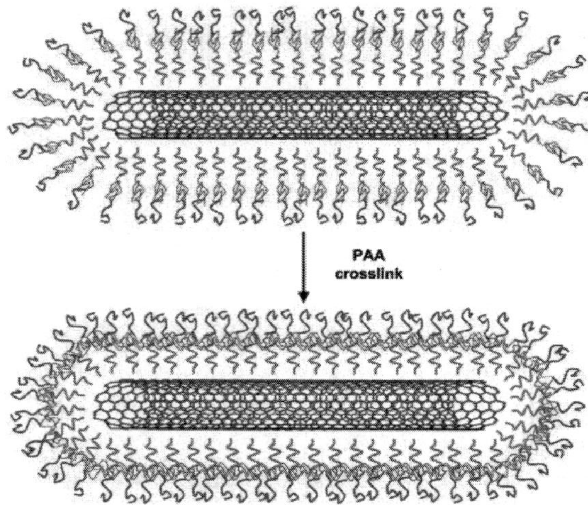

Figure 3.14 Schematic illustration of the encapsulation of a SWCNT inside a PEG-egg. Blue represents PEG segment, pink is PAA and gray is PS. Reproduced with permission of [66]. Copyright 2007, Elsevier.

R. Suckeveriene et al. [67] prepared transparent poly(styrene-b-acrylamide)(PS-b-AM)/MWCNT films by dynamic interfacial emulsion polymerization for coating applications. They used an emulsifier-free ultrasonically assisted copolymerization of acrylamide and styrene in the presence of MWCNTs, resulting in uniform dispersions stable for months. By high-resolution scanning electron microscopy (HRSEM) they observed two phases for composites with 0.275 and 0.55 wt% of MWCNTs: a porous one with MWCNTs and a smooth, uniform one. PS exhibited drop-like structure similar to porous particle phase. Since MWCNT surface is hydrophobic, MWCNTs were dispersed within the PAM fraction. That is, PAM fraction was attached to MWCNT surface. Films obtained from stable suspensions were transparent and with potential coating applications.

M. Kumar et al. [68] prepared low fouling negatively charged ultrafiltration membranes for protein separation based on poly(aryelene-b-ether sulfone) (PA-b-PES) and sulfonated (PA-b-PES) copolymer and acid-functionalized MWCNTs by nonsolvent induced phase separation (NIPS) method. They obtained membranes of around 150 μm thickness. By TEM they observed that membranes presented an anisotropic structure: a dense top skin layer, a porous sublayer and fully developed finger-like macrovoids at the bottom. The water permeabilities and increased protein fouling resistances of the membranes were dependent on the fraction of MWCNTs. The protein transmission as function of pH value and fraction of MWCNTs was studied for two model proteins (bovine serum albumin and myoglobin) and found to be controlled by size exclusion and the content of MWCNTs. The highest transmission of proteins was obtained for the membrane containing 2 wt% of MWCNTs relative to total membrane polymer. The selectivity of the hybrid membranes for the separation of the binary protein mixture could be systematically increased by increasing surface charge density by increasing fraction of MWCNTs. By changing MWCNT concentration, also porosity and hydrophobicity were adjusted. They pointed out that those membranes also could be suitable for the removal of organic matter from wastewater.

N. Yu et al. [69] prepared composites based on poly(ethylene-b-ethylene oxide) (PE-b-PEO) block copolymer and SWCNTs or MWCNTs by inducing crystallization of the copolymer on the nanotubes with the aid of supercritical CO_2. By TEM they demonstrated that the double-crystalline copolymer could be periodically decorated along the tubes, obtaining novel amphiphilic nanohybrid structure. They analyzed the effect of solvent finding that their selectivity to the block copolymer segments played a crucial role on the morphology of PE-b-PEO assembling on CNTs. With 1,2-dichlorobenzene or xylene (selective for PE), PE-b-PEO formed periodic patterns, where the CNT-induced PE crystallization was critical for the formation of the regular nanostructure. With *N,N*-dimethylacetamide, more selective for PEO, they observed only thin polymer coatings on CNTs, without any periodic pattern. They also found that the decorating degree of PE-b-PEO on the surface increased with the increase of supercritical CO_2 pressure. They showed a controllable method based on the use of supercritical CO_2 to fabricate functional CNT-based nanocomposites containing block copolymers with different micromorphologies depending on organic solvents.

Finally, nanocomposites made from CNTs and block copolymers containing a vinylpyridine block have also been synthesized by several authors. In this way, H. Shin et al. [70] prepared composites based on SWCNTs and poly(styrene-b-4-vinylpyridine) (PS-b-P4VP) block copolymer in order to create block copolymer micelles adhered to SWCNT surface. Block copolymer micelles can be used as dispersant of SWCNTs in different solvents. They used PS-b-P4VP copolymers with different molecular weights and block ratios. Composites were prepared by solution mixing with toluene as solvent. Droplets of suspension were deposited onto carbon-coated grids for preparing films and analyze their morphology by TEM. Neat copolymer spin coated film presented a morphology consisting on core/corona micelles closely packed into a pseudo-hexagonal structure. Morphologies obtained for SWCNT and (PS-b-P4VP)/SWCNT films casted from toluene can be seen in Fig. 3.15. Films obtained for SWCNT/toluene solutions showed aggregated nanotube ropes caused by the large van der Waals forces. For (PS-b-P4VP)/SWCNT films, micelles adhering to the surface of SWCNTs were found. Micelles were located in between two nanotubes (arrows in Fig. 3.15d), implying a possible de-bundling of SWCNTs by micelles. The addition of block copolymer enhanced suspension stability with the micelles overcoming the van der Waals force through steric hindrance. PS blocks (corona) wrapped themselves onto SWCNT surface.

They also modified the surface of nanotubes by nanoparticles synthesized in the P4VP core of the micelles. In addition, micelles self-assembled upon heat treatment into periodic nanostructures where a few of nanotubes were embedded in the preferable microdomains [55, 56]. They added silver acetate (AgAc) to copolymer/SWCNT solutions in toluene. AgAc was selectively coordinated to the P4VP block and was reduced obtaining Ag nanoparticles by reduction with electron beams of TEM. In that way they modified the surface of SWCNTs by nanoparticles in the P4VP core. Films were annealed at 220°C for 12 h and micelles were transformed into cylindrical P4VP nanodomains in a PS matrix with Ag nanoparticles selectively located on the P4VP block. They confined nanotubes in the cylindrical P4VP microdomains containing Ag nanoparticles.

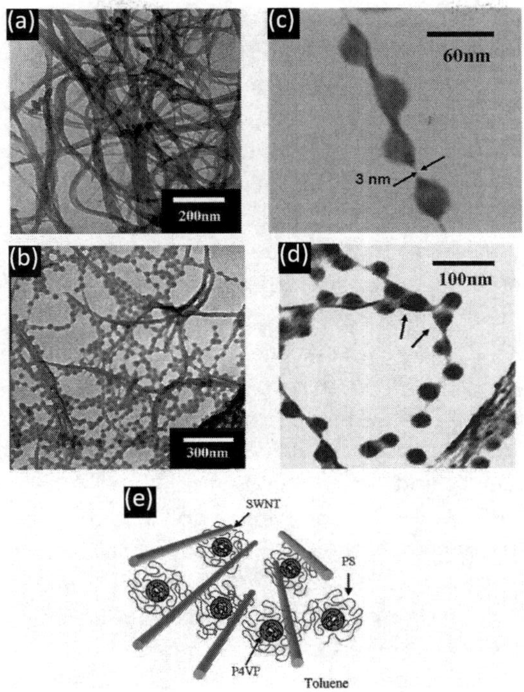

Figure 3.15 TEM images of (a) raw SWCNTs, (b), (c), and (d) SWCNTs dispersed and stabilized with block copolymer. (e) Schematic model of the nanostructure of SWCNTs and block copolymer with corona PS blocks selectively absorbed on the surface of tubes. Reproduced with permission of [70]. Copyright 2005, American Chemical Society.

Same authors [71] also used PS-b-P4VP block copolymer to prepare transparent, low-electric resistance nanocomposites with SWCNTs and Au nanoparticles for transparent electrode applications. By spin coating mixed SWCNT/toluene and (PS-b-P4VP)/toluene solutions they obtained films with block copolymer micelles presenting P4VP core and PS corona attached on SWCNT surface, overcoming van der Waals forces between them. They doped solutions with $HAuCl_4 \cdot 3H_2O$ salt (without deterioration of their stability) that went selectively to P4VP domains. They used different salt amounts to tune film conductance from 0.01 to 1 S/m. By electron beam salt was reduced to Au nanoparticles in the core regions of micelles. They analyzed morphologies finding that SWCNTs were evenly distributed with the micelles, with Au nanoparticles in their core region. This selectively doping of polar P4VP domains adhering to nanotube surface led to nanocomposites with resistance values more than two orders or magnitude lower, due to the increased number of mobile charge carriers in the nanotubes facilitated by charge transfer doping. Moreover, obtained nanocomposites were capable for nanopatterning over a large area: they successfully used them as source and drain electrodes in an organic thin film transistor (OTFT) with pentacene channel layer presenting 85% transmittance and 6000 Ω/m resistance.

C. Li et al. [72] prepared synthesized and analyzed self-assembly behavior of poly(fluorenylstyrene-b-2-vinylpyridine) block copolymers and their blends with SWCNTs. They varied the block composition, molecular weight and length of fluorene side chain in order to control morphologies. Nanocomposite films were prepared by casting from copolymer/CNT solutions in different solvents. They first analyzed the effect of fluorene units: they increased the phase segregation parameter χN and the Kuhn length of PS block. When compared with PS-b-P2VP copolymer, the phase behavior of these copolymers was understood in terms of the conformational asymmetry of block copolymers arising from the grafting of fluorene units onto the backbone of PS block. The phase diagram obtained for these block copolymer showed spherical structures, hexagonally packed cylinders, lamellar, gyroidal and hexagonally perforated lamellae, depending on their molecular conformation. They used COOH-modified SWCNTs to prepare nanocomposites with block copolymers. The hydrogen bonding between COOH and P2VP moieties effectively allowed for an improved dispersability of SWCNTs within P2VP domains, revealed by systematic shifts in the order-disorder transitions produced. The encapsulation of SWCNTs was found to be dependent on the microstructure of the copolymer: Two-dimensional lamellar morphology was more efficient than one-dimensional cylinder-based one. They also analyzed the effect of morphology on conductivity, finding that it was lower for cylinders than for lamellar. This seemed to indicate that some aggregation was present within P2VP cylinders.

References

1. Iijima, S. (1991). Helical microtubules of graphitic carbon, *Nature*, **354**, pp. 56–57.
2. Iijima, S., and Ichidashi, T. (1993). Single-shell carbon nanotubes of 1-nm diameter, *Nature*, **363**, pp. 603–604.
3. Baughman, R. H., Zakhidov, A. A., and Heer, W. A. (2002). Carbon nanotubes: The route towards application, *Science*, **297**, pp. 787–792.
4. Vigolo, B., Coulon, C., Maugey, M., Zakri, C., and Poulin, P. (2005). An experimental approach to the percolation of sticky nanotubes, *Science*, **309**, pp. 920–923.
5. Wang, Z., Liu, Q., Zhu, H., Liu, H., and Chen, Y. (2007). Dispersing multi-walled carbon nanotubes with water-soluble block copolymers and they use as support for metallic nanoparticles, *Carbon*, **45**, pp. 285–292.
6. Wang, W., Xie, X., and Ye, Y. (2010). Crystallization induced block copolymer decoration on carbon nanotubes, *Carbon*, **48**, pp. 1680–1683.
7. Costa, P., Silva, J., Anson-Casaos, A., Martinez, M. T., Abad, M. J., Viana, J., and Lanceros-Mendez, S. (2014). Effect of carbon nanotube type and functionalization on the electrical, thermal, mechanical and electromechanical of carbon nanotube/styrene-butadiene-styrene composites for large strain sensor applications, *Composites B*, **61**, pp. 136–146.
8. Costa, P., Silva, J., Sencadas, V., Simoes, R., Viana, J. C., and Lanceros-Mendez, S. (2013). Mechanical, electrical and electro-mechanical properties of thermoplastic elastomer styrene-butadiene-styrene/multiwall carbon nanotubes composites, *J. Mater. Sci.*, **48**, pp. 1172–1179.
9. Pedroni, L. G., Araujo, J. R., Felisberti, M. I., and Nogueira, A. F. (2012). Nanocomposites base don MWCNT and styrene-butadiene-styrene block copolymers: Effect of preparation method on dispersion and polymer-filler interactions, *Compos. Sci. Technol.*, **72**, pp. 1487–1492.
10. Ilcikova, M., Mrlik, M., Sedlacek, T., Chorvat, D., Krupa, I., Slouf, M., Koynov, K., and Mosnacek, J. (2014). *Polymer*, **55**, pp. 211–218.
11. Liu, X., Ly, J., Han, S., Zhang, D., Requicha, A., Thompson, M., and Zhou, C. (2005). Synthesis and electronic properties of individual single-walled nanotube/polypyrrole composite nanocables, *Adv. Mater.*, **17**, pp. 2727–2732.

12. Tchmutin, I. A., Ponomarenko, A. T., Krinichnaya, E. P., Kozub, G. I., and Efimov, O. N. (2013). Electrical properties of composites based on conjugated polymers and conductive fillers, *Carbon*, **41**, pp. 1391–1395.
13. Zou, J., Liu, L., Chen, H., Khondaker, S. L., McCullough, R. D., Huo, Q., and Zhai, L. (2008). Dispersion of pristine carbon nanotubes using conjugated block copolymers, *Adv. Mater.*, **20**, pp. 2055–2060.
14. George, A. S., Ghazinejad, M., Wang, W., Ruiz, I., Ozkan, M., and Ozkan, C. S. (2012). Block copolymer-assisted fabrication of grapheme/carbon nanotube hybrid architectures and their application in supercapacitors, *MRS Proc.*, **1440**, mrss12-1440-o09-18 doi:10.1557/opl.2012.1283.
15. Lee, D. H., Lee, W. J., and Kim, S. O. (2009). Highly efficient vertical growth of wall-number-selected, N-doped carbon nanotube arrays, *Nanoletters*, **9**, pp. 1427–1432.
16. Roerdink, M., Pragt, J., Korczagin, I., Hempenius, M. A., Stockli, T., Keles, Y., Knapp, H. F., Hinderling, C., and Vancso, J. G. (2007). Templated growth of carbon nanotubes with controlled diameters using organic/organometallic block copolymers with tailored block lengths, *J. Nanosci. Nanotech.*, **7**, pp. 1052–1058.
17. Xu, P., Ji, X., Yang, H., Qi, J., Zheng, W., Abetz, V., Jiang, S., and Shen, J. (2010). Controllable fabrication of carbon nanotubes on catalysts derived from PS-b-P2VP block copolymer template and in situ synthesis of carbon nanotube/Au nanoparticles composite materials, *Mat. Chem. Phys.*, **119**, pp. 249–253.
18. Bhaviripudi, S., Mile, E., Steiner III, S. A., Zare, A. T., Dresselhaus, M. S., Belcher, M. A., and Kong, J. (2007). CVD synthesis of single-walled carbon nanotubes from gold nanoparticle catalysts, *J. Am. Chem. Soc.*, **129**, pp. 1516–1517.
19. Tseng, Y. T., Tseng, W. H., Lin, C. H., and Ho, R. M. (2007). Fabrication of double-length-scale patterns via lithography, block copolymer templating and electrodeposition, *Adv. Mater.*, **19**, pp. 3584–3588.
20. Wang, P., Lu, J., and Zhou, O. (2008). Microwave plasma enhanced chemical vapor deposition growth of few-walled carbon nanotubes using catalysts derived from an iron-containing block copolymer precursor, *Nanotechnology*, **19**, 185605 (7 pp.).
21. Chen, J. T., Shin, K., Leiston, J. M., Zhang, M., and Russel, T. P. (2006). Amorphous carbon nanotubes with tunable properties via template wetting, *Adv. Funct. Mater.*, **16**, pp. 1476–1480.

22. Bennet, R. D., Hart, A. J., and Cohen, R. E. (2006). Controlling the morphology of carbon nanotube films by varying the areal density of catalyst nano-clusters using block copolymer micellar thin films, *Adv. Mater.*, **18**, pp. 2274–2279.
23. Bennet, R. D., Hart, A. J., and Cohen, R. E. (2006). Creating patterned carbon nanotube catalyst through the microcontact printing of block copolymer micellar thin films, *Langmuir*, **22**, pp. 8273–8276.
24. Drozdov, A. D., and Christiansen, J. (2009). Thermo-viscoplasticity of carbon black-reinforced thermoplastic elastomers, *Int. J. Solid Struct.*, **46**, pp. 2298–2308.
25. Lorenz, H., Fritzsche, J., Das, A., Stockelhuber, K. W., Jurk, R., and Heinrich, G. (2009). Advanced elastomer nano-composites based on CTN-hybrid filler system, *Compos. Sci. Technol.*, **69**, pp. 2135–2143.
26. Du, J. H., Bai, J., and Cheng, H. M. (2007). The present status and key problems of carbon nanotube based polymer composites, *Express Polym. Lett.*, **1**, pp. 253–273.
27. Holzer, S., Gamb, M., Schneider, K., Knoll, K., and Weidisch, R. (2013). Deformation mechanisms in lamellar S-S/B-S triblock copolymers, *Eur. Polym. J.*, **49**, pp. 261–269.
28. Xiao, H., Li, H., and Ou, J. (2010). Modelling of piezoresistivity of carbon black filled cement composites under multi-axial strain, *Sens. Actuators A*, **160**, pp. 87–93.
29. Wang, P., Geng, S., and Ding, T. (2010). Effect of carboxyl radical on electrical resistance of multi-walled carbon nanotube filled silicone rubber composite under pressure, *Compos. Sci. Technol.*, **70**, pp. 1571–1573.
30. Liu, Y. T., Zhang, Z. L., Zhao, W., Xie, X. M., and Ye, X. Y. (2009). Selective self-assembly of surface functionalized-carbon nanotubes in block copolymer template, *Carbon*, **47**, pp. 1883–1885.
31. Lu, L., Zhou, Z., Zhang, Y., Wang, S., and Zhang, Y. (2007). Reinforcement of styrene-butadiene-styrene triblock copolymer by multi-walled carbon nanotubes via melt mixing, *Carbon*, **45**, pp. 2621–2627.
32. Lu, L., Yu, H., Wang, S., and Zhang, Y. (2009). Thermal degradation behavior of styrene-butadiene-styrene triblock copolymer/multiwalled carbon nanotube composites, *J. Appl. Polym. Sci.*, **112**, pp. 524–531.
33. Pedroni, L. G., Soto-Oviedo, M. A., Rosolen, J. M., Felisberti, M. I., and Nogueira, A. F. (2009). Conductivity and mechanical properties of composites base don MWCNT and styrene-butadiene-styrene block copolymers, *J. Appl. Polym. Sci.*, **112**, pp. 3241–3248.

34. Wichmann, M. H. G., Buschhorn, S. T., Gehrman, J., and Schulte, K. (2009). Piezoresistive response of epoxy composites with carbon nanoparticles under tensile load, *Phys. Rev. B*, **80**, 245437 (8 pp.).

35. Ji, M., Deng, H., Yan, D., Duan, L., and Fu, Q. (2014). Selective of multi-walled carbon nanotubes in thermoplastic elastomer blends: An effective method for tunable resistivity-strain sensing behavior, *Compos. Sci. Technol.*, **92**, pp. 16–26.

36. Liu, Y. T., Zhao, W., Huang, Z. Y., Gao, Y. F., Xie, X. M., and Wang, X. H. (2006). Non-covalent surface functionalization of carbon nanotubes for solubility in organic solvents, *Carbon*, **44**, pp. 1613–1616.

37. Wu, G., Zhou, L., Ou, E., Xie, Y., Xiong, Y., and Xu, W. (2010). Preparation and properties of hydroxylated styrene-butadiene-styrene triblock copolymer/multi-walled carbon nanotubes nanocomposites via covalent bond, *Mater. Sci. Eng. A*, **527**, pp. 5280–5286.

38. Meier, J. G., Crespo, C., Pelegay, J. L., Castell, P., Sainz, R., Maser, W. K., and Benito, A. M. (2011). Processing dependency of percolation threshold of MWCNT in a thermoplastic elastomeric block copolymer, *Polymer*, **52**, pp. 1788–1796.

39. Huang, J., Zhu, Y., Jiang, W., Yin, J., Tang, Q., and Yang, X. (2014). Parallel carbon nanotube stripes in polymer thin film with remarkable conductive anisotropy, *Appl. Mater. Interfaces*, **6**, pp. 1754–1758.

40. Park, I., Lee, W., Kim, J., Park, M., and Lee, H. (2007). Selective sequestering of multi-walled carbon nanotubes in self-assembled block copolymer, *Sens. Actuat. B*, **126**, pp. 301–305.

41. Peponi, L., Valentini, L., Torre, L., Mondragon, I., and Kenny, J. M. (2009). Surfactant assisted selective confinement of carbón nanotubes functionalized with octadecylamine in a poly(styrene-b-isoprene-b-styrene) block copolymer matrix, *Carbon*, **47**, pp. 2474–2480.

42. Peponi, L., Tercjak, A., Gutierrez, J., Cardinali, M., Mondragon, I., Valentini, L., and Kenny, J. M. (2010). Mapping of carbon nanotubes in the polystyrene domain of a polystyrene-b-polyisoprene-b-polystyrene block copolymer matrix using electrostatic force microscopy, *Carbon*, **48**, pp. 2590–2595.

43. Garate, H., Fascio, M. L., Mondragon, I., D'Accorso, N. B., and Goyanes, S. (2011). Surfactant-aided dispersion of polystyrene-functionalized carbon nanotubes in a nanostructured poly(styrene-b-isoprene-b-styrene) block copolymer, *Polymer*, **52**, pp. 2214–2220.

44. Corte, L., Yamauchi, K., Court, F., Cloitre, M., Hashimoto, T., and Leibler, L. (2003). Annealing and defect trapping in lamellar phases of triblock terpolymers, *Macromolecules*, **36**, pp. 7695–7706.

45. Yamaguchi, D., Cloitre, M., Panin, P., and Leibler, L. (2005). Phase behavior and viscoelastic properties of thermoplastic elastomer gels based on ABC triblock copolymer, *Macromolecules*, **38**, pp. 7798–7806.

46. Perie, T., Brosse, A. C., Girault, S. T., and Leibler, L. (2011). Nanostructured films and composites from carbon nanotubes dispersed by ABC block terpolymers in selective solvent, *Polymer*, **52**, pp. 3065–3073.

47. Saint-Aubin, K., Poulin, P., Jaillet, C., Maugey, M., and Zakri, C. (2013). Changes of morphology and properties of block copolymer induced by carbon nanotubes, *Polymer*, **54**, pp. 2285–2291.

48. Cella, R. J. (1973). Morphology of segmented polyester thermoplastic elastomer, *J. Polym. Sci: Polym. Symp.*, **42**, pp. 727–740.

49. Szymczyk, A. (2012). Poly(trimethylene therephtalate-block-tetramethylene oxide) elastomer/single-walled carbon nanotubes composites: Synthesis, structure and properties, *J. Appl. Polym. Sci.*, **126**, pp. 796–807.

50. Glenis, S., Likodimos, V., Guskos, N., Yarmis, D., Zolnierkiewicz, G., Szymczyk, A., and Lin, C. L. (2010). Magnetic properties of carbon nanotube poly(ether-ester nanocomposites, *J. Appl. Phys.*, **108**, pp. 4464–4466.

51. Broza, G. (2010). Thermoplastic elastomers with multi-walled carbon nanotubes: Influence of dispersion methods on morphology, *Compos. Sci. Technol.*, **70**, pp. 1006–1010.

52. Brostow, W., Broza, G., Datashvili, T., Hagg, H. E., and Kopyniecka, A. (2012). Poly(butyl terephtalate)/oxytetramethylene + oxidized carbon nanotube hybrids: Mechanical and tribological behavior, *J. Mater. Res.*, **27**, pp. 1815–1823.

53. Wu, S., Li, H., Huang, G., and Wu, J. (2014). Nucleating effect of multi-walled carbon nanotubes and grapheme on the crystallization kinetics and melt behavior of olefin block copolymers, *RSC Adv.*, **4**, pp. 19024–19033.

54. Park, H. S., Choi, B. G., Hong, W. H., and Hang, S. Y. (2012). Interfacial interactions of single-walled nanotubes, conjugated block copolymer hybrids for flexible transparent conductive films, *J. Phys. Chem. C*, **116**, pp. 7962–7967.

55. Goh, R. G., Motta, N., Bell, J. M., and Waclawik, E. R. (2006). Effect of substrate curvature on the absorption of poly(3-hexyltiophene) on single-walled nanotubes, *Appl. Phys. Lett.*, **88**, 053101 (3 pp.).

56. Star, A., Stoddart, J. F., Steuerman, D., Diehl, M., Boukai, A., Wong, E., Yang, X., Chung, S. W., Choi, H., and Heath, J. R. (2001). Preparation and properties of polymer-wrapped single-walled carbon nanotubes, *Angew. Chem. Int. Ed.*, **40**, pp. 1721–1725.

57. Chen, J., Liu, H., Weimer, W. A., Halls, M. D., Waldeck, D. H., and Walker, G. C. (2002). Non-covalent engineering of carbon nanotube surfaces by rigid, functional conjugated polymers, *J. Am. Chem. Soc.*, **124**, pp. 9034–9035.

58. Zou, J., Liu, L., Chen, H., Khondaker, S. L., McCullough, R. D., Huo, Q., and Zhai, L. (2008). Dispersion of carbon nanotubes using conjugated block copolymers, *Adv. Mater.*, **20**, pp. 2055–2060.

59. Olsen, B. D., and Segalman, R. A. (2008). Self-assembly or rod-coil block copolymers, *Mater. Sci. Eng. R*, **62**, pp. 37–66.

60. Sung, J., Huh, J., Choi, J. H., Kang, S. H., Choi, Y. S., Lee, G. T., Cho, J., Myoung, J. M., and Park, C. (2010). Ultrathin electronic composite sheets of metallic/semiconducting carbon nanotubes embedded in conjugated block copolymers, *Adv. Funct. Mater.*, **20**, pp. 4305–4313.

61. Ahn, J. H., Kim, H. S., Lee, K. J., Jeon, S., Kang, S. J., Sun, Y., Nuzzo, R. G., and Rogers, J. A. (2006). Heterogeneous tridimensional electronics by use of printed semiconductor materials, *Science*, **314**, pp. 1754–1757.

62. Hwang, S. K., Choi, J. R., Bae, I., Hwang, I., Cho, S. M., Huh, J., and Park, C. (2013). High temperature operating non-volatile memory of printable single-walled carbon nanotubes self-assembled with a conjugated block copolymer, *Small*, **9**, pp. 831–837.

63. Soriano-Corral, F., Ramos-de Valle, F., Enriquez-Medrano, F. J., De Leon-Martinez, P. A., Lopez-Quintanilla, M. L., and Cabrera-Alvarez, E. N. (2012). Cast nanostructured films of poly(methyl methacrylate-b-buty lacrylate)/carbon nanotubes: Influence of poly(butyl acrylate) content on film evaporation rate, morphology and electrical resistance, *J. Nanomater.*, ID 352937 (8 pp.).

64. Wang, J., Li, F., Li, Q., Sun, J., and Chen, G. X. (2013). Polystyrene-b-poly(ethylene oxide) block copolymer thin films as templates for carbon nanotube dispersion, *Thin Solid Films*, **536**, pp. 191–195.

65. Chen, Z., Liu, Y., Yao, C., and Yang, G. (2013). Preparation and characterization of poly(trymethylene terephtalate)-poly(ethylene oxide terephtalate) segmented copolymer/multi-walled carbon nanotube composites by in situ polymerization, *Polym. Eng. Sci.*, **53**, pp. 914–922.

66. Wang, R., Cherukuri, P., Duque, J. G., Leeuw, T. K., Lackey, M. K., Moran, C. H., Moore, V. C., Conyers, J. L., Smalley, R. E., Schmidt, H. K., Weismen, R., and Engel, P. S. (2007). SWCNT PEG-eggs: Single-walled carbon nanotubes in biocompatible shell-crosslinked micelles, *Carbon*, **45**, pp. 2388–2393.

67. Suckeveriene, R. Y., Rahman, R., Ovadia, M., Szczupak, D., Mechrez, G., and Narkis, M. (2014). Synthesis of surfactant-free carbon nanotube/poly(styrene-o-acrylamide) by dynamic interfacial emulsion polymerization under sonication, *Polym. Adv. Technol.*, **25**, pp. 4–8.

68. Kumar, M., and Ulbricht, M. (2014). Low fouling negatively charged hybrid ultrafiltration membranes for protein separation from sulfonated poly(arylene ether sulfone) block copolymer and functionalized multi-walled carbon nanotubes, *Sep. Purif. Technol.*, **127**, pp. 181–191.

69. Yu, N., Zheng, X., and He, L. (2011). Controllable-induced crystallization of PE-b-PEO on carbon nanotubes with the assistance of supercritical CO_2: Effect of solvent, *Macromolecules*, **44**, pp. 3958–3965.

70. Shin, H., Min, B. G., Jeong, W., and Park, C. (2005). Amphiphilic block copolymer micelles: New dispersant for single-walled carbon nanotubes, *Macromol. Rapid Commun.*, **26**, pp. 1451–1457

71. Sung, J., Jo, P. S., Shin, H., Min, B. G., Kim, D. H., and Park, C. (2008). Transparent, low-electrical-resistance nanocomposites of self-assembled block copolymers and SWCNT, *Adv. Mater.*, **20**, pp. 1505–1510.

72. Li, C., Hsu, J. C., Sugiyama, K., Hirao, A., Chen, W. C., and Mezzenga, R. (2009). Synthesis and self-assembly behavior of poly-(fluorenylstyrene)-block-poly(2-vinylpyridine) block copolymers and their mixtures with single-walled carbon nanotubes, *Macromolecules*, **42**, pp. 5793–5801.

73. Hudson, J. L., Casavant, M. J., and Tour, J. M. (2004). Water-soluble, exfoliated, nonroping single-walled carbon nanotubes, *J. Am. Chem. Soc.*, **126**, pp. 11158–11159.

Chapter 4

Block Copolymer Assisted Sol-Gel Templating

Junkal Gutierrez

Materials + Technologies Group (GMT),
Department of Chemical and Environmental Engineering,
Polytechnic School, University of the Basque Country (UPV/EHU),
Plaza Europa 1, 20018 Donostia-San Sebastián, Spain

juncal.gutierrez@ehu.eus

Block copolymer–assisted sol-gel templating is a simple, flexible and highly versatile strategy to fabricate novel multifunctional nanostructured materials. Co-assembly processes utilizing the selective interactions of amphiphilic block copolymers with inorganic nanoparticles have applied toward structure directing of a broad range of well-defined nanostructured hybrid materials. This versatile strategy is a powerful tool to design tailor-made advanced multifunctional materials with different combinations of properties by playing with block copolymer (by changing the molecular weight, volume fraction or solution conditions) and inorganic component (size, type, amount), as well as with the ratio between both components. This opens up a large number of possibilities with widespread field of applications.

Block Copolymer Nanocomposites
Edited by Galder Kortaberria and Agnieszka Tercjak

ISBN 978-981-4669-54-2 (Hardcover), 978-981-4669-55-9 (eBook)
www.panstanford.com

In this chapter, recent advances in nanostructured hybrid materials prepared by employing PS-b-PEO, PS-b-PLLA or PS-b-PVP block copolymers as structure directing agents and sol-gel-synthesized nanoparticles are reported.

4.1 Introduction

In the past decade, design, development and characterization of novel hybrid multifunctional materials with application in different fields of nanotechnology have grown, considering the tremendous market, which opens the door for new challenges in this field. The progress in nanotechnology allows creating multifunctional nanomaterials with tunable properties controlled at the nanoscale to achieve desired requirements for future macroscale applications.

In line with this idea, the chemistry of hybrid inorganic organic materials (HIOM) is also expanding because of the wide range of possibilities that it opens for the elaboration of new materials [1–3]. This field of research bridges all the different types of chemistry (organic, inorganic, organometallic and polymer) to material science.

The advantage of hybrid inorganic/organic nanocomposites is that they combine the properties of organic and inorganic components in one material. A distinct characteristic of hybrid materials is that their properties are related not only to the chemical nature of the inorganic and organic components, but also rely strongly on their synergy.

As it is already known, nanoparticles possess unique, often size-dependent properties associated with magnetic, photonic, chemical, and conductive properties, which are different from the properties of their respective bulk materials. However, to take full advantage of these unique properties for highly integrated fabrication, precise control of nanomaterial placement and its integration are essential. Thus, in recent years increasing efforts have been made to reach a full control over morphology in multiphase materials since it is a key step in controlling the final material's properties [4].

Hence, the preparation of well-ordered hybrid inorganic/organic nanocomposites with exceptional control over size, shape, composition, and organization of nanoscale domains is crucial to the development of next generation materials and devices for energy generation and storage, microelectronic, sensors, photonic, drug delivery, packaging and other applications [5–10].

In order to build up nanoscale features, two approaches have been envisaged: top-down and bottom-up [11]. In the top-down approach, nanoparticles can be directly patterned by lithographic techniques, whereas the bottom-up approach is based on a self-assembly process. In the case of lithographic techniques, these approaches typically require additional elaborate and complicated procedures, deteriorating the merits (i.e., cost effectiveness and ability to scale up) of nanomaterials prepared by wet chemistry. On the contrary, the bottom-up approach based on polymeric nanotemplates prepared from the microphase separation of block copolymers (BCPs) [12] can be a suitable alternative platform due to their ability to form self-assemble into thermodynamically highly stable ordered nanostructures with precisely controlled shapes and domain sizes, easily and cost-effectively over a large area with high fidelity [13, 14].

Consequently, scientific research related to the use of block copolymers as templates for the development of novel multifunctional nanostructured hybrid materials has evolved into an area of considerable interest [15–17]. Several research works have focused on the development of hybrid inorganic/organic materials containing well-distributed inorganic nanoparticles located in a specific microseparated phase of nanostructured BCPs [18–23]. The self-assembly capacity of the block copolymers provides an interesting route for designing and controlling the assembly of nanoparticles into highly ordered structures, which offer possible applications in the design of chemical and biological sensors, energy storage and convertors, catalytic systems, optical devices and others [6–8, 24–27].

The diblock copolymer acts as a structure-directing agent for the functional (inorganic) component and tuning the self-assembly process enables the fabrication of multifunctional nanostructures at different dimensions and length scales. The spatial arrangement of nanoparticles can be controlled by the size and morphology of the BCP nanodomains that are adjustable by chemical nature of each block, the composition and the ambient microenvironment.

On the other hand, the important factor to determine the stable incorporation of the nanoparticles in a BCP matrix lies in the compatibility of the nanoparticles with the microstructured block copolymer [4, 28–30].

A variety of different physical and chemical methods have been established for the preparation of inorganic nanoparticles or HIOM with tunable properties [31–35].

In this chapter, we focus on HIOM derived through sol-gel processing.

Sol-gel synthesis is a versatile low-cost chemical method used to prepare different kind of inorganic nanoparticles. By definition, the sol-gel synthesis is a wet-chemical technique widely used in the fields of science and engineering. In a typical sol-gel process, a colloidal suspension, or a sol, is formed from the hydrolysis and polymerization reactions of the precursors, which are usually inorganic metal salts or metal organic compounds such as metal alkoxides (MOR) [36]. The structure, connectivity and morphology of the final inorganic network depend strongly on the relative contribution of reactions, which take place during the network formation. Depending on intrinsic (i.e., metal center features such as coordination, acidity, etc.) or extrinsic (i.e., tunable reaction conditions such as solvent, water contents, pH, catalysts, reaction time, etc.) conditions, the generated network exhibits a variety of structures, from branched arrangements to compact clusters.

Taking above into account, the combination of the self-assembly of BCPs and sol-gel chemistry results in a promising strategy to fabricate hybrid multifunctional nanostructured materials. In particular, amphiphilic block copolymers, with one hydrophilic block and another hydrophobic block, are excellent materials to be used as structure directing agents. Based on this amphiphilicity, miscibility or compatibility of sol-gel precursor and one block of the BCP can be provided by the use of secondary interactions (i.e., hydrogen bond, coordinative, acid-base). Consequently, metal oxide precursors can easily selectively locate via hydrogen bonds in the hydrophilic block of the BCP and the hydrophilic loaded block and the hydrophobic block will separate through self-assembly. This enables the fabrication of well-ordered hybrid materials with spherical, cylindrical, lamellar or gyroid microstructures (with a high particle loadings) after the solvent evaporation. Here it should be noted that the microdomains can be loaded with a suitable precursor, which can be chemically transformed into the desired nanoparticles either by addition of further reagents or simply by heating.

Thus, as shown in Fig. 4.1, the fabrication protocol is realized by incorporating inorganic nanoparticles selectively within one block of the BCP and directing the self-assembly of the hybrid material into the desired morphology.

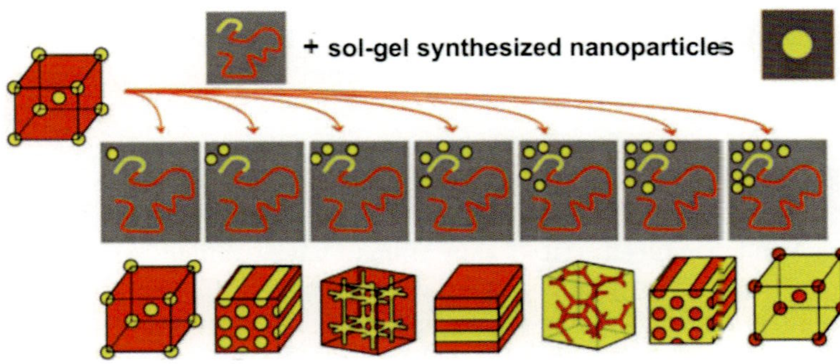

Figure 4.1 Schematic representation of the morphological behavior of BCP by varying the amount of inorganic nanoparticles. Adapted from ref. [37].

Reported approach is simple and applicable to a wide range of inorganic sol-gel-synthesized nanoparticles and different BCPs. Therefore, one can design tailor-made advanced multifunctional materials with different combinations of properties by playing with both, BCP (by changing the molecular weight and the weight ratio between the hydrophilic part and hydrophobic part) and inorganic component (size, type, or amount). This opens up numerous possibilities in the field of HIOM with a wide range of applications.

In this chapter, different HIOMs fabricated using block copolymer–assisted sol-gel templating are highlighted. This versatile strategy allows the development of novel HIOM in an extremely wide chemical composition range and shapes, therefore offering a wide range of possibilities to elaborate tailor-made advanced functional materials.

4.2 Poly(Styrene-b-Ethylene Oxide) Block Copolymer

Amphiphilic poly(styrene-b-ethylene oxide) (PS-b-PEO) block copolymer has been intensively studied as template agents, coupled with sol-gel chemistry to control the selective location and morphology of designed hybrid inorganic/organic nanocomposites, specifically systems containing titanium dioxide (TiO_2) nanoparticles as the inorganic part [38–59]. In general, one of the most common sol-gel precursors is titanium tretraisopropoxide (TTIP). TTIP can be incorporated into the hydrophilic PEO domains through coordination bonds. Moreover, the hydrolysis and condensation of TTIP turn the PEO domains chemically reactive and the titania species on the surface of the PEO domains are able to form covalent bonds with other titania species on the surface of neighboring PEO domains.

Nanostructured TiO_2 thin films have gained significant attention due to their potential applications in the fields of chemical and alternative energy industries such as photocatalysts dye-sensitized solar cells, sensors and fuel cells [60–65].

Numerous studies from different research groups attempted to develop and characterize different hybrid TiO_2/PS-b-PEO nanocomposites. In 2004, Z. Sun and J. S. Gutmann [38] reported for the first time the use of a good-poor solvent pair induced phase separation process in combination with sol–gel chemistry to fabricate titania nanostructures using PS-b-PEO block copolymer as a templating agent. Hybrid micelles were formed in the solution, and during the spin-coating process, a solvent evaporation induced arrangement of the nanostructures inside the film takes place, which might result in an ordering of the nanostructures.

Here, it should be pointed out that taking into account the versatility of this strategy, this publication established a starting point for future research works. Thus, numerous investigations were published by varying the composition of the block copolymer, changing the constituents and composition of the sol-gel solution, tuning the film deposition and post-treatment conditions, etc.

Following this work, the morphology and nanoparticle size distribution of hybrid TiO_2/PS-b-PEO nanocomposites containing different amounts of TiO_2 nanoparticles have been also reported by Z. Sun and co-workers by means of scanning force microscopy, scanning electron microscopy, and small-angle X-ray scattering analysis [39, 40].

Additionally, D. H. Kim and co-workers used atomic force microscopy (AFM) in order to study the surface morphology generated before and after thermal or ultraviolet degradation of the organic part [41, 42]. Figure 4.2 shows highly dense ordered arrays of uniformly sized titania dots (bright domains) on PS-b-PEO film. The average diameter of the titania nanoparticles was 22 nm, with average center-to-center distance between dots of 40 nm.

Figure 4.2 (a) Phase contrast AFM image of titania dot arrays on PS-b-PEO film and (b) 3D view of the height contrast AFM image of titania dot arrays on PS-b-PEO film. The images are in 1 μm × 1 μm scale. Adapted from ref. [41]. Copyright 2004, American Chemical Society.

Moreover, photoluminescence properties of hybrid TiO_2/PS-b-PEO nanocomposites as well as the morphology evolution of the nanostructures as a function of preparation conditions were also studied [43–46].

Y. J. Cheng and J. S. Gutmann [43] demonstrated that different morphologies can be obtained by adjusting the weight fractions of titanium sol-gel components, in this case, 1,4-dioxane, HCl, and TTIP. On the basis of the results, a ternary phase diagram of TiO_2 films after calcinations was mapped, which can guide the preparation of nanocomposite films with desired morphologies (Fig. 4.3).

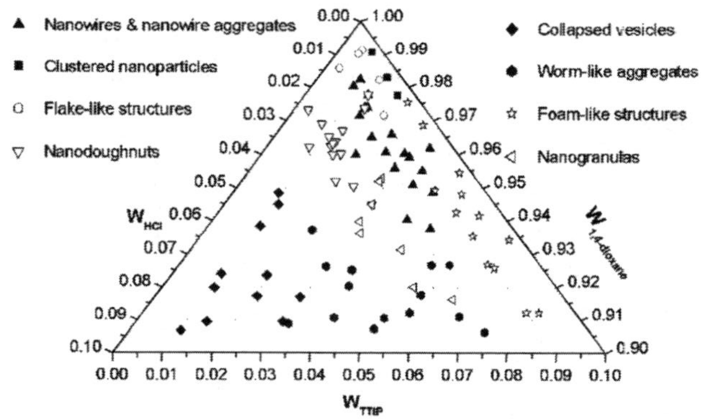

Figure 4.3 Phase diagram of TiO_2 films after calcination with varied morphologies (W, weight fraction). Reproduced with permission of [43]. Copyright 2006, American Chemical Society.

As shown in Fig. 4.3, controlling the weight ratio of the components of the sol–gel allows for a wide spread diversity of nanostructures, including nanoparticles, nanorods, nanowires, nanotubes, foam-like, flake-like, worm-like, nanodoughnuts and nanovesicles. Additionally, Fig. 4.4 shows SEM images of TiO_2 films with morphology evolution with increasing HCl weight fractions.

In light of these studied effects of varying the composition of the solution and the copolymer concentration on the morphology, J. S. Gutmann and co-workers investigated the impact of the sol-gel reaction time on the morphology evolution of titania thin films [47]. By studying the behavior and the corresponding mechanism of the influence of the sol–gel reaction time on the morphology, it is possible to gain deep insights into the templating process of the amphiphilic block copolymer. Specifically, the sol–gel reaction time was varied from 46 min to 25 h, while keeping the composition of the solution and the block copolymer concentration fixed. Obtained results demonstrated a morphology evolution from spherical mesopores to worm-shaped mesopores within TiO_2/PS-b-PEO composite films by tuning the sol-gel reaction time.

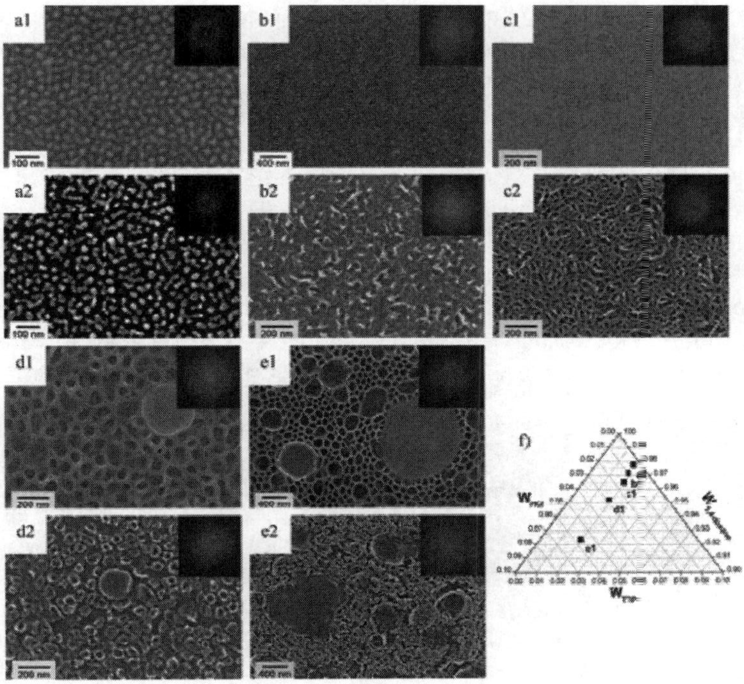

Figure 4.4 SEM images of TiO_2 films with morphology evolution with increasing HCl weight fractions. Panels a1–e1: SEM images before calcination. Panels a2–e2: SEM images after calcination. Panel f: the compositional existence regions in the phase diagram where the five structures are formed with increasing HCl weight fractions from a to e (W, weight fraction). The insets in the images are FFT patterns of the corresponding structures. Reproduced with permission of [43]. Copyright 2006, American Chemical Society.

Most recently, the same authors reported a review work related to the tailoring of titania nanostructures by the combination of sol-gel chemistry with block copolymers as structure directing agents and focused on the employment of the advanced characterization techniques which allows for the control over the structures on all size scales and also on all time scales during the synthesis [48].

On the other hand, J. Gutierrez and co-workers studied the influence of the addition of TiO_2 nanoparticles on the self-assembly of two PS-b-PEO block copolymers with different ratio between blocks varying also the sol-gel content [49]. Obtained results indicated good affinity between the PEO block and synthesized titanium sol-gel network.

This selective confinement of the inorganic part in the PEO block of the BCPs allows the generation of nanostructured systems with high nanoparticle content. Morphological study was also performed by AFM and it was observed that under the same synthesis conditions, the addition of different vol% of sol-gel in PS-b-PEO matrices provoked changes in the morphologies if compared to those detected for neat BCPs [50]. It is necessary to pointed out that well-defined nanostructured materials were obtained up to 60 vol% sol-gel content (Fig. 4.5).

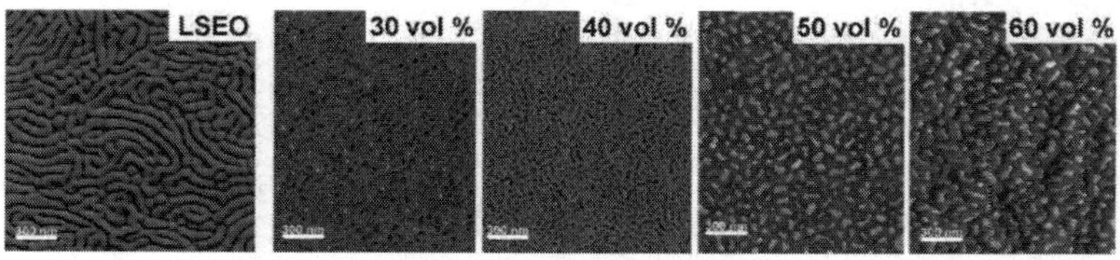

Figure 4.5 AFM phase images: block copolymer and different nanocomposites obtained by varying the amount (vol%) of sol-gel. LSEO correspond to PS-b-PEO block copolymer with $Mn^{PS} = 58600$ g mol^{-1}, $Mn^{PEO} = 31000$ g mol^{-1}, $M_w/M_n = 1.03$.

Nowadays different advanced characterization techniques are using in order to investigate the conductive properties of electronic devices at the nanoscale. Thereby, the same authors employed electrostatic force microscopy (EFM) and tunneling force microscopy (TUNA) to study the conductive properties at nanolevel of TiO_2 nanoparticles in hybrid TiO_2/PS-b-PEO nanocomposites with different inorganic content [50, 51].

EFM became a powerful tool for investigate qualitatively conductive properties of the synthesized hybrid inorganic/organic nanocomposites at the nanoscale, since EFM measurements allow to detect the electric field gradient distribution above the sample surface, which consequently allows to distinguish different conductive parts on the sample surface.

On the other hand, TUNA was employed to quantitatively determine the conductive properties at nanoscale. This technique is a versatile method to study the charge distribution and conductivity level of the synthesized hybrid inorganic/

organic nanocomposites. TUNA measurements were performed since this technique offers a unique possibility to determine local ultralow current values on hybrid TiO_2/PS-b-PEO nanocomposites at the nanometric level.

As can be seen in Fig. 4.6a, EFM results confirmed that TiO_2 nanoparticles maintain their conductive properties in the synthesized hybrid inorganic/organic nanocomposites since TiO_2 nanoparticles responded to the voltage applied to the EFM tip [51]. Moreover, TUNA results successfully demonstrated that well-dispersed TiO_2 nanoparticles had strong effect on the conductivity of the designed materials (Fig. 4.6b). It is necessary to pointed out that conductivity of these systems increased upon TiO_2 content increasing in the nanocomposites [50].

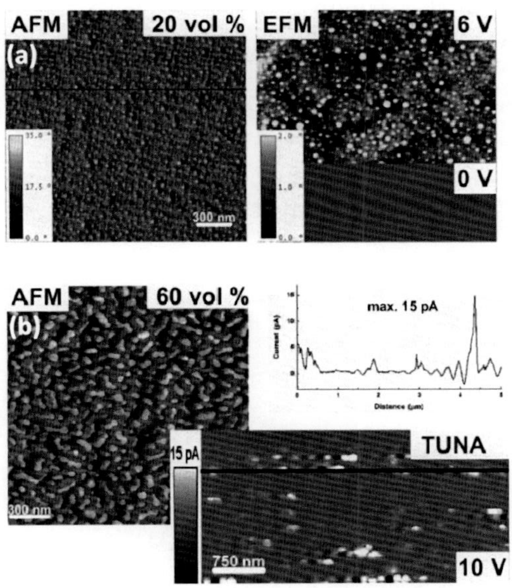

Figure 4.6 (a) AFM and EFM phase images of 20 vol% sol-gel/PS-b-PEO nanocomposite and (b) AFM and TUNA images and corresponding profile of 60 vol% sol-gel/PS-b-PEO nanocomposite.

Additionally, it is necessary to mention that highlighted hybrid materials have possible applications in the field of dye-sensitized solar cells (DSSC) and several research works were reported that focused on this idea [40, 44, 52, 53].

Regarding the multiple possibilities of the sol-gel methods, D. H. Kim and co-workers suggested to combine two different sol-gel precursors with the aim to achieve an enhancement in the final properties of designed hybrid materials [54–57]. The synergistic properties expected from the combination of two inorganic nanoparticles in a hybrid system can be controlled using block copolymer–assisted sol-gel templating strategy. In this case the inorganic precursors were selectively incorporated into PEO domains of amphiphilic PS-b-PEO due to the specific interactions.

On the one hand, hybrid Ag/TiO_2/PS-b-PEO nanocomposites were fabricated using TTIP and $AgNO_3$ as precursors [54, 55]. The organic matrix was removed by deep UV irradiation, leading to arrays of Ag/TiO_2 composite nanoparticles (Fig. 4.7). In this work, the structural evolution of the hybrid nanostructures and the internal structures of the resulting nanostructures were monitored by AFM, TEM and GISAXS studies. Moreover, as shown in Fig. 4.8, an improved photocatalytic activity of TiO_2 was achieved by the incorporation of Ag nanoparticles.

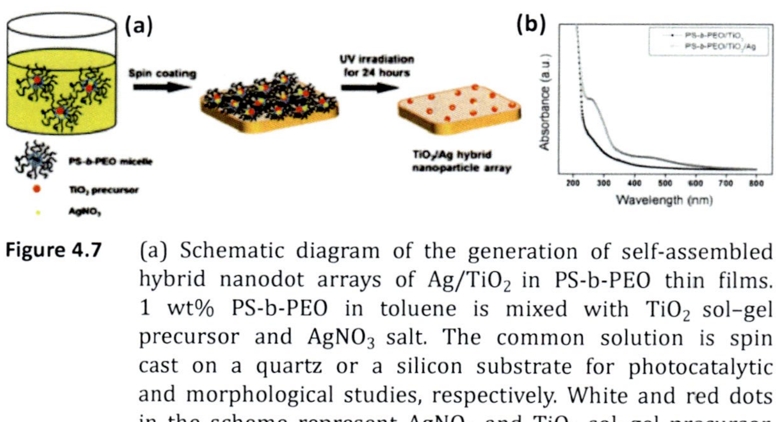

Figure 4.7 (a) Schematic diagram of the generation of self-assembled hybrid nanodot arrays of Ag/TiO_2 in PS-b-PEO thin films. 1 wt% PS-b-PEO in toluene is mixed with TiO_2 sol–gel precursor and $AgNO_3$ salt. The common solution is spin cast on a quartz or a silicon substrate for photocatalytic and morphological studies, respectively. White and red dots in the scheme represent $AgNO_3$ and TiO_2 sol–gel precursor, respectively. (b) UV-vis absorbance spectra of Ag/TiO_2/PS-b-PEO and TiO_2/PS-b-PEO thin films. The broad absorption at 400–450 nm is attributed to the surface plasmon resonance (SPR) of metallic Ag nanoparticles. The enhanced absorption below 400 nm with the addition of Ag is related to the high refractive index of TiO_2 being in contact with the Ag surface. Adapted and Reproduced with permission of [55]. Copyright 2009, Royal Society of Chemistry.

On the other hand, the same author designed composite CdS/TiO_2 nanoparticle arrays confined in nanopatterned hybrid thin films [56]. The addition of CdS quantum dots into thin films of the TiO_2/PS-b-PEO sol–gel mixture led to the morphological changes from mixed wire/hexagonal dot to well-defined, quasi-hexagonal dot arrays (Fig. 4.8). UV-vis absorption measurements showed that CdS/TiO_2/PS-b-PEO nanocomposite films exhibited enhanced and red shifted absorption behavior with increasing CdS concentration compared with the TiO_2/PS-b-PEO. An enhanced photocatalytic degradation of methylene blue was also observed in the hybrid CdS /TiO_2/PS-b-PEO nanocomposite thin film.

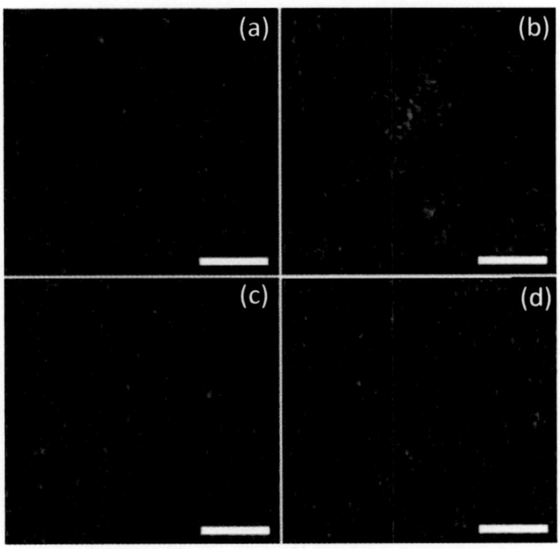

Figure 4.8 Height-contrast AFM images of CdS/TiO_2/PS-b-PEO thin films of the following compositions spin coated on silicon wafer: (a) $TiO_2$5/PS-b-PEO, (b) $TiO_2$10/PS-b-PEO, (c) CdS0.25/$TiO_2$10/PS-b-PEO and (d) CdS0.5/$TiO_2$10/ PS-b-PEO. The number following TiO_2 represents the volume percentage of TiO_2 sol–gel precursor with respect to BCP. The number following CdS represents the molar ratio of CdS to TiO_2. The scale bar represents 250 nm in length. Reproduced with permission of [56]. Copyright 2009, Royal Society of Chemistry.

Extending this methodology, hybrid Au/TiO_2 arrays were also fabricated by incorporating TTIP and tetrachloroauric (III) acid ($HAuCl_4$) inorganic precursors into PS-b-PEO templates [57]. By manipulating the relative composition of the precursors, hexagonally packed arrays of nanobowls can be obtained. Designed hybrid inorganic/organic materials and nanostructures with controlled shape and size exhibited unique photophysical properties.

Most recently, J. Gutierrez et al. synthesized hybrid nanocomposites based on PS-b-PEO block copolymer modified with a mixture of both vanadium and titanium sol-gel nanoparticles [58]. Confinement of the synthesized V:Ti nanoparticles in the PEO block of the SEO block copolymer was controlled, playing with the ratio between titanium and vanadium sol-gel precursors. AFM results indicated that increasing of the sol-gel content to 40 vol% leads to change the matrix from an organic (PS-block-rich phase) to inorganic (V:Ti nanoparticles/PEO-block-rich) one without losing high nanometric order due to confinement of the V:Ti nanoparticles in the PEO block. Quantitative nanoelectrical and nanomechanical properties were studied by means of novel powerful PeakForce (PF) tunneling atomic force microscopy (TUNA) technology and they confirmed that the employment of PeakForce TUNA is a useful innovative tool to achieve a quantitative characterization of both mechanical and electrical properties of advanced materials at the nanoscale level.

Figure 4.9 shows PF-TUNA images and corresponding profiles of 60 vol% 1:3-V:Ti/SEO nanocomposite when 10 and −10 V were applied. In both PF-TUNA current images (for 10 and −10 V) individual spherical nanoparticles can be clearly detected covering completely the substrate surface. Corresponding profiles confirmed that designed hybrid nanocomposites were able to respond to applied voltages regardless of the sign with almost the same current value (±25 pA).

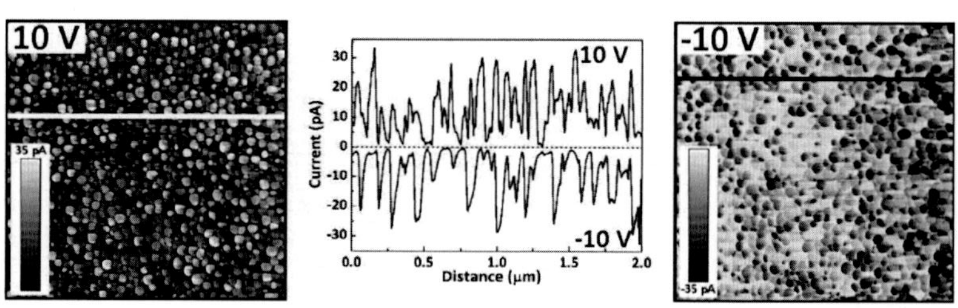

Figure 4.9 PF-TUNA current images (2 μm × 2 μm) and corresponding profiles of 60 vol% 1:3-V:Ti/SEO nanocomposite. Reproduced with permission of [58]. Copyright 2014, American Chemical Society.

Another useful route was developed by K. Sarkar and co-workers [59]. They employed diblock copolymer template assisted sol–gel approach to produce homogeneous Zn_2TiO_4 nanostructures based on spray deposition technique using zinc acetate dehydrate (ZAD) and ethylene glycol-modified titanate as precursors. This novel pathway can be better understood in following Fig. 4.10. The requirement for high surface area of the active layer for intensified dye adsorption for applications in DSSCs was met by synthesizing the sponge-like network morphology of the film. Spray deposition was also shown to allow for an upscaling in the film thickness and thereby improving the final device performance.

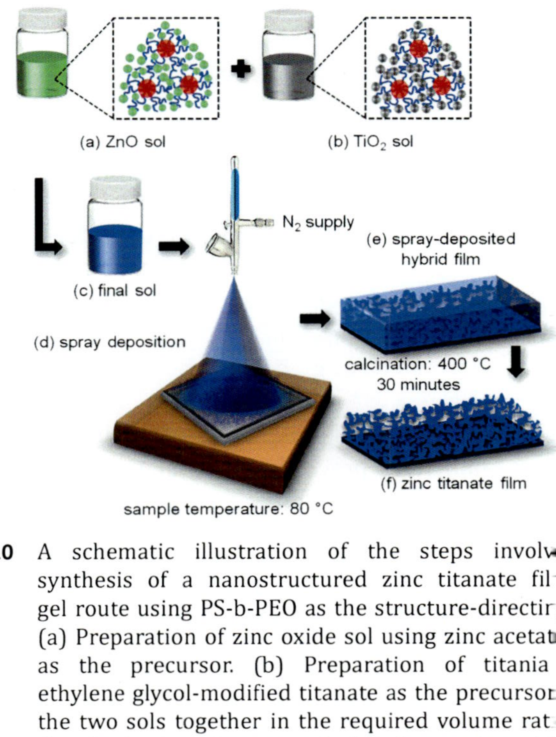

Figure 4.10 A schematic illustration of the steps involved in the synthesis of a nanostructured zinc titanate film via sol–gel route using PS-b-PEO as the structure-directing template. (a) Preparation of zinc oxide sol using zinc acetate dihydrate as the precursor. (b) Preparation of titania sol using ethylene glycol-modified titanate as the precursor. (c) Mixing the two sols together in the required volume ratio to obtain the final zinc titanate sol. (d) Spray deposition of zinc titanate sol using nitrogen as the carrier gas on a heated substrate. (e) Spray-deposited zinc titanate hybrid film. (f) Final calcination step to remove the polymer template yielding pure zinc orthotitanate nanostructured film. Reproduced with permission of [59]. Copyright 2014, Royal Society of Chemistry.

4.3 Polystyrene-b-Polylactide Block Copolymer

Degradable BCPs that can form nanoporous polymers have drawn considerable attention for use in nanofabrication. Polylactide-containing BCPs such as poly(styrene)-b-poly(lactide) (PS-b-PLA) and poly(styrene)-b-poly(L-lactide) (PS-b-PLLA) are highly suited to the fabrication of nanoporous polymers because of the unstable character of the ester group, which can be easily hydrolytically decomposed [66, 67]. Thus, by taking advantage of the degradable character of polylactide block, a porous network can be formed with tailored dimension and topology [68]. In other words, nanoporous polymers with well-defined nanochannels can be fabricated by selective removal of one of the blocks after microphase segregation.

Taking the above into account, the combination of the self-assembly of degradable BCPs and sol-gel chemistry results in a novel method for the fabrication of precisely controlled nanohybrids or highly porous inorganic networks.

Among all of the ordered nanostructures formed by the self-assembly of BCPs, the double gyroid is one of the most appealing morphologies for practical applications because of its unique geometry, comprising a matrix and two continuous but independent, interpenetrating networks in three-dimensional space [69].

H. Y. Hsueh and co-workers [70, 71] established a platform technology to fabricate precisely controlled nanohybrids and mesoporous ceramic materials using gyroid-forming polystyrene-b-poly(L-lactide) (PS-b-PLLA) degradable block copolymer as template agent for different sol-gel-synthesized inorganic nanoparticles. Figure 4.11 shows schematically the fabrication pathway.

Following this useful method, this group developed bicontinuous ceramics with high surface area using sol-gel-synthesized TiO_2 nanoparticles as can be visualized in Fig. 4.12 [71], and nanoporous SiO_2 inorganic materials with exceptionally low refractive index (as low as 1.1) and low absorbance fabricated also by templated sol-gel reaction [70, 72].

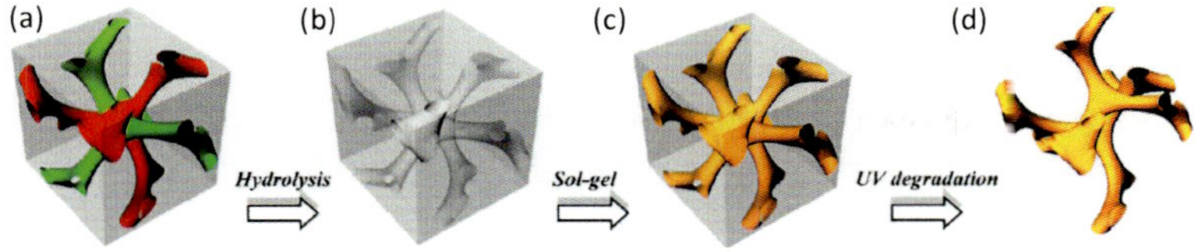

Figure 4.11 Schematic illustration for the creation of well-defined inorganic nanoporous gyroid from BCP templating. (a) PS-b-PLLA gyroid morphology (skeleton of double gyroid structure with two identical networks (green and red). (b) Gyroid-forming nanoporous PS template after the removal of minority PLLA network. (c) PS/SiO_2 gyroid nanohybrids via the templated sol-gel process. (d) Nanoporous gyroid SiO_2 after the UV removal of PS template. Reproduced with permission of [70]. Copyright 2010, American Chemical Society.

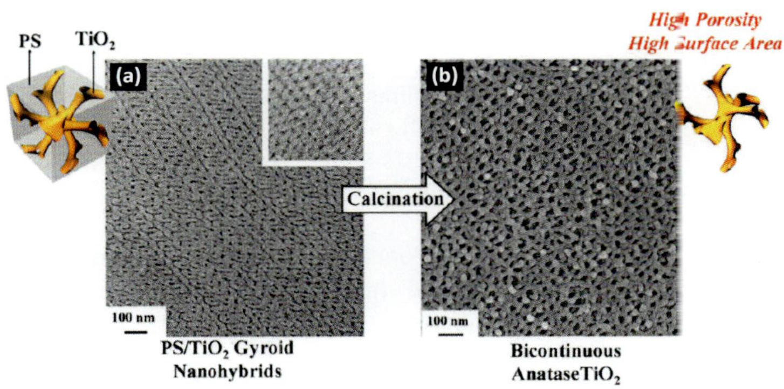

Figure 4.12 TEM micrograph of (a) the PS/TiO_2 gyroid nanohybrids without staining. The acid-catalyzed sol-gel reaction was performed using DEA (DEA/TTIP mole ratio of 0.5) as a chelating agent in methanol at 5°C. The inset shows the [220] projection of the double gyroid morphology and (b) FESEM micrograph of bicontinuous TiO_2 from the PS/TiO_2 gyroid nanohybrids after removal of the PS matrix by specific calcination process. Reproduced with permission of [71]. Copyright 2012, American Chemical Society.

T. H. Nguyen et al. employed poly(styrene)-b-poly(lactide) (PS-b-PLA) block copolymer as template for sol-gel reaction [73]. They generate silica replicas of the porous films based on a simple infiltration of the porous structure by a solution containing silica sol-gel precursors. As can be shown in Fig. 4.13 different nanostructures can be obtained using selective vapor annealing treatment and consequently, original and tunable structures can be generated.

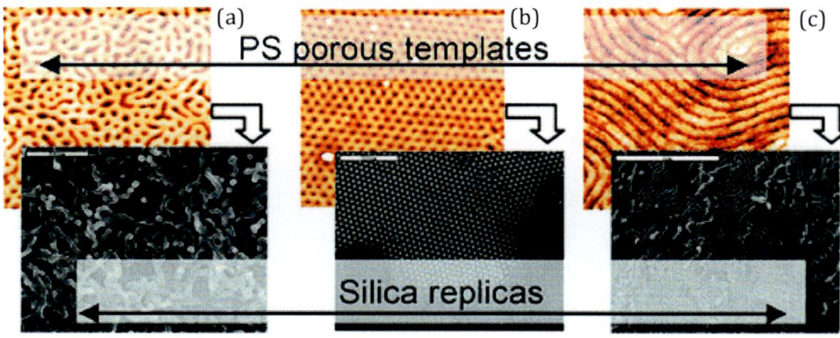

Figure 4.13 AFM (1 μm × 1 μm) images of PS porous template and SEM images of silica replica of porous network derived from PS-b-PLA thin films: (a) as spun; (b) after 4 h of exposure in THF; (c) after 8 h of exposure in THF. Reproduced with permission of [73]. Copyright 2012, American Chemical Society.

As a result, this novel approach provides a useful method to fabricate high-surface-area and high-porosity nanohybrids and mesoporous ceramic materials with self-supporting structure and controlled crystalline phase for practical applications in different fields of nanotechnology such as hybrid solar cells, batteries, fuel cells, and catalytic devices, high density systems, electronic devices, and supports for nanofluidics. Additionally, it should be noted that this pathway can be applied for different ceramic oxides including the oxides of Al, Si, Ti, Zn, and Zr which can also be fabricated from BCP templates via sol-gel process.

4.4 Polystyrene-b-Polyvynilpiridine (PS-b-PVP) Block Copolymer

Amphiphilic polystyrene-b-poly(n-vinylpyridine) (PS-b-PnVP) (n = 2, 4) is a well-known BCP that is commercially available and can be easily synthesized by sequential anionic polymerization.

PnVP block of the BCP has a broad ability to for complexes with metal ions, so patterned hybrid inorganic/organic nanostructured materials based on this BCP have been extensible studied. The amphiphilic nature of this kind of block copolymer allows the selective location of sol-gel-synthesized nanoparticles in PnVP hydrophilic block of the block copolymer by exploiting specific interactions between the PnVP block and metallic precursors. Usually, metallic precursor loaded PS-b-P4VP micelles are deposited onto a substrate by spin-coating or dip coating in order to generate hybrid inorganic/organic nanostructured thin films.

Once again the most investigated system is based on titanium dioxide nanoparticles. Several authors studied hybrid TiO_2/PS-b-PnVP nanocomposites using sol-gel process.

C.–C. Weng et al. [74] reported the synthesis of arrayed, needle-like nanostructures of the rutile phase crystal TiO_2 using ordered TiO_2 seeds prepared from a PS-b-P4VP block copolymer template and starting from titanium oxide sulfate hydrate precursor.

The preparation method can easily understand from Fig. 4.14a. Additionally, as example two SEM images of obtained TiO_2 needle film are shown in Fig. 4.14b,c.

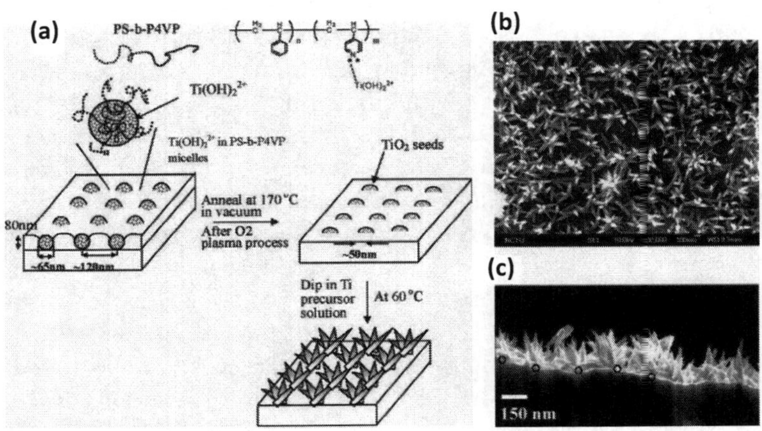

Figure 4.14 (a) Synthesis of needle-like TiO_2 nanostructures with ordered patterns, (b) SEM micrograph of TiO_2 needle film and (c) a cross-sectional profile of b. Adapted and Reproduced with permission of [74]. Copyright 2004, American Chemical Society.

These fascinating TiO_2 nanostructures can be fabricated with variable spatial positions and densities using PS-b-P4VP block copolymer templates with different molecular weights and changing the growth parameters, such as reaction times or the reaction concentration. In this research work playing with mentioned parameters, authors developed TiO_2 short and long needles with 40–50 nm and 130–150 nm in length, respectively, and with two different distances between the two needle bunches (160 and 120 nm).

The most common problems associated with the templating method are the control of the amount of titanium precursor loaded into their micellar film and the control over template morphology. With the aim to solve these difficulties, Song and co-worker [75] presented another approach named one-step or in situ reducing method using TTIP as precursor. Based on the coordination reaction between titanium isopropoxide and the P4VP block of the block copolymer, titanium isopropoxide can be loaded into the PS-b-P4VP micelle core. Thus, titanium dioxide nanoparticles can be synthesized within the PS-b-P4VP copolymer thin films. Using this micellar solution method, the final thin film morphology can be tuned using the precursor-copolymer interaction rather than just the polymer–polymer interaction or the polymer-solution interaction providing yet another way to control the morphology of nanostructures. This research work was focused on the morphological evolution of synthesized hybrid nanocomposites with different contents of titanium isopropoxide. Their AFM and transmission electron microscopy (TEM) studies indicated that the morphology of hybrid micellar films can be tuned by simply altering the metal precursor loading. A spherical to ribbon-like transition was observed at higher titanium isopropoxide to P4VP ratios. The transition could be due to fusion of the spherical micelles when titanium isopropoxide coordinates to P4VP.

Another useful route to achieve ordered metal oxide nanoparticles was reported by M. Roulet and co-workers, based on a simple impregnation of the thin films by a spin coating technique [76].

The synthesis method involves three steps: (1) deposition of PS-b-P4VP thin film by spin-coating and reorganization by selective solvent vapor annealing; (2) selective deposition of inorganic precursor on P4VP block domains of the PS-b-P4VP thin film by spin-coating; (3) synthesis of oxide nanoparticles and the elimination of the polymer scaffold by thermal annealing.

In this work, the authors demonstrated that the developed procedure works with a wide range of metallic (silicon, titanium, cerium, ruthenium, zinc and manganese) oxides.

The difference in the organization was related to the initial selective solvent vapor treatment. In the case of silica and titania nanoparticles PS-b-P4VP thin film was exposed to THF in order to have mainly P4VP cylinders perpendicular to the film surface. Comparing silica and titania SEM images with the other ones, one can clearly observed the influence of the initial annealing in the final nanoparticles array morphology. For titania and silica nanoparticles, circular shape and hexagonal organization was obtained. The difference in size of the synthesized nanoparticles was attributed to the different degree of shrinkage of precursors during the heating step. Zinc nanoparticles are the smallest ones. This phenomenon can be related with both solubility parameters and the difference of the way for the precursor's solutions to condense onto the P4VP domains. They also obtained ruthenium, cerium and manganese oxide arrays (samples prepared without selective solvent vapor treatment). The average areas, the length and the distance between nanoparticles were higher if compared with silica and titania nanoparticles which confirmed the importance of the initial vapor exposure reorganization.

Obtained results revealed the versatility of this simple and fast spin coating process. Consequently the developed protocol can be extended to the synthesis of all kinds of nano-objects which corresponding precursors are soluble in P4VP solvent confirming the.

Most recently, K. Sakar and co-workers [77] fabricated zinc oxide (ZnO) thin films with a nanogrid-like morphology via block copolymer-assisted sol–gel synthesis route. They proposed a fruitful combination of sol–gel synthesis and selective solvent vapor annealing method [78–81] to provoke a desired structural rearrangement of the hybrid nanocomposite films. They employed ZAD as ZnO precursor and hybrid nanocomposite preparation protocol can be visualized in the Fig. 4.15.

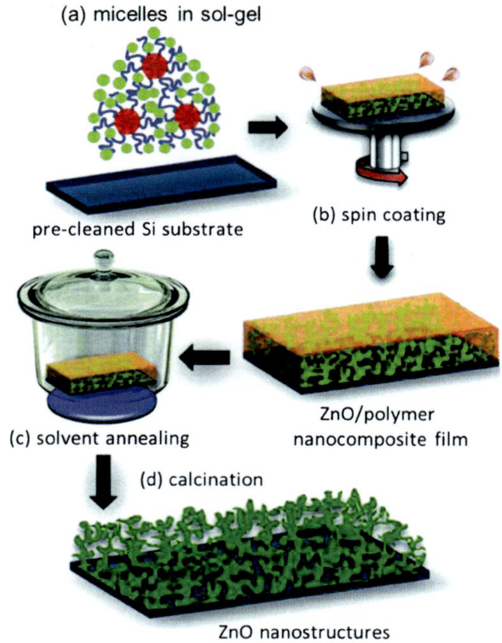

Figure 4.15 Schematic representation of the steps involved in the fabrication of ZnO composite thin films and solvent annealing: (a) preparing ZnO sol–gel solution, (b) spin-coating the polymer/ZnO nanocomposite film, (c) solvent annealing in THF vapor for different times, (d) high-temperature calcination step to obtain the nanostructured ZnO film. Reproduced with permission of [77]. Copyright 2014, Royal Society of Chemistry.

The main focus of this work was to study the final ZnO nanostructured morphology and the tunability of the length scales of these structures, for which the diblock copolymer template was removed via a high-temperature treatment after the exposure of the hybrid films to a suitable solvent vapor for a given time period.

Solvent annealing was done using tetrahydrofuran (THF) vapor and hence used as a tool to tune the pore size of the final grid-like films. During annealing of the hybrid ZnO films, the templating polymer undergoes structural rearrangements.

Generated morphology of ZnO resembling nanogrids have a large number of potential applications as active material in dye-sensitized solar cells, photonic devices, photocatalysts or chemical/gas sensors since they can be prepared on large scale samples, which is a prerequisite for successful application.

References

1. Nicole, L., Laberty-Robert, C., Rozes, L., and Sanchez, C. (2014). Hybrid materials science: A promised land for the integrative design of multifunctional materials, *Nanoscale*, **6**, pp. 6267–6292.
2. Sanchez, C., Julian, B., Belleville, P., and Popall, M. (2005). Applications of hybrid organic-inorganic nanocomposites, *J. Mater. Chem.*, **15**, pp. 3559–3592.
3. Fahmi, A., Pietsch, T., Mendoza, C., and Cheval, N. (2009). Functional hybrid materials, *Mater. Today*, **12**, pp. 44–50.
4. Mezzenga, R., and Ruokolainen, J. (2009). Nanocomposites: Nanoparticles in the right place, *Nat. Mater.*, **8**, pp. 926–923.
5. Briseno, A. L., and Yang, P. (2009). Optoelectronics: Combining chemical worlds, *Nat. Mater.*, **8**, pp. 7–8.
6. De Rosa, C., Auriemma, F., Di Girolamo, R., Pepe, G. P., Napolitano, T., and Scaldaferri, R. (2010). Enabling strategies in organic electronics using ordered block copolymer nanostructures, *Adv. Mater.*, **22**, pp. 5414–5419.
7. Lin, Y., Daga, V. K., Anderson, E. R., Gido, R. P., and Watkins, J. J. (2011). Nanoparticle-driven assembly of block copolymers: A simple route to ordered hybrid materials, *J. Am. Chem. Soc.*, **133**, pp. 6513–6516.
8. Park, S., Lee, D. H., Xu, J., Kim, B., Hong, S. W., Jeong U., Xu, T., and Russell, T. P. (2009). Macroscopic 10-terabit-per-square-inch arrays from block copolymers with lateral order, *Science*, **323**, pp. 1030–1033.
9. Rancatore, B. J., Mauldin, C. E., Tung, S. H., Wang, C., Hexemer, A., Strzalka, J., Fréchet, J. M. J., and Xu, T. (2010). Nanostructured organic semiconductors *via* directed supramolecular assembly, *ACS Nano*, **4**, pp. 2721–2729.
10. Liu, L., Shang, L., Guo, S. J., Li, D., Liu, C. Y., Qi, L., and Dong, S. J. (2009). Organic–inorganic hybrid material for the cells immobilization: Long-term viability mechanism and application in BOD sensors, *Biosens. Bioelectron.*, **25**, pp. 523–526.

11. Wan, X. J., Wang, D., and Liu, S. Y. (2010). Fluorescent pH-sensing organic/inorganic hybrid mesoporous silica nanoparticles with tunable redox-responsive release capability, *Langmuir*, **26**, pp. 15574–15579.
12. Albert, J. N. L., and Epps, T. H. III. (2010). Self-assembly of block copolymer thin films, *Mater. Today,* **13**, pp. 24–33.
13. Li, M., Coenjarts, C. A., and Ober, C. K. (2005). Patternable block copolymers, *Adv. Polym. Sci.*, **190**, pp. 183–226.
14. Li, M., and Ober, C. K. (2006). Block copolymer patterns and templates, *Mater. Today*, **9**, pp. 30–39.
15. Haryono, A., and Binder, W. H. (2006). Controlled arrangement of nanoparticle arrays in block-copolymer domains, *Small,* **2**, pp. 600–611.
16. Sarkar, B., and Alexandridis, P. (2015). Block copolymer-nanoparticle composites: Structure, functional properties, and processing, *Prog. Polym. Sci.*, **40**, pp. 33–62.
17. Lin, Y., Daga, V. K., Anderson, E. R., Gido, S. P., and Watkins, J. J. (2011). Nanoparticle-driven assembly of block copolymers: A simple route to ordered hybrid materials, *J. Am. Chem. Soc.*, **133**, pp. 6513–6516.
18. Lin, Y., Daga, V. K., Anderson, E. R., Gido, S. P., and Watkins, J. J. (2011). Nanoparticle-driven assembly of block copolymers: A simple route to ordered hybrid materials, *J. Am. Chem. Soc.*, **133**, pp. 6513–6516.
19. Taheri, S. M., Fischer, S., and Förster, S. (2011). Routes to nanoparticle-polymer superlattices, *Polymers*, **3**, pp. 662–673.
20. Vukovic, I., ten Brinke, G., and Loos, K. (2013). Block copolymer template-directed synthesis of well-ordered metallic nanostructures, *Polymer*, **54**, pp. 2591–2605.
21. Matsen, M. W., and Thompson, R. B. (2008). Particle distributions in a block copolymer nanocomposite, *Macromolecules*, **41**, pp. 1853–1860.
22. Kang, H., Detcheverry, F. A., Mangham, A. N., Stoykovich, M. P., Daoulas, K. C., Hamers, R. J., Müller, M., de Pablo, J. J., and Nealey, P. F. (2008). Hierarchical assembly of nanoparticle superstructures from block copolymer-nanoparticle composites, *Phys. Rev. Lett.*, **100**, p. 148303.
23. Gaines, M. K., Smith, S. D., Samseth, J., Bockstaller, M. R., Thompson, R. B., Rasmussen, K. O., and Spontak, R. J. (2008). Nanoparticle-regulated phase behavior of ordered block copolymers, *Soft Matter*, **4**, pp. 1609–1612.

24. Gupta, A. K., and Gupta, M. (2005). Synthesis and surface engineering of iron oxide nanoparticles for biomedical applications, *Biomaterials*, **26**, pp. 3995–4021.

25. Jain, P. K., Huang, X. H., El-Sayed, I. H., and El-Sayed, M A. (2008). Noble metals on the nanoscale: Optical and photothermal properties and some applications in imaging, sensing, biology, and medicine, *Acc. Chem. Res.*, **41**, pp. 1578–1586.

26. Stewart, M. E., Anderton, C. R., Thompson, L. B., Maria, J. Gray, S. K., Rogers, J. A., and Nuzzo, R. G. (2008). Nanostructured plasmonic Sensors, *Chem. Rev.*, **108**, pp. 494–521.

27. Arico, A. S., Bruce, B., Scrosati, B., Tarascon, J. M., and van Schalkwijk, W. (2005). Nanostructured materials for advanced energy conversion and storage devices, *Nat. Mater.*, **4**, pp. 366–377.

28. Balazs, A. C., Emrick, T., and Russell, T. P. (2006). Nanoparticle polymer composites: Where two small worlds meet, *Science*, **314**, pp. 1107–1110.

29. Bockstaller, M. R., Mickiewicz, R. A., and Thomas, E. L. (2005). Block copolymer nanocomposites: Perspectives for tailored functional materials, *Adv. Mater.*, **17**, pp. 1331–1349.

30. Sudeep, P. K., and Emrick, T. (2009). Pearls of wisdom: Stringing nanoparticles and polymers into new assemblies and materials, *ACS Nano*, **3**, pp. 2870–2875.

31. Brunet-Bruneau, A., Amenitsch, H., Albouy, P. A., and Sanchez, C. (2003). Highly porous TiO_2 anatase optical thin films with cubic mesostructure stabilized at 700°C, *Chem. Mater.*, **15**, pp. 4562–4570.

32. Dupuis, A. C. (2005). The catalyst in the CCVD of carbon nanotubes: a review, *Prog. Mater. Sci.*, **50**, pp. 929–961.

33. Grosso, D., Soler-Illia, G., Crepaldi, E. L., Cagnol, F., Sinturel, C., Bourgeois, A., Hutter, E., and Fendler, J. H. (2004). Exploitation of localized surface plasmon resonance, *Adv. Mater.*, **16**, pp. 1685–1706.

34. Murray, C. B., Kagan, C. R., and Bawendi, M. G. (2000). Synthesis and characterization of monodisperse nanocrystals and close-packed nanocrystal assemblies, *Ann. Rev. Mater. Sci.*, **30**, pp. 545–610.

35. Pardo, R., Zayat, M., and Levy, D. (2011). Photochromic organic–inorganic hybrid materials, *Chem. Soc. Rev.*, **40**, pp. 672–687.

36. Brinker, C. J., and Scherer, G. W. (1990). *Sol-Gel Science*. Academic Press, San Diego.

37. Orilall, M. C., and Wiesner, U. (2011). Block copolymer based composition and morphology control in nanostructured hybrid materials for energy conversion and storage: Solar cells, batteries, and fuel cells, *Chem. Soc. Rev.*, **40**, pp. 520–535.

38. Sun, Z., and Gutmann, J. S. (2004). Synthesis of TiO_2 nanoparticles in ultrathin block copolymer films-an integral geometry study, *Phys. A,* **339**, pp. 80–85.

39. Sun, Z., Wolkenhauer, M., Bumbu, G. G., Kim, D. H., and Gutmann, J. S. (2005), GISAXS investigation of TiO_2 nanoparticles in PS-b-PEO block copolymer films, *Phys. B,* **357**, pp. 141–143.

40. Sun, Z., Kim, D. H., Wolkenhauer, M., Bumbu, G. G., Knoll, W., and Gutmann, J. S. (2006). Synthesis and photoluminescence of titania nanoparticle arrays templated by block-copolymer thin films, *Chem Phys Chem*, **7**, pp. 370–378.

41. Kim, D. H., Kim, S. H., Lavery, K., and Russell, T. P. (2004). Inorganic nanodots from thin films of block copolymers, *Nano Lett.,* **4**, pp. 1841–1844.

42. Kim, D. H., Sun, Z. C., Russell, T. P., Knoll, W., and Gutmann, J. S. (2005). Organic–inorganic nanohybridization by block copolymer thin film *Adv. Funct. Mater.*, **15**, pp. 1160–1164.

43. Cheng, Y. J., and Gutmann, J. S. (2006). Morphology phase diagram of ultrathin anatase TiO_2 films templated by a single PS-*b*-PEO block copolymer, *J. Am. Chem. Soc.*, **128**, pp. 4658–4674.

44. Perlich, J., Kaune, G., Memesa, M., Gutmann, J. S., and Müller-Buschbaum, P. (2009). Sponge-like structures for application in photovoltaics, *Phil. Trans. R. Soc. A,* **367**, pp. 1783–1798.

45. Peng, J., Li, X., Kim, D. H., and Knoll, W. (2007). Fabrication and photocatalytic activities of morphology-controlled titania nanoobject arrays by block copolymer templates, *Macromol. Rapid Commun.*, **28**, pp. 2055–2061.

46. Li, X., Peng, J., Kang, J. H., Choy, J. H., Steinhart, M., Knoll, W., and Kim, D. H. (2008). One step route to the fabrication of arrays of TiO_2 nanobowls *via* a complementary block copolymer templating and sol-gel process, *Soft Matter,* **4**, pp. 515–521.

47. Cheng, Y. J., Zhou, S., Wolkenhauer, M., Bumbu, G. G., Lenz, S., Memesa, M., Nett, S., Emmerling, S., Steffen, W., and Gutmann, J. S. (2013). Morphology evolution in mesoporous titania block copolymer composite films with increasing sol-gel reaction time, *Eur. J. Inorg. Chem.*, pp. 1127–1133.

48. Rawolle, M., Niedermeier, M. A., Kaune, G., Perlich, J., Lellig, P., Memesa, M., Cheng, Y. J., Gutmann, J. S., and Buschbaum, P. M. (2012). Fabrication and characterization of nanostructured titania films with integrated function from inorganic-organic hybrid materials, *Chem. Soc. Rev.*, **41**, pp. 5131–5142.

49. Gutierrez, J., Tercjak, A., Garcia, I., Peponi. L., and Mondragon, I. (2008). Hybryd titanium dioxide/PS-b-PEO block copolymer nanocomposites based on sol-gel synthesis, *Nanotechnology*, **19**, pp. 155607/1–155607/8.

50. Gutierrez, J., Tercjak, A., and Mondragon, I. (2010). Conductive behavior of high TiO_2 nanoparticle content of inorganic/organic nanostructured composites, *J. Am. Chem. Soc.*, **132**, pp. 873–878.

51. Gutierrez, J., Tercjak, A., Peponi, L., and Mondragon, I. (2009). Conductive properties of inorganic and organic TiO_2/polystyrene-*block*-poly(ethylene oxide) nanocomposites, *J. Phys. Chem. C*, **113**, pp. 8601–8605.

52. Liu, Z., Li, Y., Zhao, Z., Cui, Y., Hara, K., and Miyauchi, M. (2010). Block copolymer templated nanoporous TiO_2 for quantum-dot-sensitized solar cells, *J. Mater. Chem.*, **20**, pp. 492–497.

53. Gutierrez, J., Tercjak, A., Martin, L., and Mondragon, I. (2011). Electrical properties of TiO_2/SEO nanocomposites: From macro to nano, *Electrochim. Acta*, **56**, pp. 5582–5586.

54. Cha, M. A., Shin, C., Kannaiyan, D., Jang, Y. H., Kochuveedu, S. T., Ryu, D. Y., and Kim, D. H. (2009). A versatile approach to the fabrication of TiO_2 nanostructures with reverse morphology and mesoporous Ag/TiO_2 thin films cooperative PS-b-PEO self-assembly and a sol-gel process, *J. Mater. Chem.*, **19**, pp. 7245–7250.

55. Kannaiyan, D., Cha, M. A., Jang, Y. H., Sohn, B. H., Huh, J., Park, C., and Kim, D. H. (2009). Efficient photocatalytic hybrid Ag/TiO_2 nanodot arrays integrated into nanopatterned block copolymer thin films, *New J. Chem.*, **33**, pp. 2431–2436.

56. Kannaiyan, D., Kim, E., Won, N., Kim, K. W., Jang, Y. H., Cha, M. A., Ryu, D. Y., Kim, S., and Kim, D. H. (2010). On the synergistic coupling properties of composite CdS/TiO_2 nanoparticle arrays confined in nanopatterned hybrid thin films, *J. Mater. Chem.*, **20**, pp. 677–682.

57. Li, X., Peng, J., Kang, J. H., Choy, J. H., Steinhart, M., Knoll, W., and Kim, D. H. (2008). One step rote to the fabrication of arrays of TiO_2 nanobowls *via* a complementary block copolymer templating and sol-gel process, *Soft Matter*, **4**, pp. 515–521.

58. Gutierrez, J., Mondragon, I., and Tercjak, A. (2014). Quantitative nanoelectrical and nanomechanical properties of nanostructured hybrid composites by peakforce tunneling atomic force microscopy, *J. Phys. Chem. C*, **118**, pp. 1206–1212.
59. Sarkar, K., Braden, E. V., Fröschi, T., Hüsing, N., and Buschbaum, P. M. (2014). Spray-deposited zinc titanate films obtained *via* sol-gel synthesis for application in dye-sensitized solar cells, *J. Mater. Chem. A*, **2**, pp.15008–15014.
60. Docampo, P., Guldin, S., Stefik, M., Tiwana, P., Orilall, M. C., Huttner, S., Sai, H., Wiesner, U., Steiner, U., and Snaith, H. J. (2010). Control of solid-state dye-sensitized solar cell performance by block-copolymer-directed TiO_2 synthesis, *Adv. Funct. Mater.*, **20**, pp. 1787–1796.
61. Rawolle, M., Niedermeier, M. A., Kaune, G., Perlich, J., Lellig, P., Memesa, M., Cheng, Y. J., Gutmann, J. S., and Muller-Buschbaum, P. (2012). Fabrication and characterization of nanostructured titania films with integrated function from inorganic–organic hybrid materials, *Chem. Soc. Rev.*, **41**, pp. 5131–5142.
62. Chen, X., and Mao, S. S. (2007). Titanium dioxide nanomaterials: Synthesis, properties, modifications, and applications, *Chem. Rev.*, **107**, pp. 2891–2959.
63. Kubacka, A., Fernandez-Garcia, M., and Colon, G. (2012). Advanced nanoarchitectures for solar photocatalytic applications, *Chem. Rev.*, **112**, pp. 1555–1614.
64. Chandra, D., Bekki, M., Nakamura, M., Sonezaki, S., Ohji, T., Kato, K., and Kimura, T. (2011). Dye-sensitized biosystem sensing using macroporous semiconducting metal oxide films, *J. Mater. Chem.*, **21**, pp. 5738–5744.
65. Crepaldi, E. L., Soler-Illia, G., Grosso, D., Cagnol, F., Ribot, F., and Sanchez, C. (2003). Controlled formation of highly organized mesoporous titania thin films: From mesostructured hybrids to mesoporous nanoanatase TiO_2, *J. Am. Chem. Soc.*, **125**, pp. 9770–9786.
66. Zalusky, A. S., Olayo-Valles, R., Wolf, J. H., and Hillmyer, M. A. (2002). Ordered nanoporous polymers from polystyrene-polylactide block copolymers, *J. Am. Chem. Soc.*, **124**, pp. 12761–12773.
67. Ho, R. M., Tseng, W. H., Fan, H. W., Chiang, Y. W., Lin, C. C., Ko, B. T., and Huang, B. H. (2005). Solvent-induced microdomain orientation in polystyrene-b-poly(l-lactide) diblock copolymer thin films for nanopatterning, *Polymer*, **46**, pp. 9362–9377.

68. Olson, D. A., Chen, L., and Hillmyer, M. A. (2008). Templating nanoporous polymers with ordered block copolymers, *Chem. Mater.*, **20**, pp. 869–890.

69. Hajduk, D. A., Harper, P. E., Gruner, S. M., Honeker, C. C., Kim, G., and Thomas, E. L. (1994). The gyroid: A new equilibrium morphology in weakly segregated diblock copolymers, *Macromolecules,* **27**, pp. 4063–4075.

70. Hsueh, H. Y., Chen, H. Y., She, M. S., Chen, C. K., Ho, R. M., Gwo, S., Hasegawa, H., and Thomas, E. L. (2010). Inorganic gyroid with exceptionally low refractive index from block copolymer templating, *Nano Lett.*, **10**, pp. 4994–5000.

71. Hsueh, H. Y., and Ho, R. M. (2012). Bicontinuous ceramics with high surface area from block copolymer templates, *Langmuir,* **28**, pp. 8518–8529.

72. Wang, H. F., Yu, L. H., Wang, X. B., and Ho, R. M. (2014). A facile method to fabricate double gyroid as a polymer template for nanohybrids, *Macromolecules,* **47**, pp. 7993–8001.

73. Nguyen, T. H., Vayer, M., Grosso, D., Amenitsch, H., and Sinturel, C. (2012). Using sol–gel replications to assess the porosity of block-copolymer derived thin films. *J. Phys. Chem. C,* **116**, pp. 5295–5302.

74. Weng, C. C., Hsu, K. F., and Wei, K. H. (2004). Synthesis of arrayed TiO_2 needlelike nanostructures via a polystyrene-*block*-poly(4-vinylpyridine) diblock copolymer template, *Chem. Mater.,* **16**, pp. 4080–4086.

75. Song, L., Lam, Y. M., Boothroyd, C., and Teo, P. W. (2007). One-step synthesis of titania nanoparticles from PS-P4VP diblock copolymer solution, *Nanotechnology*, **18**, p. 135605(6 pp.).

76. Roulet, M., Vayer, M., and Sinturel, C. (2013). A simple route to ordered metal oxide nanoparticle arrays using block copolymer thin films, *Eur. Polym. J.*, **49**, pp. 3897–3903.

77. Sarkar, K., Schaffer, C. J., Mosegui-Gonzalez, D., Naumann, A., Perlich, J., and Müller-Buschbaum, P. (2014). Tuning the pore size of ZnO nano-grids *via* time-dependent solvent annealing, *J. Mater. Chem. A*, **2**, pp. 6945–6951.

78. Li, X., Peng, J., Wen, Y., Kim, D. H., and Knoll, W. (2007). Morphology change of asymmetric diblock copolymer micellar films during solvent annealing, *Polymer*, **48**, pp. 2434–2443.

79. Kim, S. H., Misner, M. J., and Russell, T. P. (2004). Solvent-induced ordering in thin film diblock copolymer/homopolymer mixtures, *Adv. Mater.*, **16**, pp. 2119–2123.

80. Gutierrez, J., Tercjak, A., Garcia, I., and Mondragon, I. (2009). The effect of thermal and vapor annealing treatments on the self-assembly of TiO_2/PS-b-PMMA nanocomposites generated via the sol-gel process, *Nanotechnology*, **20**, pp. 225603/1–225603/9.

81. Kim, B., Hong, S. W., Park, S., Xu, S., Hong, S. K., and Russell, T. P. (2011). Phase transition behavior in thin films of block copolymers by use of immiscible solvent vapors, *Soft Mater*, **7**, pp. 443–447.

Chapter 5

Nanostructured Epoxy-Based Thermosetting Materials Modified with Amphiphilic Block Copolymers

Agnieszka Tercjak

Materials + Technologies Group (GMT),
Department of Chemical and Environmental Engineering,
Polytechnic School, University of the Basque Country (UPV/EHU),
Plaza Europa 1, 20018 Donostia-San Sebastián, Spain

agnieszka.tercjaks@ehu.es

This chapter is focused on the employment of different amphiphilic block copolymers for the fabrication of nanostructured thermosetting materials. Immiscible block of this type of block copolymers microphase separated from the thermoset epoxy-based matrices leading to nanostructured thermosetting materials. Moreover, the addition of amphiphilic block copolymer increases the toughness of nanostructured thermosetting materials making them potential candidate as template for dispersion and localization of inorganic and organic nanoentities.

Block Copolymer Nanocomposites
Edited by Galder Kortaberria and Agnieszka Tercjak

ISBN 978-981-4669-54-2 (Hardcover), 978-981-4669-55-9 (eBook)
www.panstanford.com

5.1 Introduction

Thermosets based on epoxy resin are one of the important classes of thermosetting polymers which show a broad range of application as high-performance materials ensuring excellent adhesion, chemical and heat resistance, good mechanical properties and very good electrical insulating properties as well as high optical transparency. The main drawback of this kind of thermosetting polymers from the point of view of their applications as adhesives, surface coatings and composites matrices for aeronautic or electronic is low fracture toughness.

Several different approaches were explored to improve the fracture toughness of epoxy thermosets. However, one of the successful pathways to achieve high improvements on the toughness of these systems is incorporation of homopolymers, both elastomers [1–5] and thermoplastics [5–9], and block copolymers (BCPs) [10–14]. Use of the BCPs not only improves the toughness of thermosetting polymers but also leads to nanostructured thermosets, which can act as templates for dispersion and selective localization of low-molecular-weight organic molecules (azobencen and liquid crystals, which will be described in more details in Chapter 7) or inorganic nanoobjects such as nanoparticles, carbon nanotubes, nanofibers and others.

One can distinguish two different pathways for generation of nanostructured thermosets based on epoxy resin (Scheme 5.1). The first one is related to employing amphiphilic block copolymers, which consist of both epoxy-miscible and epoxy-immiscible blocks. The second one is based on generation of nanostructured thermosets using the concept of chemical compatibilization, where block copolymers with reactive groups incorporated into one block can promote interaction with epoxy resin during network formation.

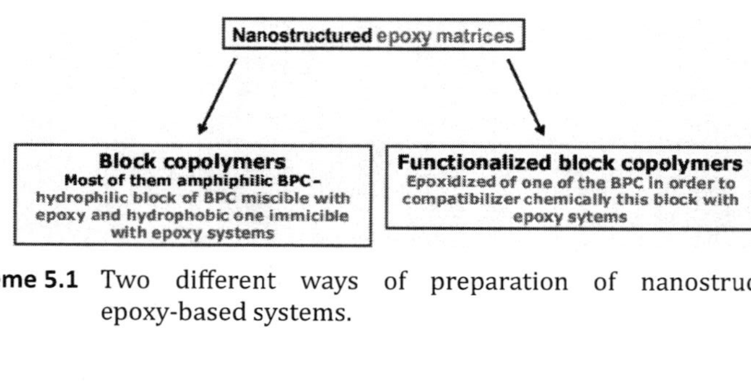

Scheme 5.1 Two different ways of preparation of nanostructured epoxy-based systems.

The employment of the second concept of fabrication of nanostructured thermosets will be described in details in Chapter 6.

5.1.1 Methods of Formation of Nanostructured Thermosetting Systems

Nanostructured thermosetting materials can be formed by following two different mechanisms. In the first one, the epoxy precursor acts as a selective solvent; consequently, microphase separation takes place before the curing reaction and the epoxy network formation process only fixes the final morphology of nanostructured thermosetting materials. Thus, the self-assembly of the immiscible, with epoxy precursors, block of block copolymer separated in BCP and epoxy precursors mixtures.

In the second pathway, the microphase separation of immiscible with epoxy precursor block of block copolymer takes place during reaction-induced phase separation (RIPS). Thus, the mixture of BCP and epoxy precursors is miscible before curing and separation takes place during network formation.

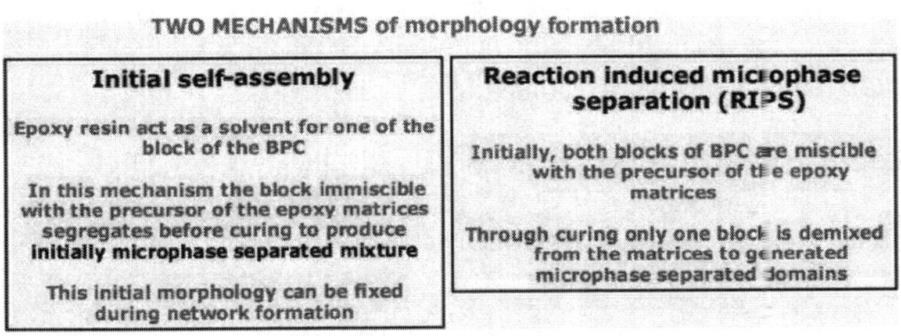

Scheme 5.2 Two mechanisms of morphology formation in nanostructured epoxy-based systems.

5.1.2 Amphiphilic Block Copolymers for Nanostructured Thermosets

One of the much known pathways for preparation of nanostructured thermosetting materials is the use of amphiphilic block copolymers with one block miscible and one immiscible with thermosetting matrices.

Table 5.1 Amphiphilic block copolymers used as nanostructured agent

Block copolymers	Abbreviation	Thermosetting precursor	References
Poly(ethylene oxide)-b-poly(ethyl ethylene)	PEO-b-PEE	DGEBA + PA	[16]
Poly(ethylene oxide)-b-poly(ethylene-alt-propylene)	PEO-b-PEP	DGEBA + MDA	[17]
Poly(ethylene oxide)-b-poly(propylene oxide)	PEO-b-PPO	DGEBA + MDA	[18]
Poly(ethylene oxide)-b-poly(propylene oxide)-b-poly(ethylene oxide)	PEO-b-PPO-b-PEO	DGEBA + MDA DGEBA + MXDA	[18, 19, 21, 24, 25, 29, 30] [14]
Polyethylene-b-poly(ethylene oxide)	PE-b-PEO	DGEBA + MDA DGEBA + MCDEA	[20] [26]
Poly(hexylene oxide)-b-poly(ethylene oxide)	PHO-b-PEO	DGEBA + PN	[12]
Poly(ethylene oxide)-b-poly(dimethylsiloxane)	PEO-b-PDMS	DGEBA + MDA	[23]
Poly(ethylene oxide)-b-poly(ε-caprolactone)	PEO-b-PCL	DGEBA + MOCA	[27]
Poly(ethylene oxide)-b-polystyrene	PEO-b-PS	DGEBA + MDA DGEBA + MXDA DGEBA + MCDEA DGEBA + DDM	[28] [32–34] [35–38] [39]
Poly(ε-caprolactone)-b-polybutadiene-b-poly(ε-caprolactone)	PCL-b-PB-b-PCL	DGEBA + MOCA	[40]
Poly(ε-caprolactone)-b-poly(n-butyl acrylate)	PCL-b-PBA	DGEBA + MOCA	[41]
Poly(heptadecafluorodecyl acrylate)-b-poly(caprolactone)	PaF-b-PCL	DGEBA + MCDEA	[42]
Polydimethylsiloxane-b-poly(ε-caprolactone)-b-polystyrene	PDMS-b-PCL-b-PS	DGEBA + MOCA	[43]
Poly(ε-caprolactone)-b-polystyrene	PCL-b-PS	DGEBA + MOCA	[44]

The microphase separation of the immiscible block of BCP leads to nanostructured thermosetting materials. In this field, many research groups [10, 12, 15–38] have effectively worked on the fabrication and characterization of nanostructured thermosetting materials employing amphiphilic block copolymers mainly containing PEO-block [10, 12, 16–39], PCL-block [27, 40–45], and PMMA-block [11, 46–49].

5.2 Self-Assembly of BCP before Curing

Hillmyer et al. [16] developed the first nanostructured epoxy resin-based thermosetting system using amphiphilic poly(ethylene oxide)-b-poly(ethyl ethylene) block copolymer (PEO-b-PEE). They employed low-molecular-weight diglycidyl ether of bisphenol A (DGEBA) as epoxy monomer and phthalic anhydride (PA) as hardener. As it is well known, before the curing reaction the PEO block of synthesized PEO-b-PEE block copolymer [50, 51] is miscible with epoxy resin, while the PEE block is immiscible. Consequently, precursors of thermosetting system act as a selective solvent for PEO-b-PEE block copolymer and lead to microphase separation of the PEE block before curing. As can be seen in Fig. 5.1, the small-angle neutron scattering (SANS) confirm the microphase separation of the PEE block before curing reaction.

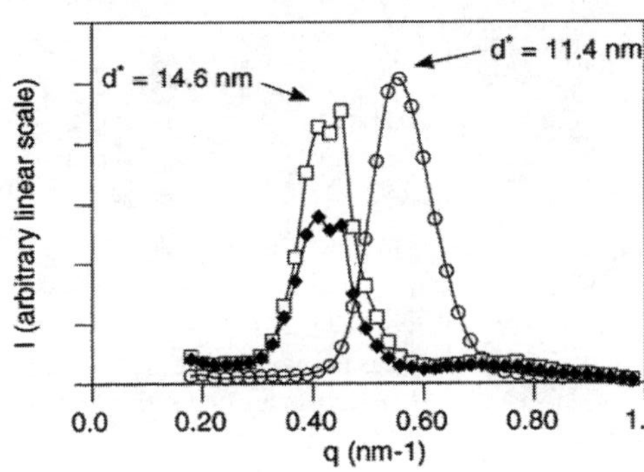

Figure 5.1 SANS profiles for PEO-b-PEE block copolymer (open circles, 75°C) and a mixture of an PEO-b-PEE-(DGEBA/PA) containing 30 wt% PEO-b-PEE block copolymer: before curing (open squares, 75°C) and cured (filled diamonds, 200°C). Reproduced with permission of [16]. Copyright 1997, American Chemical Society.

SANS profile of PEO-b-PEE block copolymers at 75°C with principal reflection at q^* equal 0.55 nm^{-1} allowed the authors to identify interplanar spacing of 11.4 nm. The 30 wt% PEO-b-PEE-(DGEBA/PA) system before curing showed interplanar spacing equal to 14.6 nm, while the same thermosetting system after curing showed 15 nm.

The hexagonally packed cylinder morphology of 25 wt% PEO-b-PEE-(DGEBA/PA) cured system was confirmed by transmission electron microscopy (TEM). In order to visualize the morphology of this system, the sample was cut and stained with RuO_4 vapor. The dark circular regions in the TEM image correspond to PEO shells of the core/shell morphology clearly identified based on TEM measurement (Fig. 5.2). Simultaneously, the white regions in TEM image correspond to epoxy immiscible PEE block.

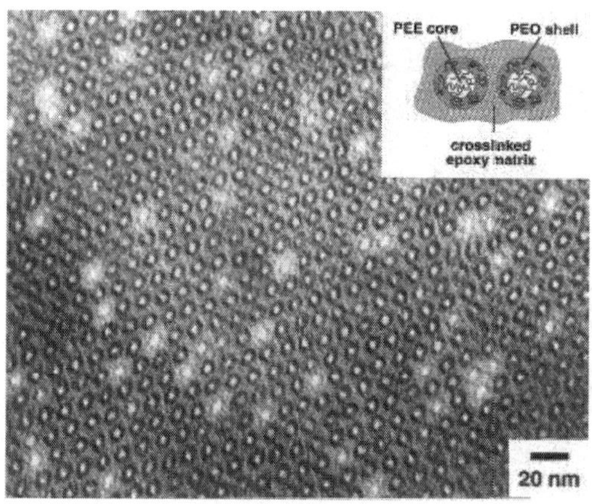

Figure 5.2 TEM of PEO-b-PEE-(DGEBA/PA) cured system (25 wt% PEO-b-PEE block copolymer). The sample was microtomed at room temperature and stained with RuO_4. The inset depicts the idealized nanostructure. Reproduced with permission of [16]. Copyright 1997, American Chemical Society.

Almost in the same time, Lipic et al. [17] from the same research group used synthesized poly(ethylene oxide)-b-poly(ethylene-alt-propylene) (PEO-b-PEP) diblock copolymer [52, 53] to nanostructured thermosetting system based on a low-molecular-weight DGEBA as epoxy monomer and methylene dianiline (MDA) as aromatic amine hardener. As shown in Fig. 5.3, mixtures of DGEBA epoxy resin modified with different PEO-b-PEP block copolymer content were nanostructured before curing. The morphologies of PEO-b-PEP/DGEBA mixtures depend strongly on the PEO-b-PEO block copolymer content and the temperature in which measurement was done. The authors reported that the SAXS measurement confirmed self-organization of PEO-b-PEP block copolymer in PEO-b-PEP/DGEBA mixtures. Their morphologies were changed from the lamellar (L), bicontinuous cubic gyroid (G), hexagonally packed cylindrical (C) to body-centered cubic packed spherical (S) structures.

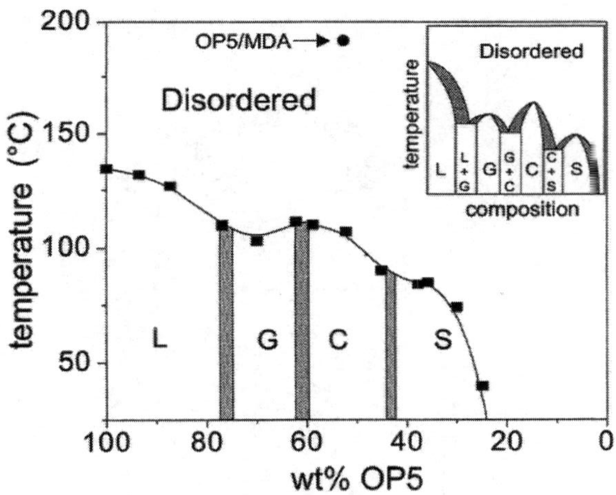

Figure 5.3 Phase diagram of PEO-b-PEP/DGEBA blends. Filled squares represent TODT's measured by dynamic mechanical spectroscopy during isochronal heating. Ordered phase morphologies (lamellar (L), gyroid (G), cylindrical (C), and spheres (S)) were identified with SAXS. The shaded regions denote two-phase coexistence. A thermodynamically proper phase diagram (i.e., one that conforms to the Gibb's phase rule) with expected coexistence windows between the ordered phases and the disordered phase is given in the inset. TODT for an PEO-b-PEP/MDA blend containing 52 wt% PEO-b-PEP block copolymer (OP5) is identified with the filled circle. Reproduced with permission of [17]. Copyright 1998, American Chemical Society.

Lipic et al. [17] reported also that the nanostructures detected in PEO-b-PEP/DGEBA mixtures are maintained in thermosetting systems modified with PEO-b-PEP block copolymer as shown in Fig. 5.4. In order to know the structure of 52 wt% PEO-b-PEP-(DGEBA/MDA) thermosetting system, the authors monitored the samples employing small-angle X-ray scattering (SAXS) during 8 h at 100°C. They proved that gelation of the curing process does not significantly affect the principal spacing, d*. Additionally, the increase in the different molecular weights of the epoxy resin in PEO-b-PEP/DGEBA mixtures leads to a 16% increase in spacing (from 9.1 to 10.6 nm for 348 and 6100 g/mol epoxy resin molecular weights, respectively).

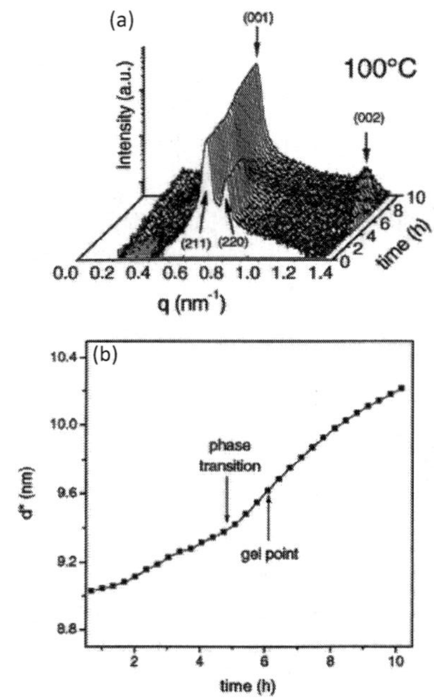

Figure 5.4 In situ SAXS data obtained during curing of 69 wt% PEO-b-PEP (DGEBA/MDA) mixture. (a) SAXS patterns collected while the sample was held at 100°C for 10 h. Extinction of the (220) reflection and growth of the (002) reflection indicates a transition from the G to the L phase at about 5 h of cure. (b) Change in principal spacing, d*, as a function of cure time. Reproduced with permission of [17]. Copyright 1998, American Chemical Society.

The authors also confirmed that the results obtained using SAXS are in good agreement with the TEM image of 52 wt% PEO-b-

PEP-(DGEBA/MDA) cured system, which clearly indicated lamellar structure, where the PEP block forms darker cores surrounded by a PEO shells well dispersed in the brighter epoxy matrix.

5.2.1 Poly(Ethylene Oxide)-b-Poly(Propylene Oxide)-b-Poly(Ethylene Oxide)

One of the first fabricated nanostructured systems using poly (ethylene oxide)-b-poly(propylene oxide)-b-poly(ethylene oxide) triblock copolymer (PEO-b-PPO-b-PEO) with different PEO block content as nanostructured agent was published by Guo et al. [19]. They used a diglycidyl ether of bisphenol A (DGEBA) and 4,4′-methylenedianiline (MDA) as epoxy precursor. The authors confirmed the nanostructuration of thermosetting materials using AFM and TEM. TEM images of PEO-b-PPO-b-PEO-(DGEBA/MDA) cured systems modified with PEO-b-PPO-b-PEO triblock copolymer containing 30 wt% PEO block are shown in Fig. 5.5 and 80 wt% PEO block in Fig. 5.6. All investigated thermosetting materials had heterogeneous morphology at the nanoscale level. The investigated system was stained with RuO_4; consequently the dark area corresponded to the separated PPO block-rich phase, while the bright area corresponded to PEO block/epoxy-rich matrices.

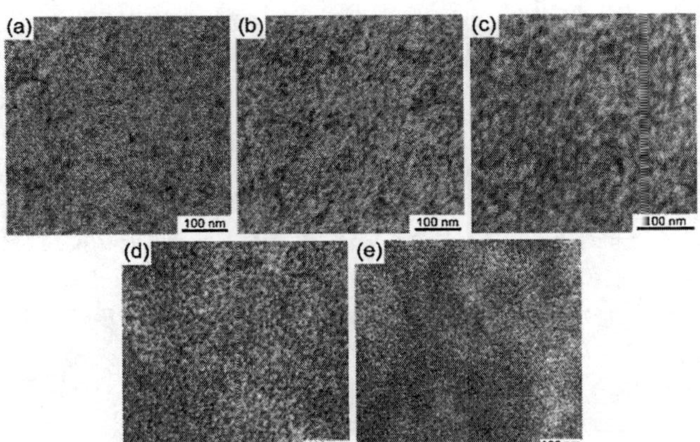

Figure 5.5 TEM images of PEO-b-PPO-b-PEO-(DGEBA/MDA) cured systems containing (a) 10, (b) 20, (c) 30, (d) 40, and (e) 50 wt% PEO-b-PPO-b-PEO block copolymer with 30 wt% PEO block. Reproduced with permission of [19]. Copyright 2002, American Chemical Society.

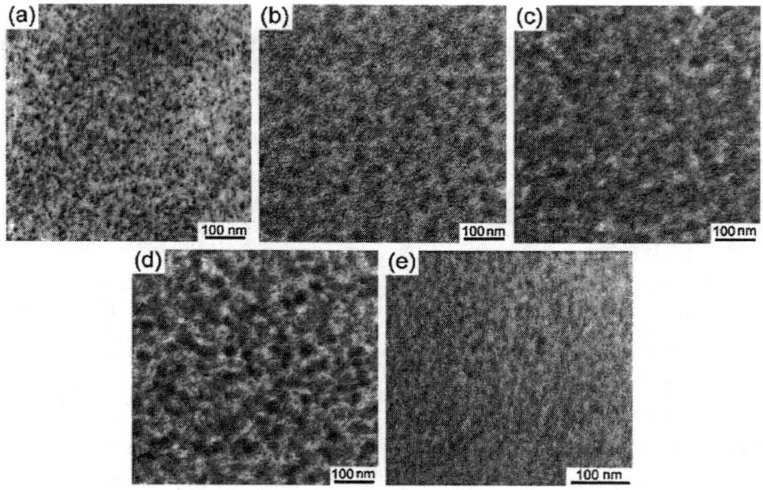

Figure 5.6 TEM images of PEO-b-PPO-b-PEO-(DGEBA/MDA) cured systems containing (a) 10, (b) 20, (c) 30, (d) 40, and (e) 50 wt% PEO-b-PPO-b-PEO block copolymer with 80 wt% PEO block. Reproduced with permission of [19]. Copyright 2002, American Chemical Society.

Thermosetting materials nanostructured with PEO-b-PPO-b-PEO triblock copolymer containing 30 wt% PEO block possessed spherical micelles morphology with separated PPO domains size equal to 10 nm up to 20 wt% PEO-b-PPO-b-PEO-(DGEBA/MDA) cured systems. For the same modifier, addition of 30 wt% BCP into DGEBA/MDA cured system led to worm-like micelles morphology, where most of spherical micelles appeared to be connected between each other forming bicontinous morphology structure for more than 30 wt% BCP. Simultaneously, PEO-b-PPO-b-PEO-(DGEBA/MDA) cured systems modified with PEO-b-PPO-b-PEO triblock copolymer containing 80 wt% PEO block showed spherical and worm-like micelles morphology. The size of microphase separated PPO block domains were around 10 nm for thermosetting system modified with 10 wt% PEO-b-PPO-b-PEO block copolymer. The increase of BCP content led to changes in the morphology just more microphase separated PEO micelles appeared as worm-like micelles finally forming bicontinous morphology. The size of microphase separated domains changed from 50 to 100 nm for thermosetting materials containing

20–50 wt% PEO-b-PPO-b-PEO block copolymer containing 80 wt% PEO block.

The research work published by Guo et al. [19] related to AFM results of PEO-b-PPO-b-PEO-(DGEBA/MDA) cured systems modified with PEO-b-PPO-b-PEO block copolymer containing both 30 and 80 wt% PEO block proved the microphase separation of PPO block in PEO block/epoxy-rich matrices. AFM results confirmed the size of separated domains. AFM results are in good agreement with TEM results described above.

Larrañaga et al. [21, 24, 25, 30] used PEO-b-PPO-b-PEO block copolymer to studied competition between macro- and microseparation in epoxy system based on a diglycidyl ether of bisphenol A (DGEBA) and 4,4′-diaminodiphenylmethane (DDM). The authors studied the influence of the curing reaction on the final morphologies. They reported that in order to design nanostructured thermosetting materials based on PEO-b-PPO-b-PEO block copolymer, LCST behavior should be taken into account. Consequently, the curing conditions could be chosen in the function of miscibility between PEO-b-PPO-b-PEO block copolymer and epoxy precursors.

Authors proved that depend on the relation between PEO and PPO blocks in PEO-b-PPO-b-PEO block copolymer precuring at lower temperature (80°C) led to nanostructured PEO-b-PPO-b-PEO-(DGEBA/DDM) cured systems while high temperature (140°C) led to macrostructured PEO-b-PPO-b-PEO-(DGEBA/DDM) cured systems. Microphase separated character of PEO-b-PPO-b-PEO-(DGEBA/DDM) cured systems with 10, 20 and 30 wt% PEO-b-PPO-b-PEO block copolymer contents was investigated using AFM as shown in Fig. 5.7. All systems showed spherical morphology with the average size around 10–20 nm in diameter. The size of microphase separated PPO block-rich phase increased with increasing the BCP content. Additionally, all investigated systems cured at lower temperature were transparent at room temperature and indicated lack of macrophase separation.

Larrañaga et al. [30] proved also that for high PEO block content in PEO-b-PPO-b-PEO block copolymer nanostructured thermosetting materials in both curing conditions (precured at 80 and 140°C) can be obtained. Thus, independently on the curing conditions, as shown in Fig. 5.7, the morphology of the PEO-b-PPO-b-PEO-(DGEBA/DDM) cured systems were separation of

PPO block-rich phase at the nanometric scale mostly as spherical micelles.

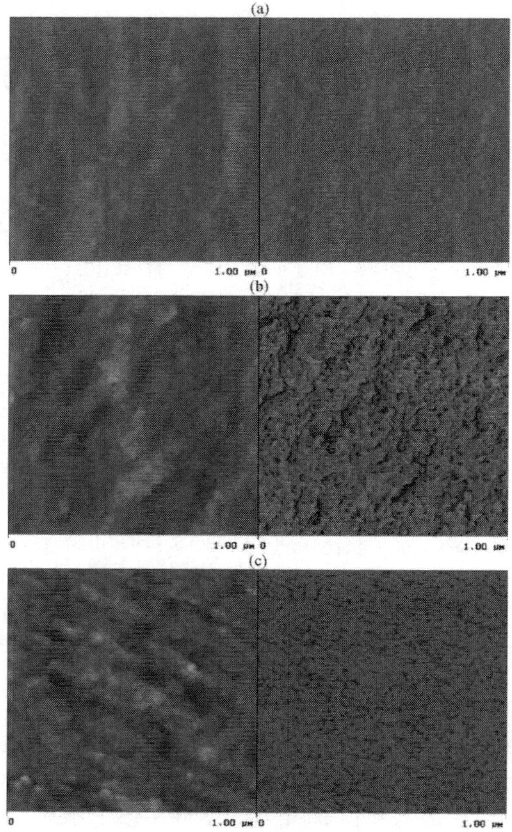

Figure 5.7 AFM images of PEO-b-PPO-b-PEO-(DGEBA/DDM) cured systems containing (a) 10, (b) 20, (c) 30 wt% PEO-b-PPO-b-PEO block copolymer cured at 80°C and postcured at 190°C. Reproduced with permission of [21]. Copyright 2005, Elsevier.

Here, it should be pointed out that for high BCP block copolymer these spherical micelles form worm-like micelles at both cure temperatures as marked in Fig. 5.8. The authors confirmed that the absence of macrophase separation of PPO block-rich phase was a consequence of the strong interactions between PEO and epoxy matrix.

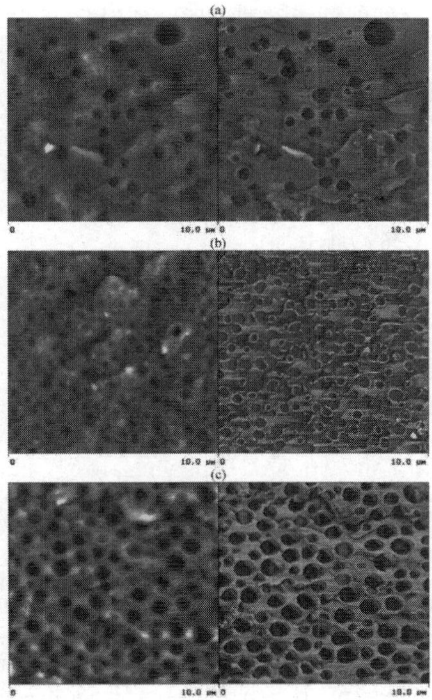

Figure 5.8 AFM images of PEO-b-PPO-b-PEO-(DGEBA/DDM) cured systems containing (a) 10, (b) 20, (c) 30 wt% PEO-b-PPO-b-PEO block copolymer cured at 80°C (1) and at 140°C (2). Reproduced with permission of [21]. Copyright 2005, Elsevier.

Thus, the high content of PEO block in the PEO-b-PPO-b-PEO block copolymer made this block copolymer highly miscible with epoxy precursors, which prevents the macrophase separation in thermosetting materials modified with PEO-b-PPO-b-PEO block copolymer. The interactions between PEO and epoxy precursors stabilized the micelles formed during the curing process at any cure temperature avoiding macrophase separation.

In order to prove the effect of the addition of PEO-b-PPO-b-PEO block copolymer on the mechanical properties of the PEO-b-PPO-b-PEO-(DGEBA/DDM) cured systems in comparison with the DGEBA/DDM cured system, Larañaga et al. [30] investigated the mechanical properties of these system cured at two different procuring temperatures. The authors confirmed that flexural modulus increased for the thermosetting materials modified with PEO-b-PPO-b-PEO block copolymer with higher PEO block contents. The results proved that mechanical properties of thermosetting materials depended strongly on the microphase separation in these systems and content of the miscible block in the block copolymer. The fracture toughness of PEO-b-PPO-b-PEO-(DGEBA/DDM) cured systems was also higher than neat DGEBA/DDM cured systems.

PEO-b-PPO-b-PEO block copolymer was also used by Cano et al. [14] for nanostructured epoxy system based on a diglycidyl ether of bisphenol A (DGEBA) and m-xylylenediamine (MXDA). In this particular case, thermosetting systems were prepared without any solvent and, taking into account their lower critical solution temperature (LCST), were precured at ambient temperature. All investigated thermosetting materials except system containing 50 wt% BCP were transparent at room temperature indicated lack of macrophase separation. The authors proved the nanostructured character of designed thermosetting materials employing AFM and TEM to study the morphology. As showed in Fig. 5.9, PEO-b-PPO-b-PEO-(DGEBA/MXDA) cured systems containing less than 25 wt% PEO-b-PPO-b-PEO block copolymer showed microphase separation while addition of 50 wt% PEO-b-PPO-b-PEO block copolymer presented macrophase separation. 5 wt% PEO-b-PPO-b-PEO-(DGEBA/MXDA) cured system showed spherical morphology where dark separated domains correspond to PPO block-rich phase with the size around 10–30 nm homogenously dispersed in brighter PEO block/epoxy-rich phase. Moreover, as proved by authors using AFM y TEM results, introduction of 15 and 25 wt% PEO-b-PPO-b-PEO block copolymer into epoxy system led to worm-like morphology. The size of the worm-like domains increased from 10 to 10–20 nm in diameter and from 60 nm to a wide range of lengths between 60 and 150 nm. Addition of 50 wt% of BCP led to macrophase separation as shown in Fig. 5.9.

Cano et al. [14] reported also mechanical properties of PEO-b-PPO-b-PEO-(DGEBA/MXDA) cured systems. As expected, due to plasticization effect of PEO-b-PPO-b-PEO block copolymer, the flexural modulus of the PEO-b-PPO-b-PEO-(DGEBA/MXDA) cured systems were lower than the flexural modulus of the neat DGEBA/MXDA cured system. Moreover, the flexural modulus of nanostructured thermosetting materials decreased with increase of BCP content.

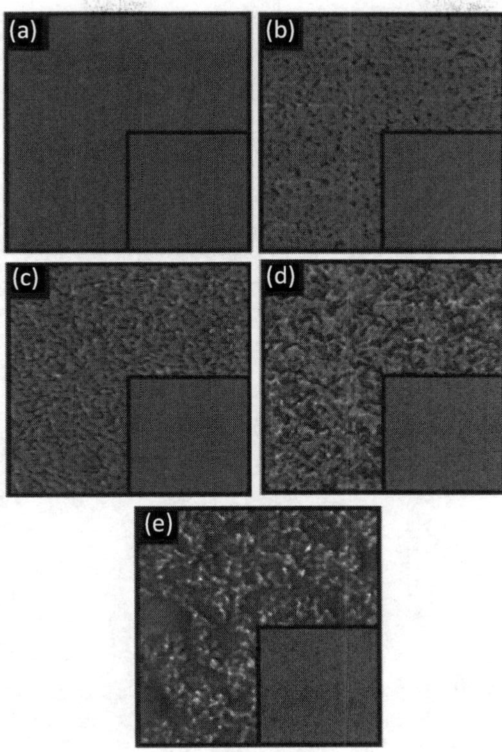

Figure 5.9 AFM phase images (1 μm × 1 μm) of PEO-b-PPO-b-PEO-(DGEBA/MXDA) containing (a) 0, (b) 5, (c) 15, (d) 25 and (e) 50 wt% of PEO-b-PPO-b-PEO block copolymer. The insets correspond to 5 μm × 5 μm AFM images. Samples were prepared using ultramicrotomy. Reproduced with permission of [14]. Copyright 2014, Elsevier.

The toughness of the PEO-b-PPO-b-PEO-(DGEBA/MXDA) cured systems increased compared with the neat DGEBA/MXDA cured system. The addition of 5 and 15 wt% of PEO-b-PPO-b-PEO block copolymer improved the toughness of 30% and 14% in the case of 5 wt% and 15 wt% PEO-b-PPO-b-PEO-(DGEBA/MXDA) cured system, respectively. The authors summarized that the addition of the PEO-b-PPO-PEO block copolymer provoked deterioration in the flexural behavior; on the other hand, it is worth considering the improvement in the toughness of investigated nanostructured thermosetting materials.

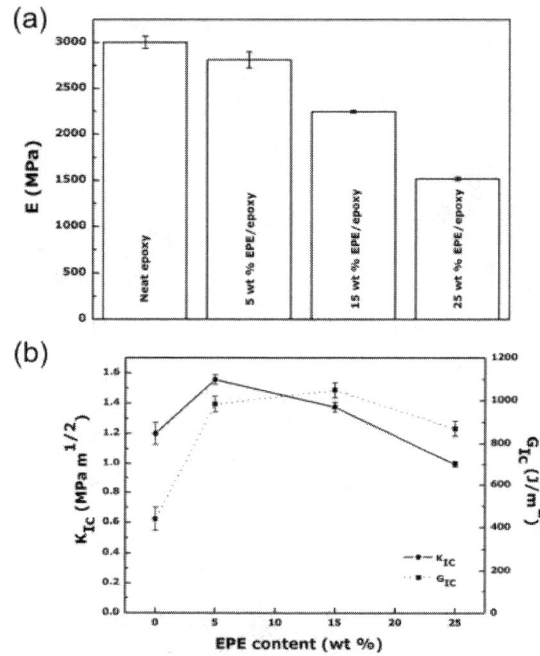

Figure 5.10 (a) Flexural modulus (*E*) and (b) critical stress intensity factor (KIC) and the critical strain energy release rate (GIC). Adapted and Reproduced with permission of [14]. Copyright 2014, Elsevier.

5.2.2 Polystyrene-b-Poly(Ethylene Oxide)

Synthesized via the atom transfer radical polymerization (ATRP) amphiphilic polystyrene-b-poly(ethylene oxide) block copolymer (PS-b-PEO) was employed as nanostructured agent for thermosetting epoxy resin-based materials by Meng et al. [28]. A DGEBA and 4,4′-methylenebis(2-chloroaniline) (MOCA) was used as thermoset precursors. All investigated mixture consist of DGEBA, MOCA and synthesized PS-b-PEO block copolymer were transparent at room temperature suggesting homogeneous and luck of macroscopic phase separation at the scale of the wavelength of visible light. The authors examined the possibility of the presence of the self-assembled nanostructures carried out SAXS measurement at room temperature and at the beginning of the curing reaction. As visualized in Fig. 5.11, SAXS profiles of 10 wt% PS-b-PEO/DGEBA mixture at room temperature presented a peak at q equal 0.03 Å^{-1} corresponding to long period ordering structure of 20.9 nm and indicated the microphase separation in this mixture.

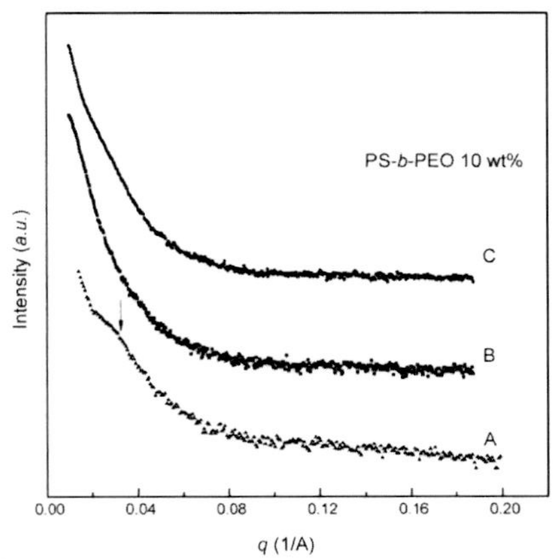

Figure 5.11 SAXS profiles of the blends of epoxy precursors with 10 wt% PS-b-PEO diblock copolymer: (A) DGEBA/PS-b-PEO blend at room temperature; (B) DGEBA + MOCA/PS-b-PEO at room temperature, (C) DGEBA + MOCA/PS-b-PEO at the beginning of curing reaction. Reproduced with permission of [28]. Copyright 2006, American Chemical Society.

As pointed out by the authors, this phenomenon can be related to the formation of the micelle structure since DGEBA can acts as selective solvent of PS-b-PEO block copolymer. Additionally, the authors proved lack of microphase separation in the same 10 wt% PS-b-PEO/DGEBA mixture when adequate stoichiometric quantity of MOCA was added. This mixture did not show any microphase separation at room temperature or at the beginning of curing reaction, which confirms the miscibility before network formation. Thus, no scattering peaks displayed in the SAXS profiles of these mixtures confirm that any nanostructures are formed before curing. This behavior can be ascribed to the change in solubility parameters resulting from the addition of MOCA to the system. Additionally, here it should be noted that results corresponding to SAXS profile of 10 wt% PS-b-PEO/DGEBA mixture at room temperature is in a good agreement with the fact that all investigated DGEBA, MOCA and synthesized PS-b-PEO block copolymer mixtures were homogenous at room temperature. From SAXS results, the authors conclude that any self-assembled nanostructures present in PS-b-PEO-(DGEBA/MOCA) cured systems must be formed by a reaction-induced phase separation and not self-assembly mechanism.

Thermosetting systems modified with synthesized PS-b-PEO block copolymer were cured at 150°C for 2 h and 180°C for 2 h to be sure that the curing reactions were completed. All investigated PS-b-PEO-(DGEBA/MOCA) cured systems were transparent confirming no macroscopic phase separation. The morphologies of 10, 20, 30 and 40 wt% PS-b-PEO-(DGEBA/MOCA) cured systems were investigated using AFM and SAXS. As shown in Fig. 5.12, all investigated thermosetting systems possessed nanostructured morphologies. Addition of 10 wt% PS-b-PEO block copolymer led to microphase separation of PS block as spherical domains homogeneously dispersed in continuous PEO block/epoxy-rich matrix. The size of microseparated PS block-rich phase was around 10–20 nm. The increase of PS-b-PEO block copolymer content provoked the increase of quantity of spherical microseparated PS block domains and some of them appeared interconnected as clearly shown in Fig. 5.12.

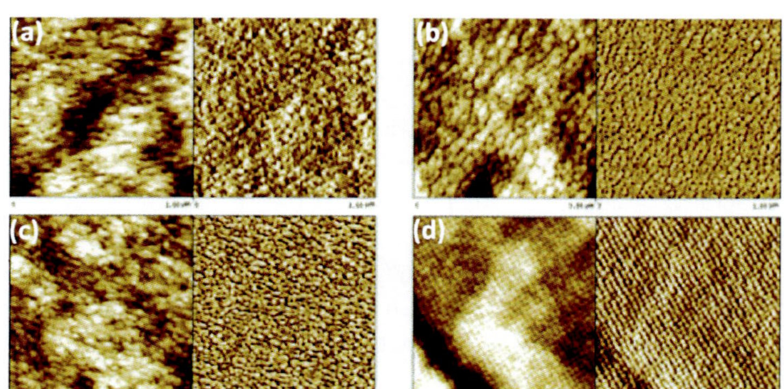

Figure 5.12 AFM images (left/right:topography/phase) of PS-b-PEO-(DGEBA/MOCA) cured systems containing (a) 10, (b) 20, (c) 30, and (d) 40 wt% of PS-b-PEO diblock copolymer. Reproduced with permission of [28]. Copyright 2006, American Chemical Society.

The size of the spherical particle remains almost unchanged. It is interesting to note that the thermosetting system showed long-range ordered nanostructured up to 40 wt% (Fig. 5.12).

The morphologies of the thermosetting systems were also investigated by SAXS measurement. As shown in Fig. 5.3, the well-defined scattering peaks were observed for all investigated systems, confirming microphase separation in thermosets systems modified with PS-b-PEO block copolymer. The SAXS profiles of all investigated PS-b-PEO-(DGEBA/MOCA) cured systems exhibited the multiple scattering maxima as denoted with the arrows in Fig. 5.3 confirming presence of long-range ordered nanostructures. Increase of the PS-b-PEO block copolymer content in thermosetting systems did not lead to significant changes in the positions of the scattering maxima. However, slight shifts to the higher q values with increase PS-b-PEO content can be clearly observed (Fig. 5.13). The scattering peaks of investigated thermosetting systems are positioned at q values of 1, 30.5, 60.5 (or 70.5), 90.5, 120.5, and 160.5 corresponding to the first-order scattering peak positions. The position of the scattering peaks identified in SAXS profiles of all thermosetting systems modified with PS-b-PEO block copolymers confirmed presence of spherical (or cylindrical) domains arranged in cubic lattices such as body-centered cubic (bcc), face-centered cubic (fcc), or simple cubic symmetries. The SAXS results were in good agreement with AFM, which indicated homogenous morphology with the domain size around 10–20 nm.

Meng et al. [28] concluded that thermosetting systems modified with synthesized PS-b-PEO block copolymer formed long-range ordered nanostructures provoked by RIPS.

The PS-b-PEO block copolymer was also used as a nanostructured agent for the fabrication of the thermosetting epoxy resin materials by Gutierrez et al. [35, 36] and Tercjak et al. [32–34, 37, 38]. They used as thermoset precursor a diglycidylether of bisphenol A epoxy monomer (DGEBA) and 4,4′-methylenebis(3-chloro 2,6-diethylaniline) (MCDEA) [35, 36, 38] or m-xylylenediamine (MXDA) [32–34, 37].

The authors employed two amphiphilic PS-b-PEP block copolymers with low (LPS-b-PEO) and high (HPS-b-PEO) PS block contents (M_n(PS) = 58,600 g/mol, M_n(PEO) = 31,000 g/mol, M_w/M_n = 1.03 for LPS-b-PEO and M_n(PS) = 125,000 g/mol,

M_n(PEO) = 16,100 g/mol, M_w/M_n = 1.4 for HPS-b-PEO). All results reported by Gutierrez at al. [35, 36] and Tercjak et al. [32–34, 37, 38] are ascribed to the bulk behavior since thermosetting systems were prepared in a 1.5 mm-thick parallelepipedic mold.

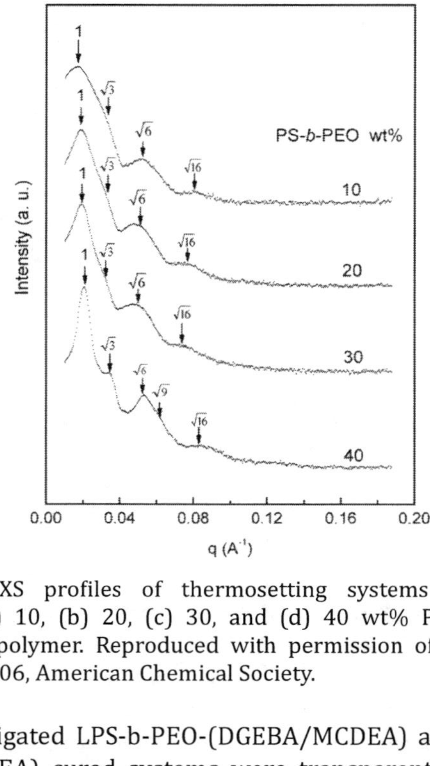

Figure 5.13 SAXS profiles of thermosetting systems modified with (a) 10, (b) 20, (c) 30, and (d) 40 wt% PS-b-PEO diblock copolymer. Reproduced with permission of [28]. Copyright 2006, American Chemical Society.

All investigated LPS-b-PEO-(DGEBA/MCDEA) and HPS-b-PEO-(DGEBA/MCDEA) cured systems were transparent indicating the lack of macrophase separation. The authors used AFM in order to study the morphologies generated in fully cured thermosetting materials. In the case of thermosetting systems modified with the PS-b-PEO block copolymer, the PEO block of the amphiphilic PS-b-PEO block copolymer is miscible with the epoxy matrix and the PS block is immiscible leading to the microphase separation of the PS block domains during curing process. Consequently, the light continuous areas in the AFM phase images correspond to PEO block/epoxy-rich matrix, whereas the dark areas correspond to PS block-rich phase. As visualized in Fig. 5.14,

LPS-b-PEO-(DGEBA/MCDEA) cured systems contain 5 and 10 wt% LPS-b-PEO showed spherical PS domains (dark domains) uniformly microphase separated in the continuous PEO block/epoxy-rich matrix.

Figure 5.14 AFM phase images (5 μm × 5 μm) of LPS-b-PEO-(DGEBA/MCDEA) cured systems. TEM image of 50 wt% LPS-b-PEO-(DGEBA/MCDEA) inset at the bottom of the corresponding AFM phase image. Reproduced with permission of [36]. Copyright 2014, Elsevier.

The size of microphase separated PS block domains was between 40 and 50 nm in diameter. The increase of LPS-b-PEO block copolymer content to 10 wt% led to the increase of the size of microseparated PS block domains and decrease of the distance between them. The addition of 20 wt% LPS-b-PEO into DGEBA/MCDEA cured system changed the morphology from spherical to hexagonal structure with cylinders of PS block arranged perpendicularly to the cut surface. Similarly, 30 wt% LPS-b-PEO-(DGEBA/MCDEA) cured system exhibited a hexagonally ordered structure; however, in this case, the cylinder of microseparated PS block was arranged perpendicularly and parallel to the cut surface. In the case of 40 wt% LPS-b-PEO-(DGEBA/MCDEA) system, the size of microseparated PS block cylinders increased and the most of them were parallel-oriented to the cut surface. The addition of 50 wt% LPS-b-PEO block copolymer led to lamellar morphology with a lamellar spacing of ~30 nm. This morphology was confirmed by TEM (bottom-right inset in the corresponding AFM-phase image). This research work published by Gutierrez et al. [35, 36] confirmed that amphiphilic PS-b-PEO block copolymer with low PS block content leads to highly transparent nanostructured materials, which can act as template to control dispersion and confinement of different nanoobjects such as nanoparticles, nanotubes and others as will be described in detail in Chapter 8.

Different morphologies were generated when high PS block content PS-b-PEO block copolymer was employed for the fabrication of the nanostructured thermosetting systems based on DGEBA/MCDEA cured system. As shown in Fig. 5.15, the introduction of 5 wt% HPS-b-PEO into DGEBA/MCDEA cured system led to the microphase separation of spherical PS-block domains with a size distribution of 50–80 nm in diameter.

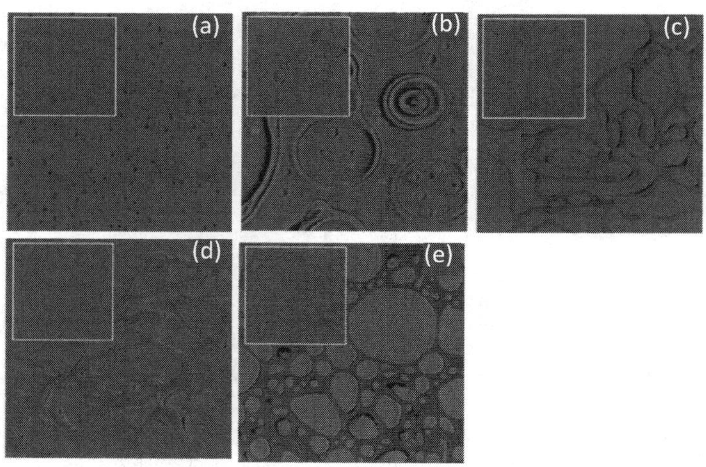

Figure 5.15 AFM phase images (5 μm × 5 μm) of different HPS-b-PEO-(DGEBA/MCDEA) cured systems containing: (a) 5 wt%, (b) 10 wt%, (c) 20 wt%, (d) 30 wt% and (e) 40 wt%. Left, top inset in each Figure corresponds to 15 μm × 15 μm. Reproduced with permission of [38]. Copyright 2014, Elsevier.

The increase of PS-b-PEO block copolymer to 10 wt% HSEO led to changes in the morphology from spherical micelles to vesicles-like morphology. This system showed several size vesicles structure with well-defined two separated phases, the dark PS block-rich phases formed outer layers and an inner layer consist of PEO block/epoxy-rich phase acted as the vesicle shells with the core being the epoxy-rich phase. For 20 and 30 wt% HPS-b-PEO block copolymer, the morphology of thermosetting materials change to long interconnected worm-like micelles of PS block-rich phase well-dispersed in PEO block/epoxy-rich matrices. Some of these long interconnected worm-like micelles formed bilayered structure means layers of the epoxy-rich phase surrounded by PS-block-rich layer. The morphology generated in 20 and 30 wt% HPS-b-PEO-(DGEBA/MCDEA) was very regular and repeatable being the wide of the long worm-like micelles around 60–110 nm.

Introduction of 40 wt% HPS-b-PEO into DGEBA/MCDEA epoxy matrix led to phase inversion. Thus, PEO block/epoxy-rich phase was macrophase separated into continuous PS block-rich phase.

5.2.3 Polyethylene-b-Poly(Ethylene Oxide)

Guo et al. [20] published fabrication procedure and characterization of nanostructured thermosetting materials modified with polyethylene-b-poly(ethylene oxide) (PE-b-PEO) block copolymer. They used as thermoset precursor a diglycidylether of bisphenol A epoxy monomer (DGEBA) and 4,4′-methylenedianiline (MDA). Thermosetting materials modified with less than 40 wt% of PE-b-PEO block copolymer content are transparent at room temperature. Introduction of 50 wt% PE-b-PEO block copolymer into DGEBA/MDA epoxy matrix leads to translucent or opaque thermosetting materials at room temperature. However, all thermosetting materials are transparent in temperature higher than melting temperature of PE-b-PEO block copolymer indicated lack of macrophase separation. The morphologies of PE-b-PEO-(DGEBA/MDA) cured systems modified with different PE-b-PEO block copolymer content were studied by TEM, AFM and SAXS. The authors proved by TEM measurement (Fig. 5.16) that PE-b-PEO-(DGEBA/MDA) cured systems possessed the heterogeneous morphology at the nanoscale. The dark areas in TEM images correspond to PEO block, which was preferentially stained with RuO_4. Addition of even 50 wt% of PE-b-PEO block copolymer leads to spherical structure with the domain size of 10 nm in diameter. The average distance between the spherical micelles decreases with increasing PE-b-PEO block copolymer content in PE-b-PEO-(DGEBA/MDA) cured systems. 40 and 50 wt% PE-b-PEO-(DGEBA/MDA) cured systems exhibited the morphology where microphase separated spherical micelles (size around 10–15 nm in diameter) aggregated to form bicontinuous microphase structures.

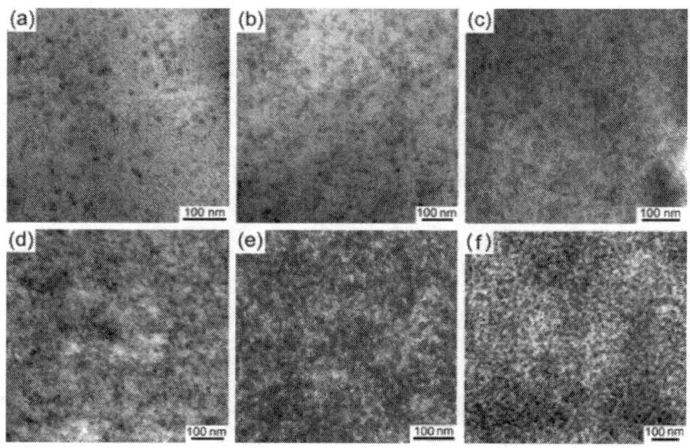

Figure 5.16 TEM images of PE-b-PEO-(DGEBA/MDA) cured systems containing (a) 5, (b) 10, (c) 20, (d) 30, (e) 40, and (f) 50 wt% of PE-b-PEO block copolymer. All investigated samples were stained with RuO_4. Reproduced with permission of [20]. Copyright 2014, Elsevier.

Guo et al. [20] reported also that the morphologies of (PE-b-PEO-DGEBA/MDA) cured systems proved by TEM measurement were in good agreement with AFM results. AFM images clearly visualized the spherical micelles with the size around 10–15 nm in diameter. Additionally, AFM results confirmed bicontinous microphase separation.

As published by Guo et al. [20], SAXS patterns of PE-b-PEO-(DGEBA/MDA) cured systems shown in Fig. 5.17 confirmed morphologies identified by AFM and TEM results for these thermosetting materials. The SAXS curves for these nanostructured thermosetting materials identified presence of the spherical micelles (or aggregated spherical micelles forming b continous structure).

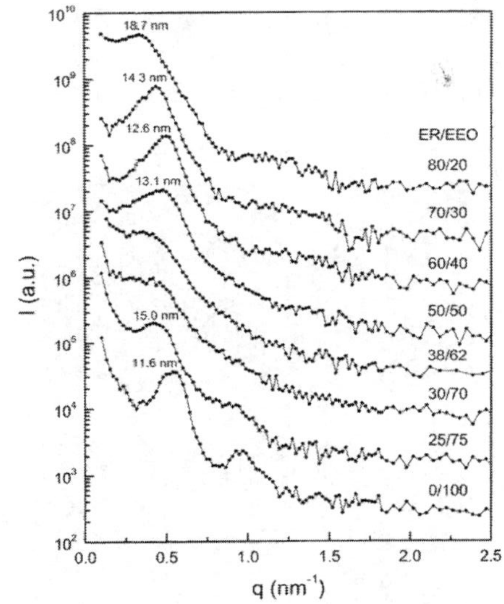

Figure 5.17 SAXS patterns of PE-b-PEO-(DGEBA/MDA) cured systems containing different PE-b-PEO block copolymer content at room temperature. Reproduced with permission of [20]. Copyright 2014, Elsevier.

The position of scattering peak corresponding to microphase separation of PE block increasing with increase of PE-b-PEO block copolymer content from 20 to 40 wt%. Thus, the value of the scattering vector, **q**, gradually increased, which proved that the long spacing in real space decreased from 18.7 to 12.6 nm. This confirms that the average distance between the neighboring spherical micelles decreases with increasing PE-b-PEO block copolymer content. Similarly, as for TEM measurement, the SAXS curve of 40 wt% PE-b-PEO-(DGEBA/MDA) cured system exhibited highly aggregated spherical micelles with distance in real space of 12.6 nm, which is the size of the microphase structure visualized in the TEM result. AFM, TEM and SAXS results confirmed the possibility of used an amphiphilic, semicrystalline PE-b-PEO block copolymer to prepared nanostructured thermosetting materials. The authors proved that adequate preparation avoids crystallization of PE-b-PEO block copolymer during RIPS and leads to nanostructuration of DGEBA/MDA cured system, where immiscible PE block microphase separated as spherical micelles.

Tercjak et al. [26] employed also the same PE-b-PEO block copolymers to nanostructured DGEBA/MCDEA cured system. As shown in Fig. 5.18, they studied the effect of curing conduction on the possibility of the fabrication of nanostructured thermosetting materials.

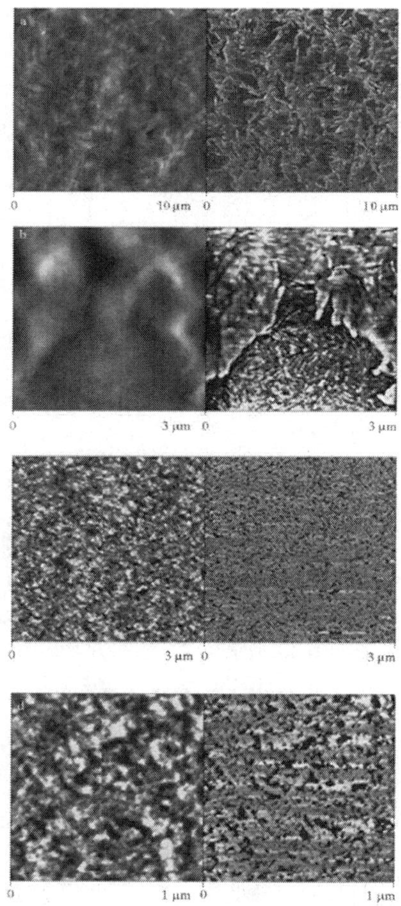

Figure 5.18 AFM images of 40 wt% PE-b-PEO-(DGEBA/MCDEA) cured system (a) precured 12 h at 120°C, (b) detail of (a), (c) precured 48 h at 120°C and (d) detail of (c). Reproduced with permission of [24]. Copyright 2014, Elsevier.

As visualized in Fig. 5.18, 40 wt% PE-b-PEO-(DGEBA/MCDEA) cured system showed spherulites of PE block separated from PEO block/epoxy-rich phase. However, if observed in more details in Fig. 5.18b, self-assembly of PE block was also clearly visualized; 40 wt% PE-b-PEO-(DGEBA/MCDEA) cured system precured 48 h at 120°C showed spherical micelles of PE block of PE-b-PEO block copolymer microphase separated from PEO block/epoxy matrices. The authors proved that the fabrication of nanostructured thermosetting materials depends strongly on the curing conditions.

References

1. Iijima, T., Tomoi, M., Yamasaki, J., Kakiuchi, H. (1999). Toughening of epoxy resins by modification with acrylic elastomers containing pendant epoxy groups, *Eur. Polym. J.*, **26**, pp. 145–151.
2. Imanaka, M., Nakamura, Y., Nishimura, A., Iida, T. (2003). Fracture toughness of rubber-modified epoxy adhesives: Effect of plastic deformability of the matrix phase, *Comput. Sci. Technol.*, **63**, pp. 41–51.
3. Tripathy, R., Ojha, U., Faust, R. (2011). Polyisobutylene modified bisphenol A diglycidyl ether based epoxy resins possessing improved mechanical properties, *Macromolecules*, **44**, pp. 6800–6809.
4. Kinloch, A. J., Lee, S. H., Taylor, A. C. (2014). Improving the fracture toughness and the cyclic-fatigue resistance of epoxy-polymer blends, *Polymer*, **55**, pp. 6325–6334.
5. Sha, Y., Hui, C. Y., Kramer, E. J., Hahn, S. F., Berglund, C. A. (1996). Fracture toughness and failure mechanisms of epoxy/rubber-modified polystyrene (HIPS) interfaces reinforced by grafted chains, *Macromolecules*, **29**, pp. 4728–4736.
6. Di Pasquale, G., Motta, O., Recca, A., Carter, J. T., McGrail, P. T., Acierno, D. (1997). New high-performance thermoplastic toughened epoxy thermosets, *Polymer*, **38**, pp. 4345–4348.
7. Ha, Ch.-S., Kim, Y., Lee, W.-K., Cho, W.-J., Kim, Y. (1998). Fracture toughness and properties of plasticized PVC and thermoplastic polyurethane blends, *Polymer*, **39**, pp. 4765–4772.
8. Gan, W., Yu, Y., Wang, M., Tao, Q., Li, S. (2003). Viscoelastic effects on the phase separation in thermoplastics-modified epoxy resin, *Macromolecules*, **36**, 7746–7751.

9. Johnsen, B. B., Kinloch, A. J., Taylor, A. C. (2005). Toughness of syndiotactic polystyrene/epoxy polymer blends: Microstructure and toughening mechanisms, *Polymer*, **46**, pp. 7352–7369.
10. Dean, J. M., Grubbs, R. B., Saad, W., Cook, R. F., Bates, F. S. (2003). Mechanical properties of block copolymer vesicle and micelle modified epoxies, *J. Polym. Sci. Part B: Polym. Phys.*, **41**, pp. 2444–2456.
11. Rebizant, V., Venet, A.-S., Tournilhac, F., Girard-Reydet, E., Navarro, Ch., Pascault, J.-P., Leibler, L. (2004). Chemistry and mechanical properties of epoxy-based thermosets reinforced by reactive and nonreactive SBMX block copolymers, *Macromolecules*, **34**, pp. 8017–8027.
12. Thio, Y. S., Wu, J., Bates, F. S. (2006). Epoxy toughening using low molecular weight poly(hexylene oxide)-poly(ethylene oxide) diblock copolymers, *Macromolecules*, **39**, pp. 7187–7189.
13. Liu, J. D., Thompson, Z. J., Sue, H.-J., Bates, F. S., Hillmyer, M. A., Dettloff, M., Jacob, G., Verghese, N., Pham, H. (2010). Toughening of epoxies with block copolymer micelles of wormlike morphology, *Macromolecules*, **43**, pp. 7238–7243.
14. Cano, L., Builes, D. H., Tercjak, A. (2014). Morphological and mechanical study of nanostructured epoxy systems modified with amphiphilic poly(ethylene oxide-b-propylene oxide-b-ethylene oxide)triblock copolymer, *Polymer*, **55**, pp. 738–745.
15. Ruiz-Perez, L., Royston, G. J., Fairclough, J. P. A., Ryan, A. J. (2008). Toughening by nanostructure, *Polymer*, **49**, pp. 4475–4488.
16. Hillmyer, M. A., Lipic, P. M., Hajduk, D. A., Almdal, K., Bates, F. S. (1997). Self-assembly and polymerization of epoxy resin-amphiphilic block copolymer nanocomposites, *J. Am. Chem. Soc.*, **119**, pp. 2749–2750.
17. Lipic, P. M., Bates, F. S., Hillmyer, M. A. (1998). Nanostructured thermosets from self-assembled amphiphilic block copolymer/epoxy resin mixtures, *J. Am. Chem. Soc.*, **120**, pp. 8963–8970.
18. Mijovic, J., Shen, M., Sy, J. W., Mondragon, I. (2000). Dynamics and morphology in nanostructured thermoset network/block copolymer blends during network formation. *Macromolecules*, **33**, pp. 5235–5244.
19. Guo, Q., Thomann, R., Gronski, W., Thurn-Albrecht, T. (2002). Phase behavior, crystallization, and hierarchical nanostructures in self-organized thermoset blends of epoxy resin and amphiphilic poly(ethyleneoxide)-block-poly(propyleneoxide)-block-poly(ethylene oxide) triblock copolymers, *Macromolecules*, **35**, pp. 3133–3144.
20. Guo, Q., Thomann, R., Gronski, W., Staneva, R., Ivanova, R., Stühn, B. (2003). Nanostructures, semicrytalline morphology, and nanoscale confinement effect on the crystallization kinetics in self-

organized block copolymer/thermoset blends, *Macromolecules*, **36**, pp. 3635–3645.

21. Larrañaga, M., Gabilondo, N., Kortaberria, G., Serrano, E., Remiro, P., Riccardi, C. C., Mondragon, I. (2005). Micro- or nanoseparated phases in thermoset blends of an epoxy resin and PEO–PPO–PEO triblock copolymer, *Polymer*, **46**, pp. 7082–7093.
22. Guo, Q., Wang, K., Chen, L., Zheng, S., Halley, P. J. (2006). Phase behavior, crystallization, and nanostructures in thermoset blends of epoxy resin and amphiphilic star-shaped block copolymers, *J. Polym. Sci. Part B: Polym. Phys.*, **44**, pp. 975–985.
23. Guo, Q., Chen, F., Wang, K., Chen, L. (2006). Nanostructured thermoset epoxy resin templated by an amphiphilic poly(ethylene oxide)-block-poly(dimethylsiloxane) diblock copolymer, *J. Polym. Sci. Part B: Polym. Phys.*, **44**, pp. 3042–3052.
24. Larrañaga, M., Martin, M. D., Gabilondo, N., Kortaberria, G., Eceiza, A., Riccardi, C. C., Mondragon, I. (2006). Toward microphase separation in epoxy systems containing PEO–PPO–PEO block copolymers by controlling cure conditions and molar ratios between blocks. Part 1. Cure kinetics, *Colloid Polym. Sci.*, **284**, pp. 1403–1410.
25. Larrañaga, M., Arruti, P., Serrano, E., de la Caba, K., Remiro, P., Riccardi, C. C., Mondragon, I. (2006). Towards microphase separation in epoxy systems containing PEO/PPO/PEO block copolymers by controlling cure conditions and molar ratios between blocks. Part 2. Structural characterization, *Colloid Polym. Sci.*, **284**, pp. 1419–1430.
26. Tercjak, A., Larrañaga, M., Martin, M. D., Mondragon, I. (2006). Thermally reversible nanostructured thermosetting blends modified with poly(ethylene-b-ethylene oxide) diblock copolymer, *J. Therm. Anal. Calorim.*, **86**, 663–668.
27. Meng, F., Zheng, S., Liu, T. (2006). Epoxy resin containing poly(ethylene oxide)-block-poly(ε-caprolactone) diblock copolymer: Effect of curing agents on nanostructures, *Polymer*, **47**, pp. 7590–7600.
28. Meng, F., Zheng, S., Li, H., Liang, Q., Liu, T. (2006). Formation of ordered nanostructures in epoxy thermosets: A mechanism of reaction-induced microphase separation, *Macromolecules*, **39**, pp. 5072–5080.
29. Sun, P., Dang, Q., Li, B., Chen, T., Wang, Y., Lin, H., Jin, Q., Ding, D. (2005). Mobility, miscibility, and microdomain structure in nanostructured thermoset blends of epoxy resin and amphiphilic poly(ethylene oxide)-block-poly(propylene oxide)-block-poly(ethylene oxide) triblock copolymers characterized by solid-state NMR, *Macromolecules*, **38**, pp. 5654–5667.

30. Larrañaga, M., Serrano, E., Martin, M. D., Tercjak, A., Kortaberria, G., de la Caba, K., Riccardi, C. C., Mondragon, I. (2007). Mechanical properties-morphology relationships in nano-/microstructured epoxy matrices modified with PEO-PPO-PEO block copolymers, *Polym. Int.*, **56**, pp. 1392–1403.

31. Liu, J. D., Thompson, Z. J., Sue, H.-J., Bates, F. S., Hillmyer, M. A., Dettloff, M., Jacob, G., Verghese, N., Pham, H. (2010). Toughening of epoxies with block copolymer micelles of wormlike morphology, *Macromolecules*, **43**, pp. 7238–7243.

32. Tercjak, A., Serrano, E., Garcia, I., Mondragon, I. (2008). Thermoresponsive meso/nanostructured thermosetting materials based on PS-b-PEO block copolymer-dispersed liquid crystal: Curing behavior and morphological variation, *Acta Mater.*, **56**, pp. 5112–5122.

33. Tercjak, A., Mondragon, I. (2008). Relationships between the morphology and thermoresponsive behavior in micro/nanostructured thermosetting matrixes containing a 4′-(hexyloxy)-4-biphenylcarbonitrile liquid crystal, *Langmuir*, **24**, pp. 11216–11224.

34. Tercjak, A., Gutierrez, J., Peponi, L., Rueda, L., Mondragon, I. (2009). Arrangement of conductive TiO_2 nanoparticles in hybrid inorganic/organic thermosetting materials using liquid crystal, *Macromolecules*, **42**, pp. 3386–3390.

35. Gutierrez, J., Tercjak, A., Mondragon, I. (2010). Transparent nanostructured thermoset composites containing well-dispersed TiO_2 nanoparticles, *J. Phys. Chem. C*, **114**, pp. 22424–22430.

36. Gutierrez, J., Mondragon, I., Tercjak, A. (2011). Morphological and optical behavior of thermoset matrix composites varying both polystyrene-block-poly(ethylene oxide) and TiO_2 nanoparticle content, *Polymer*, **52**, pp. 5699–5707.

37. Tercjak, A., Gutierrez, J., Mondragon, I. (2012). Conductive properties of inorganic/organic nanostructured systems based on block copolymers, *Mater. Sci. Forum*, **714**, pp. 153–158.

38. Tercjak, A., Gutierrez, J., Martin, M. D., Mondragon, I. (2012). Transparent titanium dioxide/block copolymer modified epoxy-based systems in the long scale microphase separation threshold, *Eur. Polym. J.*, **48**, pp. 16–25.

39. Francis, R., Baby, D. K. (2014). Toughening of epoxy thermoset with polystyrene-block-polyglycolic acid star copolymer: Nanostructure-mechanical property correlation, *Ind. Eng. Chem. Res.*, **53**, pp. 17945–17951.

40. Meng, F., Zheng, S., Zhang, W., Li, H., Liang, Q. (2006). Nanostructured thermosetting blends of epoxy resin and amphiphilic poly(ε-caprolactone)-block-polybutadiene-block-poly(ε-caprolactone) triblock copolymer, *Macromolecules*, **39**, pp. 711–719.

41. Xu, Z., Zheng, S. (2007). Reaction-induced microphase separation in epoxy thermosets containing poly(ε-caprolactone)-block-poly(n-butyl acrylate) diblock copolymer, *Macromolecules*, **40**, pp. 2548–2558.

42. Ocando, C., Serrano, E., Tercjak, A., Peña, C., Kortaberria, G., Calberg, C., Grignard, B., Jerome, R., Carrasco, P. M., Mecerreyes, D., Mondragon, I. (2007). Structure and properties of a semifluorinated diblock copolymer modified epoxy blend, *Macromolecules*, **40**, pp. 4068–4074.

43. Fan, W., Wang, L., Zheng, S. (2009). Nanostructures in thermosetting blends of epoxy resin with polydimethylsiloxane-block-poly(ε-caprolactone)-block-polystyrene ABC triblock copolymer, *Macromolecules*, **42**, pp. 327–336.

44. Zhu, L., Zhang, Ch., Han, J., Zheng, S., Li, X. (2012). Formation of nanophases in epoxy thermosets containing an organic–inorganic macrocyclic molecular brush with poly(ε-caprolactone)-block-polystyrene side chains, *Soft Matter*, **8**, pp. 7062–7072.

45. Hu, D., Zhang, Ch., Yu, R., Wang, L., Zheng, S. (2010). Self-organized thermosets involving epoxy and poly(ε-caprolactone)-block-poly(ethylene-co-ethylethylene)-block-poly(ε-caprolactone) amphiphilic triblock copolymer, *Polymer*, **51**, pp. 6047–6057.

46. Ritzenthaler, S., Court, F., David, L., Girard-Reydet, E., Leibler, L., Pascault, J. P. (2002). ABC triblock copolymers/epoxy-diamine blends. 1. Keys to achieve nanostructured thermosets, *Macromolecules*, **35**, pp. 6245–6254.

47. Maiez-Tribut, S., Pascault, J. P., Soule, E. R., Borrajo, J., Williams, R. J. J. (2007). Nanostructured epoxies based on the self-assembly of block copolymers: A new miscible block that can be tailored to different epoxy formulations, *Macromolecules*, **40**, pp. 1268–1273.

48. Fan, W., Zheng, S. (2008). Reaction-induced microphase separation in thermosetting blends of epoxy resin with poly(methyl methacrylate)-block-polystyrene block copolymers: Effect of topologies of block copolymers on morphological structures, *Polymer*, **49**, 3157–3167.

49. Blanco, M., Lopez, M., Kortaberria, G., Mondragon, I. (2009). Nanostructured thermosets from self-assembled amphiphilic block copolymer/epoxy resin mixtures: Effect of copolymer content on nanostructures, *Polym. Int.*, **59**, pp. 523–528.

50. Luo, X., Zheng, S., Zhang, N., Ma, D. (1994). Miscibility of epoxy resins/poly(ethylene oxide) blends cured with phthalic anhydride, *Polymer*, **35**, pp. 2619–2623.

51. Zheng, S., Zhang, N., Luo, X., Ma, D. (1995). Epoxy resin/poly(ethylene oxide) blends cured with aromatic amine, *Polymer*, **36**, pp. 3609–3613.

52. Hillmyer, M. A., Bates, F. S. (1996). Synthesis and characterization of model polyalkane-poly(ethylene oxide) block copolymers, *Macromolecules*, **29**, pp. 6994–7002.

53. Hillmyer, M. A., Bates, F. S., Almdal, K., Mortensen, K., Ryan, A. J., Fairclough, J. P. A. (1996). Complex phase behavior in solvent-free nonionic surfactants, *Science*, **276**, pp. 976–978.

Chapter 6

Chemically Functionalized Block Copolymers as Reactive Modifiers for Nanostructuring and Toughening Epoxy Thermosetting Materials

Connie Ocando

Instituto de Tecnología en Polímeros y Nanotecnología (ITPN)
Facultad de Ingeniería, Universidad de Buenos Aires,
Las Heras 2214, Buenos Aires, 1127AAR/Capital Federal, Argentina

connie_ocando@hotmail.com

In addition to the employment of amphiphilic block copolymers as modifiers for the generation of nanostructured thermosetting materials, described in chapter 5, this chapter will cover a review about the synthesis and employment of functionalized block copolymers, incorporating a certain degree of epoxy functionality, as reactive component for the development of well-defined ordered nanostructures in blends with epoxy systems, doing particular emphasis on showing the most relevant outcomes obtained in the group of Professor Iñaki Mondragon on this challenging subject. This chapter discusses the synthesis, miscibility,

Block Copolymer Nanocomposites
Edited by Galder Kortaberria and Agnieszka Tercjak

ISBN 978-981-4669-54-2 (Hardcover), 978-981-4669-55-9 (eBook)
www.panstanford.com

reaction, morphology and properties of functionalized block copolymer-modified thermosetting materials aiming to establish a comprehensive understanding of the structure-property relationship of reactive blends with nanostructured inclusions. The outcomes suggested that the use of reactive functionalized block copolymers instead of nonreactive ones as modifier of thermosetting systems permits the use of a wider range of curing agents and conditions to produces nanostructures and better control of morphology formation within highly crosslinked epoxy systems as well as lead to a superior degree of toughening and thermomechanical properties. Pursuing the improvement of these last properties in epoxy systems has been the main motivation for the use of this kind of modifiers.

6.1 Introduction

Nowadays, the development of novel functional polymers is interesting, from the point of view of permitting a better and wider use of conventional ones. The chemical functionalization of block copolymers is a post-polymerization process that takes places in order to introduce desirable functional groups in one of their block chains and tuning their chemico-physical properties [1–7]. The tunable thermodynamic interactions of these block copolymers induced by the chemical functionalization have been exploit as templating agents to examine the phase separation processes involve in blends with epoxy resins [8–24]. Even more interesting, this functionalization approach not only provides an epoxyphilic behavior between one block of the block copolymer and the uncured and cured states of the thermosetting system, a crucial issue for the successful development of nanostructured materials, but also could generate several reactive groups into the block copolymer chains to promote covalent linkages with the thermosetting formulation (epoxy resin or crosslinker) during the polymerization of the epoxy network. The occurrence of these covalent linkages has been widely studied in order to find a correlation between them on interfacial adhesion, size and shape of the dispersed structures, fact that could be translated into further improvement in the thermal and mechanical properties of resulting nanostructured thermosetting materials [8–10, 13–16, 23–25]. Likewise, nanostructured epoxy blends based on amphiphilic blocks copolymers having different intermolecular interactions with the epoxy resin [26], it was noticed that the use of reactive block copolymers permitted to finely control the interphase and morphology formation of the thermosetting blend [15, 17, 26].

Recently, inspired by the concept of chemical functionalization to promote favorable interactions between epoxy network and block copolymer modifiers, Wu et al. [24] introduced sulfonic acid groups into the nonpolar styrene block chains of a polystyrene-b-poly(ethylene-co-butylene)-b-polystyrene (SEBS) block copolymer with 67 wt% of PS block in order to prepare novel block ionomers and increased the solubility parameter of PS. The employment of sulfonated polystyrene-b-poly(ethylene-co-butylene)-b-polystyrene (SSEBS) block copolymer as modifier produced nanometer- or micrometer-scale structures dispersed within the crosslinked epoxy matrix, where the dimensions were determined by the sulfonation degree of SSEBS. The interactions between sulfonic acid groups of SSEBS and epoxy were established as the main driving force to avoid macroscopic phase separation in these blends. For higher sulfonation degrees, spherical core–shell microdomains with long-range ordered nanostructures were determined by TEM and SAXS analyses, where the cores were attributed to the rubbery EB (poly(ethylene-ran-butylene)) blocks, whereas the shells were assigned to the partially microphase-separated SPS (sulfonated polystyrene) blocks. It is possible to emphasize that the EB blocks are immiscible with epoxy precursor and thereby forming spherical cores prior to curing. In the case of SPS blocks, it was not established if the phase separation of SPS blocks takes place before curing or if it is induced by the curing. It is well know that the blends of PS homopolymer and epoxy resins often display the upper critical solution temperature behavior (UCST), where the maximum critical solution temperature is 80°C [27–29]. Subsequently, during the course of polymerization the PS undergoes reaction induced phase separation (RIPS) [30]. In particular in epoxy/SSEBS blends the sulfonate groups can form ionic interaction or hydrogen bonding with the epoxy precursors. Thus, the sulfonation degree of SSEBS had a great impact on the morphology because the shell thickness depends on the extent of the phase separation of SPS blocks, as well as the remaining free sulfonic acid groups in the block ionomer can react with the amine curing agent, resulting in better miscibility between SPS blocks and the cured epoxy network. Interestingly, Wu et al. [23] also found that the introduction of epoxy-miscible poly(ε-caprolactone) side chains in a SSEBS block ionomer with low content of PS to produce the ionically linked block ionomer complex SSEBS-c-PCL was successfully employed as an alternative to the covalently bonded PCL-based block copolymers to achieve nanostructures in epoxy blends. All epoxy/SSEBS and epoxy/SSEBS-c-PCL blends showed an increased fracture toughness compared to neat epoxy. Rebizant et al. [17] synthesized a polystyrene-b-polybutadiene-b-poly(methyl methacrylate)-b-poly(glycidyl methacrylate) (SBMG) block copolymer containing reactive poly(glycidyl methacrylate) block to be employed as modifier for thermosetting materials. It was found that the obtained pre-cured nanostructures were retained after curing reaction. These results demonstrated the advantages of the presence of functional reactive groups in SBMG compared to that with nonreactive (polystyrene-b-polybutadiene-b-poly(methyl methacrylate)) (SBM) block copolymers, which were expelled from the epoxy network during its formation. Thus, when the block copolymer was fitted with a certain number of reactive groups, the formation of covalent bonds between the block copolymer and the epoxy system at the early stage of curing permit locked the morphology already present before cure until the gel point and allowing a finely control of the final morphology of the blend. Additionally, the same research group synthesized polystyrene-b-polybutadiene-b-poly((methyl methacrylate)-*stat*-(methacrylic acid)) (SBMA) block copolymers incorporating carboxylic acid reactive functionalities in the last block capable of reacting with epoxide groups and with some curing agents 16. The nonreactive or slowly reacting polystyrene-b-polybutadiene-b-poly((methyl methacrylate)-*stat*-(*tert*-butyl methacrylate)) (SBMT) block copolymer were found flocculate into macroscopic structures when the epoxy network was formed with most curing agents, while the SBMA block copolymer preserved the nanoscale micellar structure. The materials performances were analyzed in terms of transparency, glass transition temperature, T_g, and linear elastic mechanics at break (critical intensity factor, K_{Ic}).

Grubbs et al. [6, 20] introduced epoxy functional groups into the polyisoprene block of a polyisoprene-b-polybutadiene block copolymer to increased its polarity and further be used as template agent of an epoxy system. Higher epoxidation degrees were necessary to ensure the miscibility of epoxidized PI block with the epoxy formulation [20]. The morphology of the blends, evidenced by SAXS and TEM, were micelles, packed spheres, hexagonal and lamellar structures as the content of block copolymer was increased in the blend. Moreover, it was found that the reactivity of oxirane groups of epoxidized PI block with the curing agent was slower than those in the resin, the higher curing temperature being necessary to complete the reaction. It was also established that a partial expulsion of epoxidized PI blocks from the epoxy-rich phase occurred. Similar behavior was observed in poly(ethylene oxide)–poly(ethylene-*alt*-propylene) (PEO–PEP) and poly(ethylene oxide)-b-poly(propylene oxide)-b-poly(ethylene oxide) (PEO-PPO-PEO) nonreactive block copolymer-modified epoxy system (Fig. 6.1a) [31, 32]. Nevertheless, this expulsion can be followed by an interfacial reaction of the majority reactive epoxy groups of block copolymer with the epoxy resin (Fig. 6.1c). In the opposite, it was found that poly(methyl acrylate-*co*-glycidyl acrylate)-b-polyisoprene reactive block copolymer had similar reactivity to the epoxy resin, permitting that the copolymer chains cure within the epoxy matrix (Fig. 6.1b) [20]. The effect of these two types of copolymer at low concentration on epoxy toughening was examined [25].

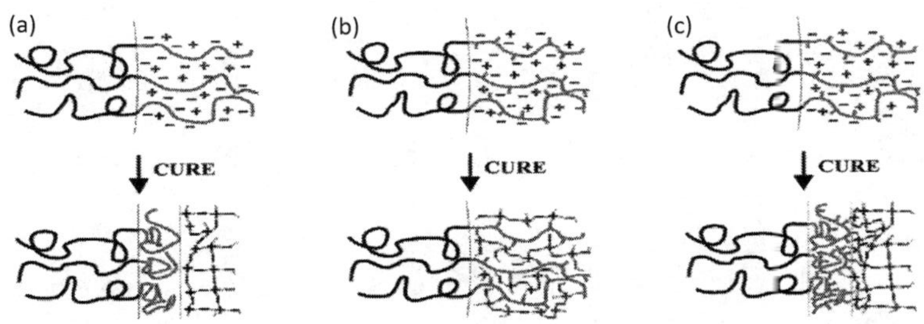

Figure 6.1 Comparison of microphase-separation processes during cure of nonreactive block copolymers (a) and reactive block copolymers: cured within epoxy matrix (b) or cured interfacially after expulsion from epoxy phase (c) Reproduced with permission of [20]. Copyright 2000, American Chemical Society.

Styrene-diene block copolymers such as poly(styrene-b-butadiene-b-styrene) (SBS) or poly(styrene-b-isoprene-b-styrene) (SIS) are the most widely used thermoplastic elastomers (TPE) [33, 34]. The evaluation of styrene-diene block copolymers as modifiers for the generation of nanostructured thermosetting materials has been of great industrial interest because they are commercially available at considerable low costs. For this purpose, the epoxidation reaction of polybutadiene block of SBS block copolymer or polyisoprene block of SIS block copolymer was used as an effective approach to promote miscibility and reaction with the thermosetting formulation during curing reaction [4, 5, 8–15, 18, 19]. Furthermore, the potential reactivity of the oxirane groups introduced in the SIS block copolymer was also exploited for the preparation of block copolymers with new amine functionalities [15]. The study of the different parameters that determine the mechanisms involved during morphology formation in thermosetting matrices modified with styrene-diene block copolymers epoxidized at several degrees, as well as a comprehensive understanding of the effect of chemical functionalization of one block, that permit the development of controlled nanostructures with enhanced matrix-domains interface interactions, on thermomechanical and fracture toughness properties of thermosetting blends will be addressed in this chapter. We also discuss the gradient on block copolymer concentration through film thickness and the effects of casting solvents used on the nanostructuring of a thermosetting epoxy coating. Finally, the stiffening of a TPE block copolymer is presented.

6.2 Functionalization of Styrene-Diene Block Copolymers

It is well studied the chemical functionalization of polydiene materials, where the epoxidation reaction has been found the most promising and advantageous method in the laboratory as well as in chemical industry [35–43]. The different procedures of epoxidation reaction of styrene-diene block copolymers that have been the goal of various investigations, for its further use as modifier in the generation of novel nanostructured thermosetting materials, will be described. In general, the polydiene epoxidation procedures used for this intention has been by phase-transfer catalysis/hydrogen peroxide and dioxiranes. The occurrence of side reaction such as ring opening and crosslinking will be also addressed.

6.2.1 Epoxidation Reaction of Styrene-Diene Block Copolymers Using Hydrogen Peroxide under Phase-Transfer Catalysis

The use of hydrogen peroxide as the oxidant component and Venturello–Ishii phase-transfer catalytic system for the functionalization of alkenes to epoxides have attracted much attention [37, 44–46]. Most of the work has been focused on SBS block copolymers [2, 5], polybutadiene [42] and hydroxyl terminated polybutadiene [36, 37]. Serrano et al. [5] investigated the epoxidation of a styrene-butadiene star block copolymer by means of this procedure, the epoxidation of polybutadiene units in the block copolymer was carried out in a biphasic system in the presence of a the phase-transfer catalyst. The phase transfer catalytic system, methyltrioctylammoniumtetrakis(dipero xotungstate)phosphate, was formed in situ from sodium tungstate dihydrate, phosphoric acid and methyltrioctylammonium chloride as a quaternary ammonium salt under acidic conditions. In the Venturello–Ishii catalytic system, the quaternary ammonium cation is the species bearing the active oxygen which are transported from the organic phase to the aqueous phase in order to enhance the reaction rate [36, 47]. Figure 6.2, shows the scheme of this epoxidation procedure.

Jian and Hay [2] studied the influences of reaction variables, such as the concentration of polymer, the concentration of catalyst, the concentration of hydrogen peroxide, the volume ratio of organic phase to aqueous phase, the type of organic solvent, the time and temperature reaction, on the conversion of double bonds of SBS block copolymer. The control of temperature and reaction time permitted the preparation of block copolymers with different degrees of functionality. The unepoxidized and epoxidized styrene-diene block copolymers containing different microstructures were characterized by spectroscopic and chromatographic techniques [5, 13]. After epoxidation reaction, the ^{1}H NMR spectra of the epoxidized copolymer showed two new signals attributed to the protons attached to the trans- and cis-epoxy groups. The order of reactivity of the isomeric units of double bonds in the polybutadiene block toward epoxidation was: 1,4- > 1,2-vinyl C=C bonds. In addition, thermal analysis by DSC showed an increase of T_g values of PB block, confirming the introduction of rigid oxirane rings in its chains. The appearance of a single T_g value for the epoxidized PB block established that the epoxidation reaction occurs at random positions along the polymer chains. This approach of epoxidation was also applied for the functionalization of SIS block copolymer [4].

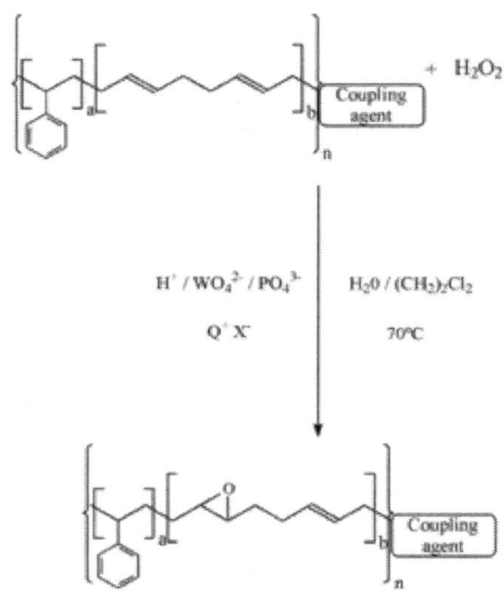

Figure 6.2 Schematic representation of epoxidation procedure using hydrogen peroxide under phase-transfer catalysis of a styrene-butadiene star block copolymer. Reproduced with permission of [5]. Copyright 2004, John Wiley and Sons.

6.2.2 Epoxidation Reaction of Styrene-Diene Block Copolymers Using Dimethyl Dioxirane without Using Phase-Transfer Catalysis

Dimethyl dioxiranes (DMD) are oxidizing agents that are usually prepared in situ from a mixture of acetone and OXONE (potassium monopersulfate). DMD has been efficiently used as stereoselective oxidants for functionalization applications, especially for epoxidations of polydienes [38, 48–51]. This method is eco-friendly and appropriate due to oxone is soluble in water and nontoxic, as well as the reaction conditions are quite simple. The epoxidation ability of block copolymers containing polyisoprene, by means of DMD without using a phase transfer catalyst has been studied in detail by varying the amount of OXONE [6] and reaction times at room temperature [4].

The successful epoxidations products of epoxidized SIS block copolymer were identified using ^{1}H-NMR, ^{13}C NMR, FTIR and SEC [4]. Regarding the selectivity of epoxidation reaction, depending of possible chain microstructures configurations of isoprene additions in the polyisoprene, it was found that 1,4-isoprene units are more reactive than 3,4-isoprene units. Figure 6.3 shows the epoxidation degree (determined from ^{1}H NMR analysis) versus reaction time of a commercial SIS using this functionalization approach. As can be seen, around 75 h of synthesis allowed the obtaining of highly epoxidized block copolymer in which 87 mol% of the polyisoprene repeat units were epoxidized.

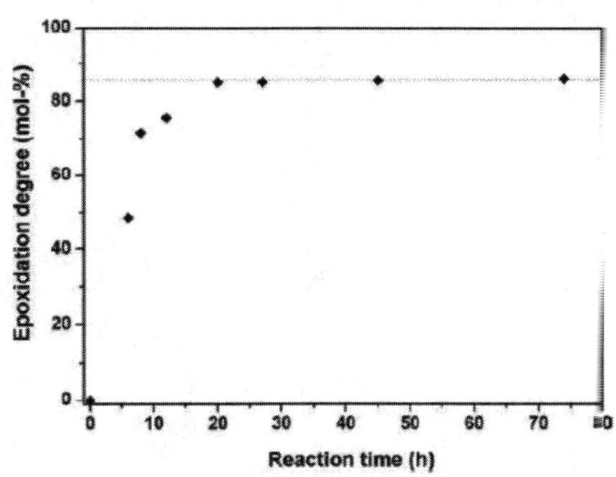

Figure 6.3 Epoxidation degree as function of reaction time for SIS block copolymer using DMD approach at 25°C. Reproduced with permission of [4]. Copyright 2011, John Wiley and Sons.

6.2.3 Introduction of Amino Reactive Groups into Epoxidized Styrene-Diene Block Copolymers

Recently and by the first time, in order to increase the crosslinking reactivity of polydiene block copolymers with the thermosetting formulation during the curing process and to study its successful application as modifier to obtain nanostructured epoxy thermosetting materials with controlled morphology and enhanced matrix-domain interactions, Garate et al. [15] evaluated the incorporation of amino reactive groups into the epoxidized polyisoprene chains of SIS block copolymer by partial oxirane ring-opening reaction by aminolysis to give a novel reactive block copolymer (eSIS-AEP). The methodology used for this purpose was similar than the used by Bonnaud et al. [52].

Figure 6.4 shows the scheme of this functionalization procedure. The amino branch groups were introduced to the oxirane ring in the epoxidized polyisoprene block chains using 1-(2-aminoethyl)piperazine (AEP) as the nucleophile to give the corresponding β-amino alcohols. This reaction was performed using triethylamine (TEA) as highly efficient catalyst for the ring-opening reactions of epoxy groups with amines [53]. The epoxidation of SIS block copolymer was carried out by the DMD oxidizing approach described in Section 6.2.2. The successful grafting reaction of amine groups into epoxidized PI chains and the different generated microstructures were determined by FTIR, ^{1}H NMR, ^{13}C NMR and SEC [15].

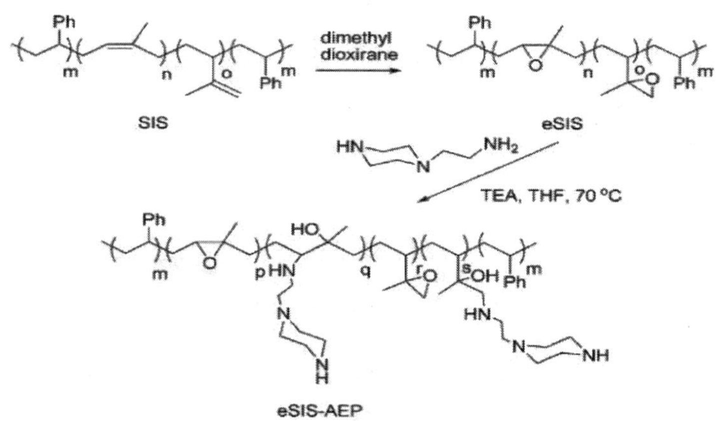

Figure 6.4 Schematic representation of reaction procedure for the introduction of epoxy and amino groups into SIS block copolymer. Reproduced with permission of [15]. Copyright 2014, American Chemical Society.

6.2.4 Evaluation of Side Reactions Depending on Functionalization Conditions and Methodology

A comprehensive study about the possible occurrence of side reactions by FTIR, ^{1}HNMR, ^{13}CNMR, and SEC, such as overoxidation, ring opening, chains scission and crosslinking, during the epoxidation procedures described above, was carried out by Garate et al. [4]. It was found that at high reaction temperature and time, the formation of side product increased in all the methodologies, which could negatively affect the solubility of epoxidized block copolymers. For the case of hydrogen peroxide under phase-transfer catalysis approach the occurrence of ring opening of oxirane groups and overoxidation was detected at higher epoxidation degrees (60 mol%), due to the identification of hydroxyl as well as carbonyl groups in the epoxidized products by spectroscopic. The acidic pH of this approach could promote oxirane ring opening reaction to give OH groups. On the other hand, SEC analyses showed a broadening in the molecular weight distribution of epoxidized block copolymers, indicating the formation of lower and higher molecular weight chains, fact that proves that crosslinking and chain scission reactions occurred [5]. Less crosslinking and scission reactions were detected by reducing the temperature. By DMD approach, side reactions were minimized and the molecular weight distribution of the parent block copolymer was almost retained. The pH value of this approach avoided the ring opening of epoxy groups and subsequent crosslinking reactions that were detected in the hydrogen peroxide phase-transfer catalysis methodology. In this sense, by the employment of DMD approach was possible to obtain higher degrees of epoxidation with a low occurrence of side reactions.

6.3 Self-Assembling Behavior of Epoxidized Styrene-Diene Block Copolymers

It is expected that the epoxidation of block copolymers can have a significant impact on the self-assembled morphologies due to the modification on thermodynamic interactions between dissimilar blocks produced by the chemical changes induced on the polydiene structure.

In this sense, Kim et al. studied the consequences on bulk thermodynamics properties at controlled degrees of epoxidation (xn) of polyisoprene block in a symmetric poly(styrene-b-isoprene) (PS-PIxn) block copolymer [54]. They established a nonlinear relation for the effective segment-segment interaction parameter (χ_{eff}) between the epoxidized polyisoprene and polystyrene block chains with the degree of epoxidation (Fig. 6.5); χ_{eff} depends on temperature, composition, architecture and molecular weight [55–58].

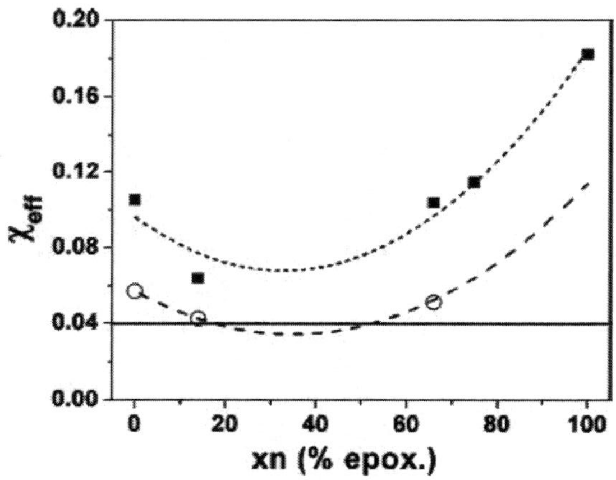

Figure 6.5 Effective interaction parameter (χ_{eff}) between poly(styrene) and partially epoxidized poly(isoprene) blocks in PS-PIxn block copolymer, determined from SAXS measurements (filled symbols) and from DMS data (empty symbols). Reproduced with permission of [54]. Copyright 2012, American Chemical Society.

The correlation of χ_{eff} between PS and the epoxidized PI with the degree of epoxidation (Fig. 6.5) was estimated by two methodologies, combining theoretical and experimental approaches: by applying the mean-field theory to the self-assembled block copolymer structures identified by small-angle X-ray scattering (SAXS) and to the order-to-disorder transition temperatures yielded by the block copolymers determined by dynamic mechanical spectroscopy (DMS).

The first approach involves the experimental determination, using SAXS analyses, of the structural scattering vector, q_1, associated to the domain periodicity or average distance between polymer phases developed by the self-assembled morphology of analyzed block copolymer, in this case a lamellar periodicity, $d_1 = 2\pi/q_1$ (Bragg scattering equation), and the employment of the mean-field theory in the strong segregation limit (Eq. 6.1) [59]. This theory is well known in the fields of polymer and colloid science [60]. To do so, they established for each experimental q_1 value of PS-PIxn block copolymer its corresponding χ_{eff} from Eq. 6.1.

$$d_1 = \frac{2\pi}{q_1} = 1.10aN^{2/3}\chi_{\text{eff}}^{1/6}, \tag{6.1}$$

where a is the statistical segment length and N is the degree of polymerization.

The dash curves in Fig. 6.5 were the best fit curve calculated from experimental data. The χ_{eff} values for PS-PIxn extracted from SAXS measurements were fitted using the binary interaction model equation, principally used for homopolymer/copolymer blends (Eq. 6.2).

$$\chi_{\text{eff}} = \frac{\text{xn}}{100}\chi_{\text{SEI}} + \left(1 - \frac{\text{xn}}{100}\right)\chi_{\text{SI}} - \frac{\text{xn}}{100}\left(1 - \frac{\text{xn}}{100}\right)\chi_{\text{IEI}}, \tag{6.2}$$

where S, I, and EI refer to polystyrene, polyisoprene, and epoxidized polyisoprene block chains, respectively.

The fitting values of the three pairwise interaction parameters were $\chi_{\text{SI}} \approx 0.096$, $\chi_{\text{SEI}} \approx 0.19$, and $\chi_{\text{IEI}} \approx 0.26$. This model was previously employed by Ren et al. [61] to understand the dependence of the χ_{eff} on the chemical modification extent in a quantitative way and extracted the three pairwise interaction parameters of difluorocarbene-modified PS-PI block copolymers.

Therefore, it was established that the miscibility between PIxn and PS at intermediate epoxidation degree is driven by the minimization of strongly unfavorable intramolecular interactions within the statistical copolymer relative to intermolecular interactions, that is, $\chi_{\text{IEI}} > \chi_{\text{SI}}, \chi_{\text{SEI}}$. Similar correlation of χ_{eff} with the degree of epoxidation was obtained using order-disorder transition temperatures (T_{ODT}) results from DMS [54].

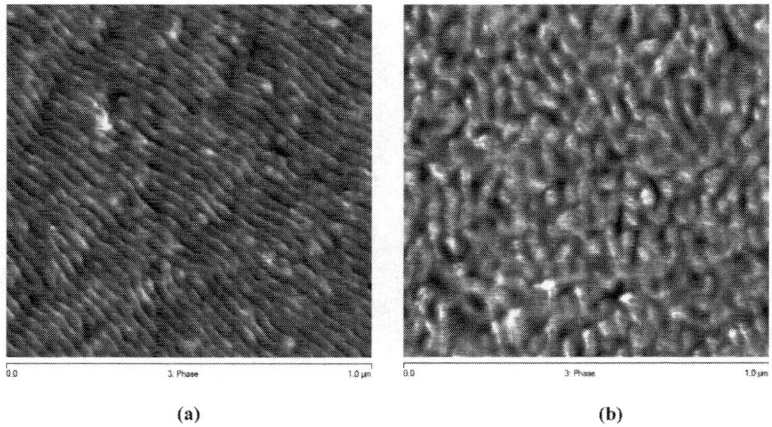

Figure 6.6 TM-AFM phase images of (a) SBS and (b) SBSep46 in bulk after annealing at 110°C for 12 h. PB and PBep phase appears dark in AFM images. Reproduced with permission of [19]. Copyright 2013, American Chemical Society.

Ocando et al. [19] studied the influence of chemical functionalization on bulk self-assembled morphologies of SBS block copolymer. Figure 6.6 shows the analysis of morphological behavior by TM-AFM for the parent SBS block and its counterpart epoxidized 46 mol% (SBSep46). It was noticed that the parent SBS block copolymer self-assembled into a well-defined lamellar nanostructure, where the lamellae were oriented perpendicular to the free surface with an interlamellar period of around 30 nm. Upon functionalization, a change in the morphology was observed; the lamellar morphology disappeared forming worm-like self-assembled morphology without long range order and specific value of periodicity distance around 40 nm (Fig. 6.6b). Similar morphological behaviors were observed by other authors in epoxidized block copolymer thin films [5]. The observed differences in domain periodicity between the SBS and SBSep46 block copolymers were related to the changes on polydispersity index. Other studies showed a clear dependence between the polydispersity index and the self-assembled morphology in block copolymers [62–66]. Moreover, Lynd and Hillmyer [63, 66] found that the increment on the polydispersity index of majority block could frustrate the chain packing into long-range order self-assembled morphology.

6.4 Miscibility and Curing Reaction Behavior of Epoxidized Styrene-Diene Block Copolymers as Template for the Generation of Nanostructured Thermosetting Materials

It is well established that the initial miscibility of at least one block of the templating block copolymer with the epoxy precursors before reaction is the key factor for inducing order on the nanometer scale in both the uncured and cured states of epoxy blends [31]. Guo et al. [67] observed macrophase separation in a poly(propylene glycol) type epoxy resin modified with polystyrene-b-polybutadiene block copolymer. In this sense, the evaluation of the miscibility, by the increment of polarity of the styrene-diene block copolymer by the epoxidation, as well as the potential reactivity for the crosslinking of oxirane groups introduced into polydiene chains with the epoxy precursors is addressed.

Serrano et al. [5] reported the synthesis and characterization of epoxidized star polystyrene–b-polybutadiene (SBep) block copolymers as templates for nanostructured thermosetting materials. On the basis of optical microscopic observation, the cloud point curve of uncured 30 wt% SBep/epoxy resin blends was obtained. It was found a systematic increase on miscibility of the blend at higher degrees of epoxidation. Moreover, Fig. 6.7 shows TM-AFM phase images at room temperature for a film prepared by solvent casting of SBep (68 mol% of epoxidation) block copolymer blended with 50 wt% DGEBA. It was noticed that the blend at uncured state was able to self-assemble into a poorly ordered morphology at nanometer scale, fact that confirmed the miscibility between the epoxidized block copolymer and the epoxy resin. On the contrary the film of parent star SB block copolymer blended with 10 wt% DGEBA analyzed by AFM showed macrophase separation.

Aiming to study the capability of epoxidized block copolymer to act as reactive modifier for the development of nanostructured thermosetting materials, dynamic DSC experiments of unreacted blends were performed for DGEBA/MCDEA, ternary SBep46/DGEBA/MCDEA and binary SBep46/MCDEA blends (Fig. 6.8) [8]. First, the slight decrease of the initial T_g in ternary blends corroborated the miscibility of the epoxidized PB block with the epoxy system, the T_g corresponding to the PS block was also detected. Due to the demonstrated UCST behavior of PS in the DGEBA system [28], at temperatures of curing higher than 80°C, usually employed for DGEBA/MCDEA system, it was expected that the curing reaction at higher epoxidation degrees started from a homogeneous blend. Meng et al. [68] demonstrated this last behavior by using a poly(ε-caprolactone)-b-polybutadiene-b-poly(ε-caprolactone) as modifier of a DGEBA/MOCA system, where the formation of the nanostructures, starting form a complete miscible blend, was judged to be via RIPS of PB chains, whereas the PCL chains remained mixed within the epoxy networks.

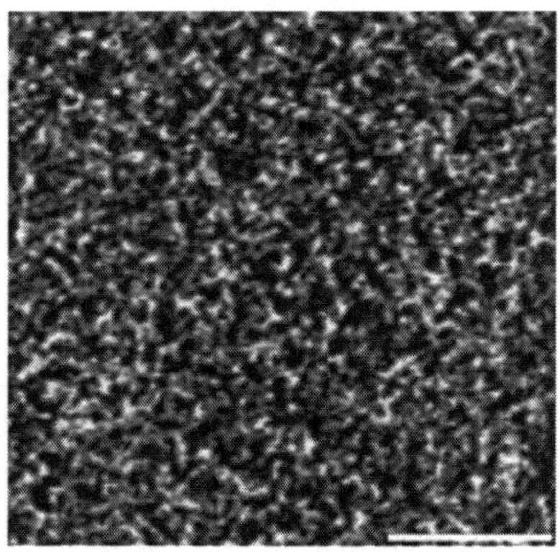

Figure 6.7 TM-AFM phase images of uncured system containing 68 mol% epoxidized copolymer blended with 50 wt% DGEBA annealed at 80°C in vacuum for 3 h. Scale bar size = 1 μm. Reproduced with permission of [5]. Copyright 2004, John Wiley and Sons.

The occurrence of crosslinking reactions between the epoxidized PB chains and the amine hardener was confirmed by an increment on the T_g associated with the epoxidized PB block of the SBep46/MCDEA binary blends after the second DSC scan (Fig. 6.8). However, it was also evidenced that epoxidized block copolymer was able to react with the hardener but at a lower rate than the corresponding for the crosslinking reaction between the DGEBA and the hardener, thereby it was most probably that this reaction occurred in the post-curing process. An increasing of reactivity with epoxidation degree was also established [8].

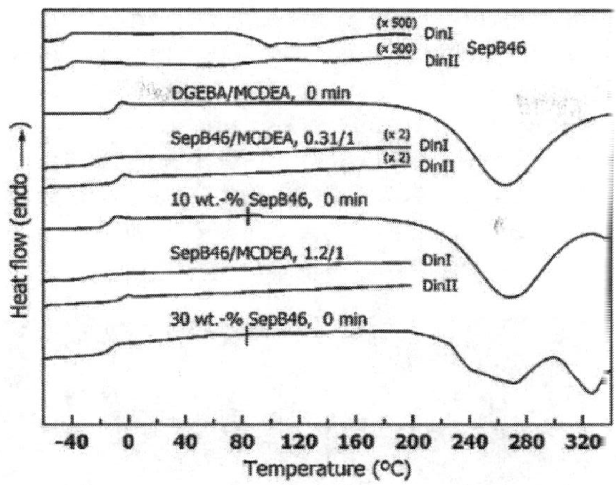

Figure 6.8 Dynamic DSC thermograms for the DGEBA/MCDEA, SBep46/MCDEA binary and SBep46/DGEBA/MCDEA ternary blends (10 and 30 wt% SBep46), curves are normalized to amine equivalents. Reproduced with permission of [8], Copyright 2007 John Wiley and Sons.

As a result, in addition to the demonstrated improvement in the miscibility of the blends, where the microphase separation process in modified systems is expected to occur through the epoxy miscible epoxidized PB block mechanism, at higher epoxidation degree the presence of the epoxidized PB block in the epoxy matrix after curing can be also due to the reaction of epoxidized PB block with the hardener through the epoxidized units. DSC results were corroborated by FT-IR analysis of epoxidized block copolymer-modified epoxy system with and without post-curing [8]. A displacement of the associated hydroxyl group band for epoxidized block copolymer-modified blends was noticed, which suggested that the miscibility can be due to the formation of intramolecular hydrogen bonds associated with unreacted oxirane groups of the epoxidized block copolymers and the OH groups developed through cure reactions of epoxy network. Similar behavior was observed in amphiphilic block copolymers containing miscible PEO as modifier of thermosetting systems, where hydrogen-bonding interactions between the components of these blends lead to a miscible system [69].

In order to precisely and systematically investigate the miscibility and the evolution of the morphology during curing process by means of TM-AFM, a novel DGEBA/Ancamine system modified with highly epoxidized poly(styrene-b-isoprene-b-styrene) block copolymer was analyzed by Garate et al. [14]. Figure 6.9 shows the TM-AFM phase image of the epoxy system modified with 23 wt% of SIS epoxidized 85 mol%, at three different stages of curing process: before curing (stage one); after curing at 80°C for 100 min (stage two); after curing at 80°C for 180 min (stage three).

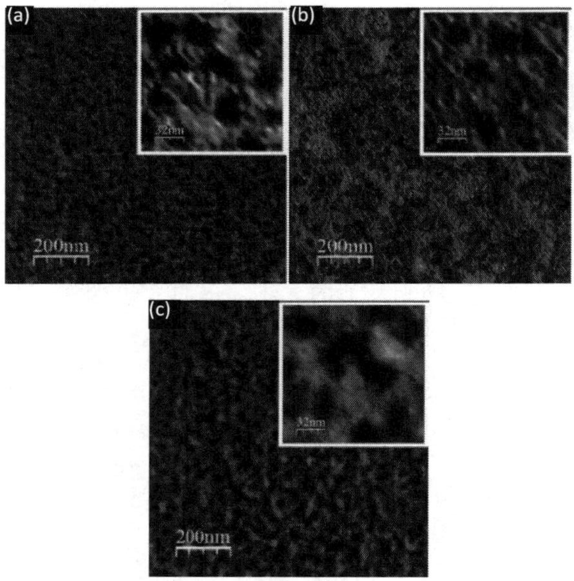

Figure 6.9 TM-AFM phase image of epoxy system/eSIS (23 wt%) blend for: (a) stage one, (b) stage two and (c) stage three. PS block appears dark in AFM images. Reproduced with permission of [14]. Copyright 2013 American Chemical Society.

As can be seen from Fig. 6.9a, at stage one the blend revealed a pre-cured structure based on sphere-like nanodomains dispersed in the epoxy matrix. This pre-cured structured was due to self-assembling of PS chains. Conversely to that was established for SBSep/DGEBA/MCDEA system, it was found that DGEBA/Ancamine/PS blend was cloudy at 80°C, suggesting that PS blocks of eSIS were not initially miscible with the epoxy precursors at the curing temperature. Further, at stage two poorly ordered sphere-like nanostructures were observed (Fig. 6.9b). Finally, these nanodomains underwent a change to bigger size and even less ordered morphology at stage three (Fig. 6.9c). The observed morphological evolution in epoxy system modified with eSIS provided an important insight that a microphase separation process was developed as crosslinking proceeds. The initially miscible epoxidized PI block could undergo a partial expulsion, where the PS nanodomains served as the template for the demixing of epoxidized PI chains, leading to a widening and loss of regularity in the microphase-separated nanodomains.

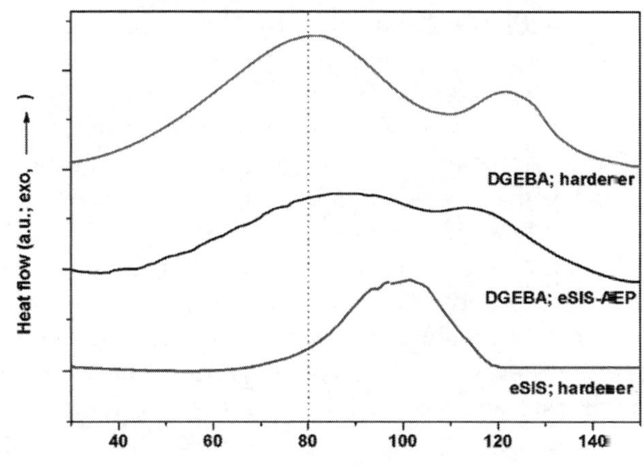

Figure 6.10 Dynamic DSC from 20 to 150°C for DGEBA-hardener, eSIS-AEP/DGEBA, and eSIS/hardener. Reproduced with permission of [15]. Copyright 2014, American Chemical Society.

Otherwise, as mentioned before, the improvement of the capability to crosslink with the epoxy precursors during curing reaction of functionalized SIS block copolymer that contains oxirane and –NH side groups (eSIS-AEP), was also corroborated by dynamic DSC experiments (Fig. 6.10) [15]. The expected increment of the crosslinking reactivity is because this amine functionalized block copolymer could react toward the primary oxirane rings of epoxy resin, fact that suppressed the lower reactivity of triply substituted epoxy groups of epoxidized PI chains respect to the terminal oxirane rings of epoxy resin with the curing agent.

The results demonstrated that at the curing temperature for these systems, the ePI-AEP chains can participate in the epoxy/amine condensation reaction. Furthermore, the thermogram for eSIS/hardener showed a higher T_{max} of exothermic peak than that for eSIS-AEP/DGEBA, corroborating that eSIS-AEP was more reactive than eSIS toward the epoxy precursors at the curing temperature. Consequently, by combining an epoxy miscible and highly reactive block (ePI-AEP) with an epoxy immiscible block (PS) for the modification of a DGEBA/Ancamine system, it was expected that the final morphology was only due to a self-assembly mechanism before curing, where the capability of eSIS-AEP to react with DGEBA during the curing process permitted that PI block cure within the epoxy matrix, without a representative change in the morphology before and after curing reaction.

It was found a strongly relationship between rheological behavior and viscoelastic phase separation in thermoplastic-modified epoxy materials, the phase separation during the cure reaction results in sudden changes in viscosity and modulus and that can be used as criteria for the identification of microphase separation during network formation [70–72]. Choe et al. [73] found in epoxy/thermoplastic blends that the viscosity begins to increase at the onset of phase separation. At this point, the cure reaction rate was readily increased in the thermoplastic-free phase. They also found a rise of G' and G'' at the onset of phase separation.

Serrano et al. [8] investigated the evolution of the complex viscosity, storage modulus (G'), and loss modulus (G''), during isothermal curing at 140°C for the neat DGEBA/MCDEA system and its blends containing 10 and 30 wt% of epoxidized SB and 30 wt% of partent SB, the rheokinetics profiles are shown in Fig. 6.11. Regarding the complex viscosity behavior during isothermal curing at 140°C (Fig. 6.11a), it was noticed that the gelation process was slightly faster when parent SB block copolymer was used as a modifier. Additionally, it was observed an increase in the magnitude of complex viscosity (inset in Fig. 6.11a) before gelation, this phenomenon was associated with SB macrophase separation. The acceleration of the epoxy polymerization rate in the modified epoxy system, in comparison with the neat epoxy system, and also the increment of the viscosity upon phase separation was related to the fact that the block copolymer was macrophase separated in the blend forming an inverted morphology, where epoxy-cured particles were dispersed in a continuous SB matrix [27, 74].

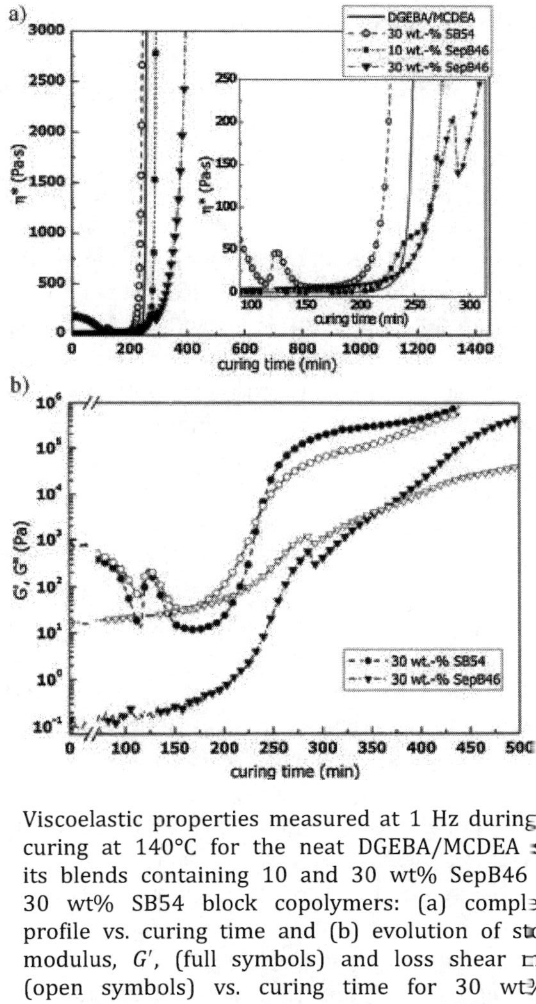

Figure 6.11 Viscoelastic properties measured at 1 Hz during isothermal curing at 140°C for the neat DGEBA/MCDEA system and its blends containing 10 and 30 wt% SepB46 as well as 30 wt% SB54 block copolymers: (a) complex viscosity profile vs. curing time and (b) evolution of storage shear modulus, *G*′, (full symbols) and loss shear modulus, *G*″, (open symbols) vs. curing time for 30 wt% modified systems. Reproduced with permission of [8]. Copyright 2007, John Wiley and Sons.

On the other hand, for the epoxidized SB-modified epoxy blends, it was also observed an increment of the magnitude of both complex viscosity and moduli just before gelation. This result suggested that the formation of nanostructures in the DEGBA/MCDEA system modified with epoxidized styrene-butadiene block copolymers occurs through reaction-induced microphase separation (RIPS). Subsequently, the phase microphase-separated morphology formed before gelation can be fixed during crosslinking reaction. The driving force for the microphase separation of PS block during curing was established; the increment on average molar mass with the occurrence of crosslinking reactions produces a decrease in the entropic contribution to the free energy of mixing, while after gelation, the presence of a crosslinked network produces an elastic contribution that limitthe amount of modifier in the growing network [30].

Similar behavior in the complex viscosity was observed in PS homopolymer-modified epoxy system, which revealed an inverted morphology [27, 74]. Nevertheless, Serrano et al. [8] proposed that in the case of nanostructured thermosetting materials based on block copolymer, the increment on complex viscosity was related to the fact that, at the onset of phase separation, the compatibilization effect of the epoxidized PB block was, at the same time, a competition between the PS and epoxy chains, due to their affinity with the epoxidized PB chains which could reduce the mobility of the epoxy chains. Indeed, after the microphase separation of PS block took place, which is the component in the blend with higher T_g, the overall viscosity decrease since the matrix was formed by a epoxidized PB block and a polymerized epoxy with conversion lower than 0.3. Therefore, the delay in the curing reaction and phase-separation process, which occurs close to the gelation process, for the PS block in epoxidized SB-modified blends, in comparison to PS homopolymer-modified blends, was related with the miscibility between the epoxy and the block copolymer due to the epoxidized PB block.

In another paper, Serrano et al. [11] confirmed by dielectric relaxation spectroscopy the retention of epoxidized PB block and microphase separation of PS block in the epoxy matrix. They analyzed the molecular dynamics of a DGEBA/MCEDA epoxy system modified with an epoxidized SB linear block copolymer during the curing and microphase separation processes.

6.5 Morphological Behavior of Cured Thermosetting Materials Modified with Epoxidized Styrene-Diene Block Copolymers

The effect of functionalization degree of styrene-diene block copolymers as well as the amount of these modifiers in blends with epoxy system on the formation mechanics of nanostructures and final microphase-separated morphology after curing reaction is addressed.

The miscibility of the polydiene chains of functionalized block copolymer with the epoxy precursors depends on the epoxidation degree, therefore it was expected that a range of epoxidation degree leads to clear differences in the formation mechanics of nanostructures and thereby in the morphology after curing reaction. The microphase-separated morphology of fully reacted blends depending on epoxidation degree of functionalized block copolymer as modifier was analyzed by TM-AFM and TEM [9, 10, 13, 18].

TM-AFM and TEM images for DGEBA/MCDEA cured blends containing 30 wt% star SB block copolymers functionalized at a range of epoxidation degrees are shown in Fig. 6.12. It was noticed that the epoxy system modified with star SB block copolymer with epoxidation degrees higher than 40 mol% revealed microphase-separated morphologies, where long-range order nanostructures formed by PS cylinders arranged in the epoxy matrix were obtained [9]. Higher magnification of TEM images of these systems (Fig. 6.12e–f) showed the presence of PB chains close to the PS cylinders, denoted by the presence of dark domains. Fact that suggested a partial demixing of non-epoxidized chains [10]. It is worth noting that for the epoxy system modified with the lower degree of epoxidation, less-ordered structure and broader size distribution for PS cylinders in comparison with higher epoxidation degrees were observed. An increase of epoxidation degree produced a decrease in domains size, probably due to a smaller amount of PB chains expulsed from the matrix around PS domains. This last behavior could be also related to the demonstrated reactivity differences of epoxidized block copolymers with the curing agent depending on the degree of functionalization [8].

Similar results were obtained by Ocando et al. [13]. In this study, the authors also established the minimum epoxidation degree needed to switch from a macrophase separated to a nanostructured epoxy system by modification with SBS block copolymers epoxidized to several degrees.

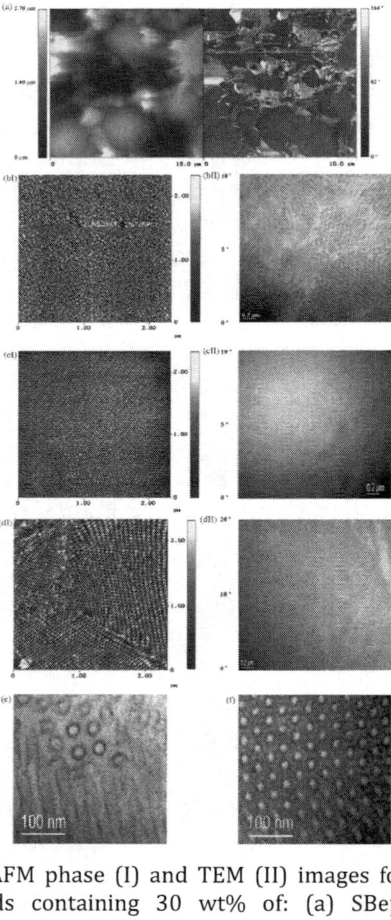

Figure 6.12 TM-AFM phase (I) and TEM (II) images for DGEBA/MCDEA blends containing 30 wt% of: (a) SBep15, (b) SBep40, (c) SBep61, and (d) SBep76 copolymers. (e and f) Figures show higher magnification of TEM images for SBep40- and SBep61-modified systems, respectively. Reproduced with permission of [9]. Copyright 2006, American Chemical Society.

For this purpose, in order to make a more accurate study, the functionalization reaction of this block copolymer was designed in order to obtain different degrees of epoxidation with less difference between each one. A minimum of 27 mol% of epoxidation, was established to ensure nanostructuring of final blends.

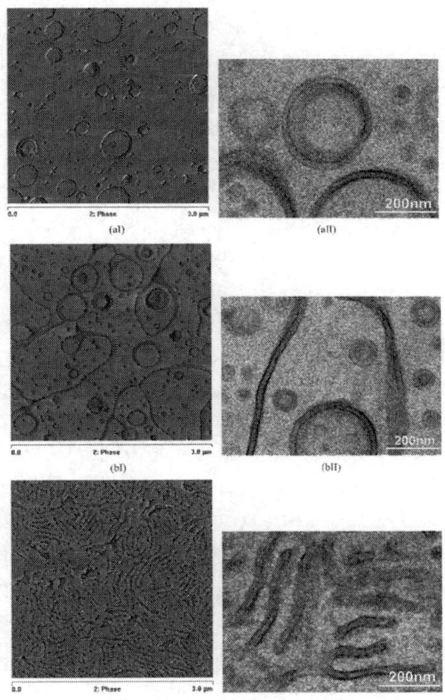

Figure 6.13 TM-AFM phase (I) and TEM (II) images for epoxy mixtures containing different amounts of SB50ep22: (a) 10 wt%, (b) 20 wt%, and (c) 30 wt%. PS block appears dark in AFM images and black in TEM images. Reproduced with permission of [18]. Copyright 2009, American Chemical Society.

Interestingly, it was found that the employment of epoxidation degrees close to the minimum threshold established to achieve nanostructuring in epoxy system modified with epoxidized SB block copolymers [18], produced microphase separation through combining two mechanics of morphology formation, first by the self-assembly of epoxidized PB chains before curing and followed by fixing these nanostructures via microphase separation of PS chains through curing reaction. New bilayered structures were developed, leading to vesicles or long worm-like micelles depending on the amount of block copolymer in the overall blend (Fig. 6.13).

This morphological behavior showed that an epoxidation degree close to the minimum established for nanostructuring is not high enough to make compatible the entire PB block with the epoxy resin before curing. Consequentially, the epoxidized PB block was capable of self-assembly into the epoxy resin thanks to the initially PS miscible block. As can be noticed from Fig. 6.13a the microphase-separated morphology was composed of well dispersed vesicles encasing within them epoxy system in a continuous epoxy matrix. The bilayered structure was corroborated because two phases were discernible in each vesicle domain. The outer layers of vesicle shells were ascribed to PS and the inner layer to epoxidized PB, being the core related to epoxy system. When the content of block copolymer was increased up to 20 wt% in the blend, the morphology revealed a transition to interconnected long worm-like micelles where layers of epoxidized PB where surrounded by PS layered domains (Fig. 6.13b). Finally, the blend containing 30 wt% block copolymer (Fig. 6.13c) revealed an entirely long worm-like morphology.

Similar morphologies were obtained in self-assembled block copolymer in solution, where one block is miscible in the solvent and the other immiscible (typically water) [75, 76]. This fact confirmed that the phase separation of amphiphilic block copolymers in polar epoxy resins can mimic the phase behavior of amphiphilic block copolymers in water.

The schematic representations about the possible organization of the block chains during the formation of nanostructures in thermosetting materials modified with different degrees of epoxidized SB block copolymer before and after epoxy curing reaction are shown in Fig. 6.14. Figures 6.14(I) and 6.14(II) show the morphology development when the degree of epoxidation of the block copolymer was close to the minimum established for nanostructuring. Figure 6.14 III shows the morphology development mechanism when the degree of epoxidation is higher than the minimum established for nanostructuring. The different mechanisms involved in the morphology development mainly depend on the content of epoxidized polybutadiene in the epoxy mixture. Thus, it is reasonable to expect that the blend containing a minor number of epoxidized PB units forms vesicular assemblies with a less interfacial curvature with the epoxy resin than does the blend with a larger number of epoxidized PB units, which forms spherical micelles [25].

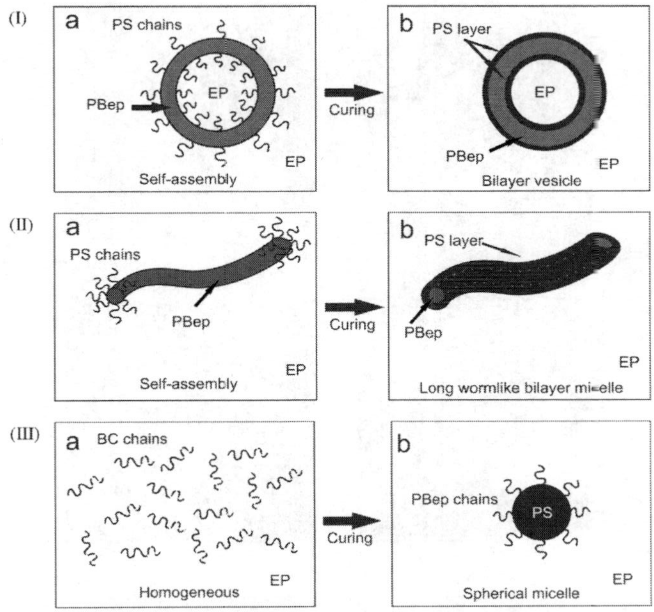

Figure 6.14 Schematic representations of blocks organization for the thermosetting blends containing (I) 10 wt%, (II) 30 wt% of SB epoxidized at values of epoxidation close to the minimum established, and (III) 10 wt% SB epoxidized at higher values than the minimum established for nanostructuring, before (a) and after curing (b). Reproduced with permission of [18]. Copyright 2009, American Chemical Society.

Regarding the effect of the amount of highly epoxidized SBS block copolymer as modifier of epoxy system on final morphology after curing reaction, it was found that the morphology was mainly depends on the PS content in the blend [13]. At contents of around 16–20 wt% PS in the overall blend, TM-AFM images

(Fig. 6.15) showed hexagonally packed cylinder morphology, where cylinders of PS were arranged parallel and perpendicularly to the cut surface, (Fig. 6.15e). As the amount of block copolymer in the epoxy system was increased, the obtained morphologies underwent a transition from micelles to interconnected spherical micelles, followed by worm-like micelles to finally cylinders arranged hexagonally.

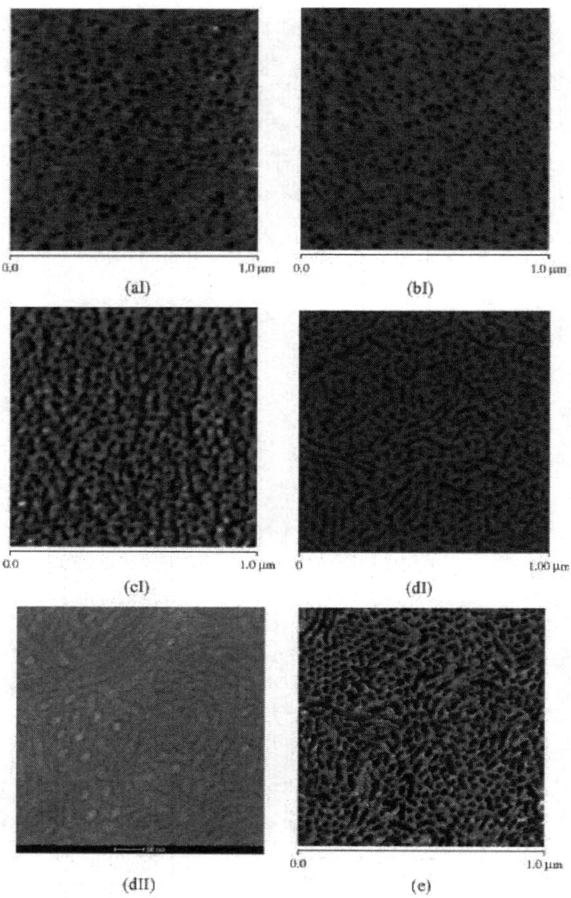

Figure 6.15 TM-AFM phase (I) and TEM (II) images for epoxy blends containing different contents of SBSep44: (a) 10, (b) 20, (c) 30, (d) 40 and (e) 50 wt%. Scale bar: 50 nm for TEM image. Reproduced with permission of [13]. Copyright 2008, John Wiley and Sons.

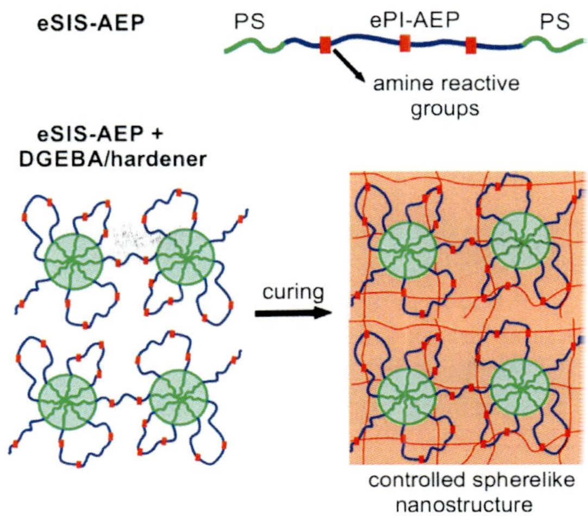

Figure 6.16 Schematic illustration of the formation of sphere-like nanodomains in the eSIS-AEP/DGEBA/Ancamine system. Reproduced with permission of [15]. Copyright 2014, American Chemical Society.

Interestingly, Garate et al. [15] attempted to control the mechanism of the formation of the nanostructures by the design of an epoxidized SIS block copolymer incorporating amine functionalities which present a demonstrated enhanced reactivity with the epoxy precursors. This functionalization approach allowed the formation of nanostructured thermosetting materials with perfect sphere-like nanodomain morphology, instead of the distorted sphere-like nanodomains with some interconnections between domains obtained for the epoxy system modified with epoxidized SIS block copolymer. Therefore the initially obtained sphere-like nanodomains were preserved, indicating that no reaction-induced microphase separation of ePI-AEP chains occurred. The mechanism of formation of nanostructures for the DGEBA/Ancamine system modified with epoxidized SIS block copolymer incorporating amine functionalities is shown in Fig. 6.16. These results showed that the reactivity of the functionalized block copolymer affected the interphase between adjacent phases, which is an important factor in determining the final morphology and properties in such materials. It was a great challenge to control the evolution of interphase thickness at nanometer scale.

6.6 Thermomechanical Properties of Cured Thermosetting Materials Modified with Epoxidized Styrene-Diene Block Copolymers

The study of the damping characteristics of nano-structured thermosetting materials based on block copolymer is one of the appropriate methodologies to provide evidence of the effective retention of one block into the crosslinked epoxy-rich phase by miscibility or reaction, as well as to study the final phase behavior and thermal properties representative of bulk after curing [23, 24, 77–80].

Figure 6.17 shows the dynamic mechanical spectra for the fully cured DGEBA/MCDEA system and its blends containing 30 wt% SB and its counterpart epoxidized at different degrees [9]. It was noticed that the neat epoxy system showed the two characteristics relaxations process at low and high temperature associated to the β and α relaxation modes, respectively. A complete different behavior was observed for 30 wt% SB and SBep15-modified systems, exhibiting three main relaxation processes. These relaxations were attributed to the α relaxation mode of epoxidized PB block at low temperatures, which is overlapping with the β relaxation of the epoxy-rich phase, and the α relaxation temperature for PS block at around 90°C. Relaxations associated with glass transition temperatures for each block in SB and SBep15 were practically unaffected by the epoxy network, thus confirming macrophase separation between these modifiers and epoxy network. It is worth noting that the liquid-like behavior observed at temperatures higher than the T_g of PS block phase for SB and SBep15-modified systems revealed that the block copolymers form the matrix in these blends. When the epoxidation degree increased up to 40 mol%, only one process, corresponding to the β relaxation of the epoxy-rich phase, was observed in the low-temperature range. This fact suggests that epoxidized PB chains were not phase separated from the epoxy matrix.

Moreover, the corresponding α relaxation process of PS chains appeared separated from the epoxy-rich phase, thus corroborating the microphase separation in cured blends. The decrease of the α relaxation of epoxy-rich phase in epoxy system-modified epoxidized block copolymer up to 40% mol of epoxidation was related to a plasticization effect induced by the incorporation of the epoxidized PB block at segmental level in the epoxy network.

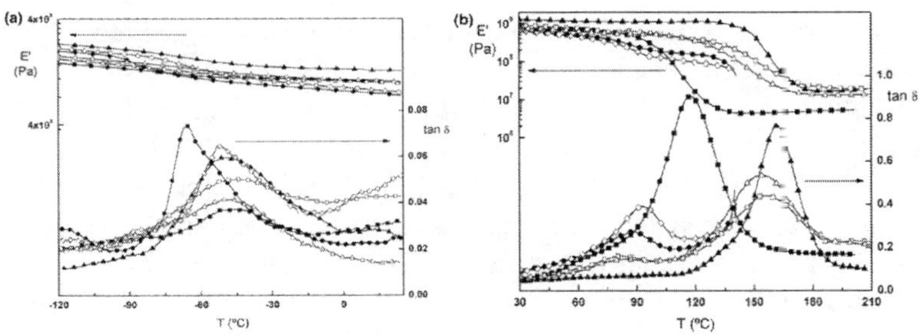

Figure 6.17 Dynamic mechanical spectra obtained at 1 Hz for neat DGEBA/MCDEA system (–2–) and its blends containing 30 wt% of: (–b–) SB, (–]–) SBep15, (–9–) SBep40, (–0–) SBep61, and (–4–) SBep76 copolymers: (a) low-temperature range and (b) high-temperature range. Reproduced with permission of [9]. Copyright 2006, American Chemical Society.

Similar thermomechanical behavior of fully cured epoxy systems modified with epoxidized SBS block copolymers at several degrees was reported by Ocando et al. [13]. The demixing of non-epoxidated PB chains surrounding PS nanodomains was corroborated by the widening of the relaxation process at low temperatures of modified epoxy systems. This behavior was less pronounced at higher epoxidation degrees because a larger incorporation of PB chains into the epoxy matrix was possible. In addition, the wide temperature range over which α relaxation process of epoxy-rich phase was affected reflects a gradient in composition in the epoxy matrix at the vicinity of the nanodomains. It is worth noting that after the α relaxation of the epoxy-rich phase, the systems modified with highly epoxidized block copolymer have an elastomer-like behavior. It was also noticed that the crosslink density (proportional to the plateau storage modulus at temperature above T_g) decreased with increasing block copolymer amount; this was because the phase-separated domains can occupy positions in between the reaction sites and thereby hinder the curing reaction at some particular sites [24].

Interestingly, it was found that it is possible to overcome the plasticization effect on T_g of resulting thermosetting blends observed in the epoxy systems modified with epoxidized block copolymers by means of the employment of block copolymers with higher degrees of epoxidation, around 80% mol, and incorporating amine functional groups as well [15]. Figure 6.18 shows the DSC thermograms obtained for the epoxy system modified with eSIS and eSIS-AEP. The eSIS/DGEBA-hardener (ER) system showed a T_g at 27°C, associated with the partial microphase separation of epoxidized PI. Conversely, such behavior was not observed for eSIS-AEP/ER system, which indicated that ePI-AEP chains remained within to the crosslinked matrix. This result corroborated the reactivity of ePI-AEP block toward DGEBA during curing. It is worth highlighting that these modified epoxy systems revealed an increment in the T_g of epoxy-rich phase, in the opposite with the system modified with low epoxidized SBS [13]. Fact that could be explain because the ePI-AEP block remained linked with the matrix and thereby affected the T_g of epoxy-rich phase as a consequence of a reduction of the free volume of the polymeric matrix.

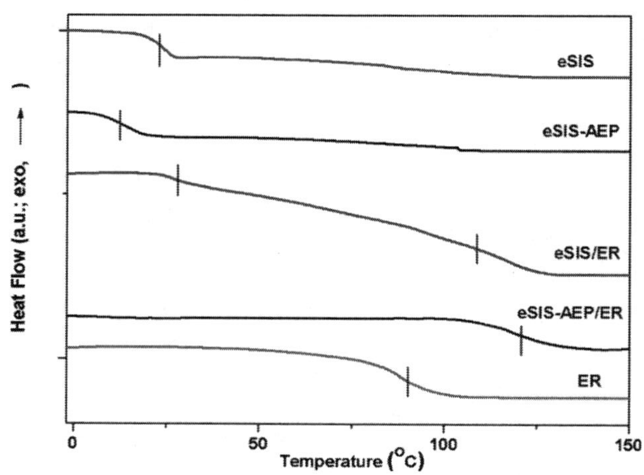

Figure 6.18 DSC curves of the second heating scan for eSIS, eSIS-AEP, and eSIS/ER, and eSIS-AEP/ER cured blends. Reproduced with permission of [15]. Copyright 2014, American Chemical Society.

6.7 Fracture Toughness Behavior of Nanostructured Epoxy Thermosetting Materials Based on Epoxidized Styrene-Diene Block Copolymers

The mechanical properties of thermosetting blends strongly depend on phase-separated morphologies [81]. To date, a considerable amount of research work has been focus on studying the fundamentals of morphology-property relationships in nanostructured thermosetting materials modified with block copolymers [16, 25, 77–80, 82–86]. Mechanical tests revealed that considerable toughness improvements can be attained at low contents of block copolymer blended with epoxy system, and the degree of this improvement depends on the shape, phase-separated domain size and matrix-domains interactions (related to interfacial adhesion) of generated morphology [23]. There exist a significant theoretical toughening mechanisms during crack growth proposed to explain the improvement of toughness for modified epoxy systems forming multiphase morphology, such as shear yielding, plastic deformation, particle bridging, crack-pinning, microcracking [87].

Recently, Liu et al. [77, 80] systematically studied micro-mechanical deformation mechanisms of energy-absorbing process in amphiphilic poly(ethylene-*alt*-propylene)-b-poly(ethyleneoxide) block copolymer modified epoxy by means of the observation of the survived crack tip damage zone of the double notch four-point-bending specimens using OM and TEM. They found experimental evidence that the primary toughening mechanism started from the nanocavitation of nanosize spherical micelles followed by shear banding of the matrix. Moreover, crack tip blunting and the formation of dilatation bands around the crack tip region were also established as energy-absorbing mechanism to contribute to the toughening.

Dean et al. [25, 86] reported the mechanical properties of reactive and nonreactive block copolymer-modified epoxy system in order to investigate the effect of morphology and interfacial adhesion on the fracture properties. For nonreactive block copolymers modifiers at low contents on epoxy system, it was found that spherical micelles had less improvement on fracture toughness than a vesicular morphology [86]. Increases in G_c of 177% were obtained in the vesicle-modified samples. Indeed, the vesicular morphology revealed higher increment on the fracture toughness even at half of block copolymer loading of the micellar morphology system. This last behavior was due to the vesicle construction. Vesicles are spherical domains consisting of a thin bilayer membrane that encases epoxy system. Therefore, as the microphase-separated block copolymer forms only the shell and the total volume of the vesicle phase consists of both the shell and the encapsulated epoxy, acting as a microparticle, a lower amount of the block copolymer is required to achieve a similar degree of toughening as that reached by the micellar morphology system. A subsequent study by Dean et al. [25] revealed that nonreactive vesicles provided poorer toughness than vesicles where one of the blocks was chemically bonded to the epoxy system. They also found a systematically increment of the fracture toughness (calculated from the stress-intensity factor, K_{Ic}) as the relation D_i/D_p (where D_i is the average interparticle distance and D_p is the average vesicle diameter) decreases. At high values of D_i/D_p, there is a weak dependence of the degree of adhesion on the toughening enhancement. Debonding and subsequent matrix plastic deformation were the toughening mechanism proposed for these blends. In addition, from this study it was found that tensile modulus decreases with an increasing vesicle volume fraction independent of the degree of interfacial adhesion between the vesicles and epoxy matrix.

Rebizant et al. [16] evaluated a nonreactive block copolymer and its counterpart incorporating carboxylic acid reactive functionalities as modifiers for different types of epoxy/curing agent blends. It was found that both copolymers lead to toughened materials, independently of the nature of the hardener and the curing process. Moreover, when the curing reaction was performed using 4,4′-diaminodiphenyl sulfone (DDS) and 4,4′-methylenedianiline (MDA), the improvement induced by the reactive block copolymer was more evident. This result was related to the change of morphology induced by reactive blending with a reduction of size as well as an enhanced of interfacial adhesion between the two phases.

Wu et al. [23, 24] studied the role of phase domain size on toughening thermosetting materials with block ionomers, in particular SSEBS. It was found that the toughening depends on the phase domain size, where the epoxy blend with submicrometer phase domains yielded the maximum toughness. This behavior was related to the fact that very small micelles could not induce enough plastic deformation. By the first time they established a quantitative correlation between the nanostructure parameters, such as core radius (R_c), effective hard-sphere radius (R_{hs}) and shell thickness (T_s) (obtained by fitting the SAXS data with a core–shell model) with the fracture toughness of the epoxy blends exhibited well-dispersed nanoscale spherical domains. The fracture toughness was found to increase with increasing R_c and R_{hs} but decrease with T_s. The blend containing the largest core and hard spheres, and the thinnest shell, displayed the highest fracture toughness values. For the epoxy blend modified with SSEBS-c-PCL, there is a large amount of free sulfonic acid groups in the block ionomer complex, which can react with MDA leading to enhanced domain-matrix adhesion. Therefore, the blend with a thinner shell has more SPS units interpenetrated into the epoxy network, leading to greater adhesion. SEM images of fracture surfaces indicated that the toughening mechanism of these systems were probably due to interfacial debonding of spherical microdomains, plastic void expansion, and coalescence, followed by small-scale matrix shear deformation.

Declet-Perez et al. [83] found that a localized epoxy network disruption, such as plasticized and damaged network, produced in the epoxy/block copolymer interface (core) of micelle-modified epoxy systems plays a crucial role in toughening. For this study, they prepared spherical micelle forming block copolymers with rubbery cores (able to cavitate) and glassy cores (unable to cavitate). Interestingly, both systems were capable of improving the fracture toughness of the epoxy system, but the degree of this improvement was greater for the rubbery core micelles than the glassy core micelles, fact related to a more easy local deformation of the matrix induced by the rubbery micelle. This finding revealed that the network disruption in the interface had a similar effect for toughening than matrix shear yielding mechanism. In this sense, it was established that the plasticized shell cavitation is a toughening mechanism that is a direct consequence of the interaction at the interphase between the miscible block of the block copolymer and the epoxy network. This novel interpretation of epoxy toughening could also explain the reason why the worm-like micelles often present higher improvement in fracture toughness than spherical micelles [79], because the high-aspect-ratio worm-like micelles link and modify larger volumes of the crosslinked network than spherical ones.

It was also found that the toughening was strongly influenced by the crosslink density of the epoxy matrix, with higher improve of fracture toughness for lower crosslink density epoxies [78].

Regarding the toughening of nanostructured epoxy materials based on epoxidized block copolymer, Ocando et al. [18] evaluated the dependence of the mechanical properties respect to the epoxidation degree, which governed the morphology, domain-matrix adhesion and size. Table 6.1 shows the fracture toughness values reported in terms of linear elastic mechanism at break, critical stress intensity factor (K_{Ic}) and strain energy release rate (G_{Ic}). It is worth highlighting that the blends containing SB65ep28 that presented nanometer-sized spherical micelles revealed a lower contribution to fracture toughness improvement than the blends containing SB50(2)ep22 that exhibited micrometer-size vesicles or long worm-like micelles with a bilayered structure. The lower fracture toughness values were for the macroscopically phase-separated blends with similar amounts of block copolymer. These results implied that the domain size of phase-separated systems was not the only factor contributing to toughening but the extent of interactions at the interface between the epoxy matrix and phase-separated domains as well as their domain shape can also be relevant factors. Contrary to the expectation, modulus decrease was less for the blends containing the block copolymer with higher amount of PB block, demonstrating that obtaining a long-range ordered morphology produced a synergistic effect on this mechanical property due to a higher extent of interactions between phases. In fact, the macrophase-separated system yielded the lower modulus value, which seemed to be related to the different miscibility extents of these systems. Similar results were reported by Serrano et al. [8]

and an interesting finding was the establishment of the higher value of toughening for the hexagonally ordered nanostructured thermosetting blends, with a flexural modulus value similar to that for neat epoxy matrix. This trend was related to the core (epoxy/block copolymer interface) formed by partially reacted epoxidized butadiene blocks, located around the PS cylinders.

Table 6.1 Summary of K_{Ic} and G_{Ic} results for epoxy systems modified with epoxidized SB block copolymers

System	Weight (%)	E (MPa)	K_{Ic} (Mpa m$^{1/2}$)	G_{Ic} (J/m^2)	G_{Ic}/G_{Ic}, epoxy
Epoxy matrix	—	2700 ± 45	0.80 ± 0.02	208 ± 10	1
SB50(2)ep15	10	2200 ± 55	1.01 ± 0.09	405 ± 70	1.96
SB50(2)ep15	30	935 ± 40	0.77 ± 0.05	560 ± 75	2.69
SB50(2)ep22	10	2150 ± 80	1.00 ± 0.03	410 ± 30	1.97
SB50(2)ep22	20	1700 ± 80	0.90 ± 0.04	415 ± 40	1.99
SB50(2)ep22	30	1570 ± 60	1.31 ± 0.06	955 ± 90	4.59
SB65ep28	10	2370 ± 85	0.87 ± 0.07	285 ± 50	1.36
SB65ep28	20	2170 ± 110	0.96 ± 0.03	370 ± 20	1.78
SB65ep28	30	2060 ± 60	1.25 ± 0.07	670 ± 80	3.22

Source: Reproduced with permission of [18]. Copyright 2009, American Chemical Society.

6.8 Nanostructuring of Thin Epoxy Coatings Modified with Epoxidized Styrene-Diene Block Copolymer

One of recent studies evaluated the gradient of block copolymer segregation and the influence of solvent evaporation on the morphology through film thickness of nanostructured thermosetting coatings based on a commercial epoxy formulation consisting of DGEBA and a low-temperature fast curing amine modified with several amounts of epoxidized SBS block copolymer by means of TM-AFM and attenuated total reflectance infrared spectroscopy [12].

Figure 6.19 shows the evolution of the microphase separate morphology analyzed at different regions through film thickness of an epoxy coating modified with 10 wt% of SBS epoxidized at 46 mol% with constant thickness prepared using different casting solvents. A differential block copolymer segregation through the entire thickness, from the air/polymer interface to the polymer/substrate interface, was noticed.

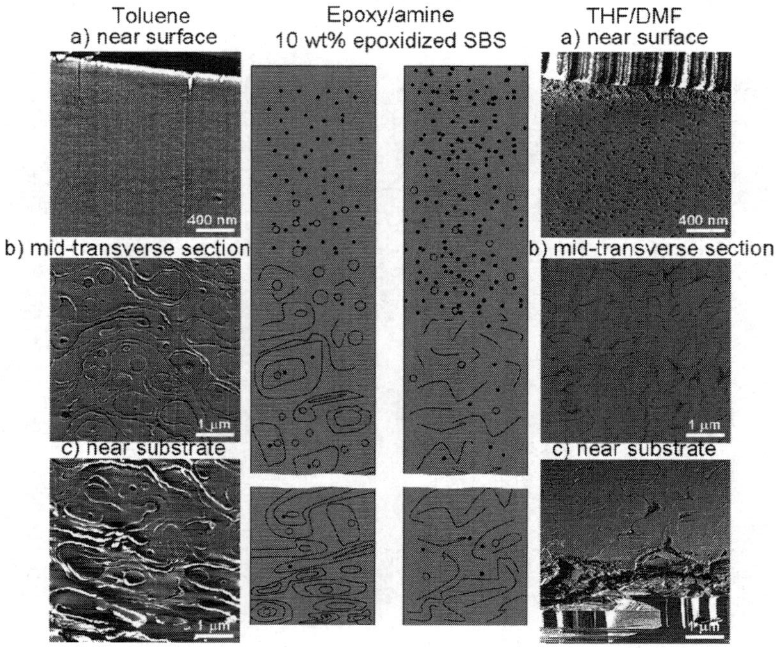

Figure 6.19 TM-AFM phase images for epoxy blends modified with 10 wt% of SBSep46 block copolymer cast in toluene and THF/DMF 1:1 mixture as a function of the location through a 240 μm film thickness: (a) near the air/polymer interface, images 2 μm × 2 μm; (b) in the midtransverse section of the film, images 5 μm × 5 μm; (c) near the polymer/substrate interface, images 5 μm × 5 μm. Reproduced with permission of [12]. Copyright 2012, American Chemical Society.

In addition, by attenuated total reflectance infrared spectroscopy, the absence of block copolymer near the air/polymer interface in the system cast with toluene observed by TM-AFM analyses was verified.

The scheme in Fig. 6.19 simplifies the sequence of TM-AFM images obtained along the film thickness. For both solvents, it was suggested that the morphology gradient could be related to the increase of block copolymer concentration from the air/polymer interface to the substrate/polymer interface by the solvent evaporation. Therefore, the progressive increment of block copolymer concentration provoked a morphological transition from growing of small micelles by coalescence, leading to the formation of short worm-like micelles which rapidly transformed into small vesicles until that larger micellar domains were achieved through the film thickness. At longer distances in depth from the air/polymer interface, longer time in contact with the solvent and longer the structures that could be achieved by coalescence. These structures were frozen before reaching the equilibrium morphological state because the fast curing rate of the epoxy/amine system used. Therefore, these morphological transitions occurred early in the curing process while localized mobility was possible (before gelation took place). Therefore, it was suggested that the whole process of morphology formation through film thickness seemed to be a competition of curing reaction, solvent evaporation, and coalescence of micelles. Clear morphological evidence was also found that the blend of THF/DMF was a better solvent with respect to toluene for these thin epoxy coatings because of the favored polymer-solvent interactions.

A similar study about the influence of casting solvent quality on the templated morphology achieved in the thermosetting matrix modified with poly(ethylene oxide)-b-poly-(ethylene-alt-propylene) (PEO–PEP) was carried out by Hermel-Davidock et al. [88]. A transition from the spherical micelle structure to the worm-like micelle structure, as a consequence of the curing process, was observed. In addition, it was proposed that this last morphological behavior is similar to the phase transition behavior that occurs in dilute block copolymer solutions. Zhang and Eisenberg [89] proposed that the transition from spherical to worm-like micelle structure was due to the adhesive collision and fusion of the nanodomains. In the case of dilute concentrations of block copolymer in an epoxy matrix, it was also proposed that as the curing progresses, the unstable, distorted spherical micelles adhesively collide with neighboring micelles and form a more stable worm-like micelle structure [88].

6.9 Self-Assembled Epoxidized Styrene-Diene Block Copolymer Modified with Epoxy System

It was established that the final morphology of the blends could be affected by factors such as the homopolymer chemistry, chain length and its local segregation or reaction with a selective phase of block copolymer [90–92]. On the other hand, it was also noticed that the shape and the range of order of self-assembled morphology could greatly affect the mechanical properties in TPE materials [93–97]. In this sense, a recently study proposed the development of bulk nanostructured TPE materials stiffened by the use of an epoxy system as modifier at low contents, in order to evaluate the effect of this thermosetting modifier on the morphological, thermomechanical and tensile properties of TPE materials [19].

For this study, Ocando et al. [19] examined blends of epoxidized SBS block copolymer (46% mol of epoxidation degree) modified with 10, 20, and 30 wt% DGEBA/MCDEA system. Via TEM and TM-AFM imaging, they observed microphase separation in all blends after curing of epoxy system (Fig. 6.20). Initially, a worm-like morphology was observed for the neat epoxidized SBS block copolymer. Subsequently, a bicontinuous morphology for the blend containing 10 wt% epoxy system was observed (Fig. 6.20a), followed by a lamellar morphology with poor-range order in the blend with 20 wt% epoxy system where some parts of the lamellae are not completely formed (Fig. 6.20b), revealing the coexistence with a bicontinuous phase. The bicontinuous morphology was reported as an intermediate morphology between the cylindrical and lamellar structure in block copolymers and blends [98–100]. Finally the blend containing 30 wt% epoxy system achieved a well-ordered lamellar morphology (Fig. 6.20c). This morphological behavior indicated that the swelling of epoxidized PB block domains with low amounts of epoxy system induced microphase separation from a poorly to well ordered self-assembly structures. Similar results were obtained by other authors [101].

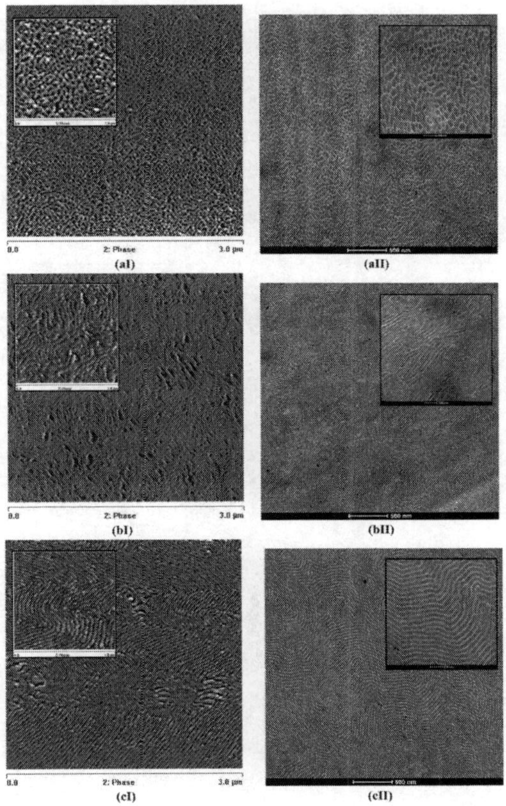

Figure 6.20 TM-AFM phase (I) and TEM (II) images of (a) 90SBSep46/10epoxy, (b) 80SBSep46/20epoxy, and (c) 70SBSep46/30epoxy systems in bulk after cured at 140°C for 72 h and 165°C for 2 h. PBep+epoxy phase appears dark in AFM images and black in TEM images. Reproduced with permission of [19]. Copyright 2013, American Chemical Society.

Tensile modulus, yield strength, and elongation at break were evaluated (Table 6.2). From these results, it was established that these blends can be still considered as semiductile materials. It was also noticed that the blending produced a significant increase on strength and stiffness being more relevant in the system with 30 wt% epoxy and well-ordered lamellar morphology. This last behavior was attributed to the fact that this system is below the T_g of the epoxidized PB-rich phase at the temperature of tensile test (room temperature), contrasting with the epoxidized SBS block copolymer modified with 10 and 20 wt% system.

Table 6.2 Tensile properties of neat SBS, SBSep46, 90SBSep46/10epoxy, 80SBSep46/20epoxy, and 70SBSep46/30epoxy systems

System	Tensile modulus (MPa)	Yield stress (MPa)	Elongation at break (%)
SBS	38 ± 6	3.8 ± 0.3	2364 ± 259
SBSep46	109 ± 19	8.0 ± 0.4	509 ± 25
90SBSep46/10epoxy	141 ± 5	11.2 ± 0.7	325 ± 34
80SBSep46/20epoxy	159 ± 8	13.4 ± 0.6	372 ± 30
70SBSep46/30epoxy	372 ± 19	26.6 ± 1.5	307 ± 21

Source: Reproduced with permission of [19], Copyright 2013, American Chemical Society.

In addition, a synergic effect between the increment of rigid chains and the development of a highly ordered morphology on mechanical properties was proposed. It is worth highlighting that the blends showed a thermoplastic-like behavior at higher temperatures, which means that the BC/epoxy blends could be processed at higher temperatures.

References

1. Wang, S. M., and Tsiang, R. C. C. (1996). Epoxidation of partially hydrogenated styrene-butadiene block copolymers using peracetic acid in a cyclohexane/water heterogeneous system, *J. Polym. Sci. Part A: Polym. Chem.*, **34,** pp. 1483–1491.
2. Jian, X., and Hay, A. S. (1991). Catalytic epoxidation of styrene–butadiene triblock copolymer with hydrogen peroxide, *J. Polym. Sci. Part A: Polym. Chem.*, **29**, pp. 1183–1189.
3. Jian, X., and Hay, A. S. (1991). Catalytic epoxidation of polyisobutylene-co-isoprene with hydrogen peroxide, *J. Polym. Sci. Part A: Polym. Chem.*, **29**, pp. 547–553.
4. Garate, H., Mondragon, I., Goyanes, S., and D'Accorso, N. B. (2011). Controlled epoxidation of poly (styrene-b-isoprene-b-styrene) block copolymer for the development of nanostructured epoxy thermosets, *J. Polym. Sci. Part A: Polym. Chem.*, **49**, pp. 4505–4513.

5. Serrano, E., Larrañaga, M., Remiro, P. M., Mondragon, I., Carrasco, P. M., Pomposo, J. A., and Mecerreyes, D. (2004). Synthesis and characterization of epoxidized styrene-butadiene block copolymers as templates for nanostructured thermosets, *Macromol. Chem. Phys*, **205**, pp. 987–996.
6. Grubbs, R. B., Broz, M. E., Dean, J. M., and Bates, F. S. (2000). Selectively epoxidized polyisoprene-polybutadiene block copolymers, *Macromolecules*, **33**, pp. 2308–2310.
7. Wu, S., Guo, Q., Zhang, T., and Mai, Y.-W. (2013). Phase behavior and nanomechanical mapping of block ionomer complexes, *Soft Matter*, **9**, pp. 2662–2672.
8. Serrano, E., Tercjak, A., Ocando, C., Larrañaga, M. Parellada, M. D., Corona-Galván, S., Mecerreyes, D., Zafeiropoulos, N. E., Stamm, M., and Mondragon, I. (2007). Curing behavior and final properties of nanostructured Thermosetting systems modified with epoxidized styrene-butadiene linear diblock copolymers, *Macromol. Chem. Phys.*, **208**, pp. 2281–2292.
9. Serrano, E., Tercjak, A., Kortaberria, G., Pomposo, J. A., Mecerreyes, D., Zafeiropoulos, N. E., Stamm, M., and Mondragon, I. (2006). Nanostructured thermosetting systems by modification with epoxidized styrene-butadiene star block copolymers. Effect of epoxidation degree, *Macromolecules*, **39**, pp. 2254–2261.
10. Serrano, E., Martin, M. D., Tercjak, A., Pomposo, J. A., Mecerreyes, D., and Mondragon, I. (2005). Nanostructured thermosetting systems from epoxidized styrene butadiene block copolymers, *Macromol. Rapid Commun.*, **26**, pp. 982–985.
11. Serrano, E., Kortaberria, G., Arruti, P., Tercjak, A., and Mondragon, I. (2009). Molecular dynamics of an epoxy resin modified with an epoxidized poly (styrene–butadiene) linear block copolymer during cure and microphase separation processes, *Eur. Polym. J.*, **45**, pp. 1046–1057.
12. Ramos, J. A., Espoìsito, L. H., Fernaìndez, R., Zalakain, I., Goyanes, S., Avgeropoulos, A., Zafeiropoulos, N. E., Kortaberria, G., and Mondragon, I. (2012). Block copolymer concentration gradient and solvent effects on nanostructuring of thin epoxy coatings modified with epoxidized styrene–butadiene–styrene block copolymers, *Macromolecules*, **45**, pp. 1483–1491.
13. Ocando, C., Tercjak, A., Serrano, E., Ramos, J. A., Corona-Galván, S., Parellada, M. D., Fernández-Berridi, M. J., and Mondragon, I. (2008). Micro-and macrophase separation of thermosetting

systems modified with epoxidized styrene-block-butadiene-block-styrene linear triblock copolymers and their influence on final mechanical properties, *Polym. Int.*, **57**, pp. 1333–1342.

14. Garate, H., Mondragon, I., D'Accorso, N. B., and Goyanes, S. (2013). Exploring microphase separation behavior of epoxidized poly (styrene-b-isoprene-b-styrene) block copolymer inside thin epoxy coatings, *Macromolecules*, **46**, pp. 2182–2187.
15. Garate, H., Goyanes, S., and D'Accorso, N. B. (2014). Controlling nanodomain Morphology of epoxy thermosets modified with reactive amine-containing epoxidized poly (styrene-b-isoprene-b-styrene) block copolymer, *Macromolecules*, **47**, pp. 7416–7423.
16. Rebizant, V., Venet, A.-S., Tournilhac, F., Girard-Reydet, E., Navarro, C., Pascault, J.-P., and Leibler, L. (2004). Chemistry and mechanical properties of epoxy-based thermosets reinforced by reactive and nonreactive SBMX block copolymers, *Macromolecules*, **37**, pp. 8017–8027.
17. Rebizant, V., Abetz, V., Tournilhac, F., Court, F., and Leibler, L., Reactive tetrablock copolymers containing glycidyl methacrylate. Synthesis and morphology control in epoxy-amine networks, *Macromolecules*, **36**, pp. 9889–9896.
18. Ocando, C., Tercjak, A., Martín, M. D., Ramos, J. A., Campo, M., and Mondragon, I. (2009). Morphology development in thermosetting mixtures through the variation on chemical functionalization degree of poly(styrene-b-butadiene) diblock copolymer modifiers. Thermomechanical properties, *Macromolecules*, **42**, pp. 6215–6224.
19. Ocando, C., Fernaìndez, R., Tercjak, A., Mondragon, I., and Eceiza, A. (2013). Nanostructured thermoplastic elastomers based on SBS triblock copolymer stiffening with low contents of epoxy system. Morphological behavior and mechanical properties, *Macromolecules*, **46**, pp. 3444–3451.
20. Grubbs, R. B., Dean, J. M., Broz, M. E., and Bates, F. S. (2000). Reactive block copolymers for modification of thermosetting epoxy, *Macromolecules*, **33**, pp. 9522–9534.
21. Grubbs, R. B., Dean, J. M., and Bates, F. S. (2001). Methacrylic block copolymers through metal-mediated living free radical polymerization for modification of thermosetting epoxy, *Macromolecules*, **34**, pp. 8593–8595.
22. Wu, S., Peng, S., Hameed, N., Guo, Q., and Mai, Y.-W. (2012). A new route to nanostructured thermosets with block ionomer complexes, *Soft Matter*, **8**, pp. 688–698.

23. Wu, S., Guo, Q., Peng, S., Hameed, N., Kraska, M., Stühn, B., and Mai, Y.-W. (2012). Toughening epoxy thermosets with block ionomer complexes: A nanostructure–mechanical property correlation, *Macromolecules*, **45**, pp. 3829–3840.

24. Wu, S., Guo, Q., Kraska, M., Stühn, B., and Mai, Y.-W. Toughening epoxy thermosets with block ionomers: The role of phase domain size, *Macromolecules*, **46**, pp. 8190–8202.

25. Dean, J. M., Grubbs, R. B., Saad, W., Cook, R. F., and Bates, F. S. (2003). Mechanical properties of block copolymer vesicle and micelle modified epoxies, *J. Polym. Sci. Part B: Polym. Phys.*, **41**, pp. 2444–2456.

26. He, X., Liu, Y., Zhang, R., Wu, Q., Chen, T., Sun, P., Wang, X., and Xue, G. (2014). Unique interphase and cross-linked network controlled by different miscible blocks in nanostructured epoxy/block copolymer blends characterized by solid-state NMR, *J. Phys. Chem. C*, **118**, pp. 13285–13299.

27. Meynie, L., Fenouillot, F., and Pascault, J.-P. (2004). Polymerization of a thermoset system into a thermoplastic matrix. Effect of the shear, *Polymer*, **45**, pp. 1867–1877.

28. Ritzenthaler, S., Court, F., David, L., Girard-Reydet, E., Leibler, L., and Pascault, J. (2002). ABC triblock copolymers/epoxy-diamine blends. 1. Keys to achieve nanostructured thermosets, *Macromolecules*, **35**, pp. 6245–6254.

29. Girard-Reydet, E., Pascault, J.-P., and Brown, H. R. (2001). Splat: A nonequilibrium morphology on the way to a microemulsion, *Macromolecules*, **34**, pp. 5349–5353.

30. Williams, R. J., Rozenberg, B. A., and Pascault, J.-P. (1997). Reaction-induced phase separation in modified thermosetting polymers, in: *Polymer Analysis Polymer Physics,* Chapter 3 (Springer, Berlin Heidelberg), **128**, pp. 95–156.

31. Lipic, P. M., Bates, F. S., and Hillmyer, M. A. (1998). Nanostructured thermosets from self-assembled amphiphilic block copolymer/epoxy resin mixtures, *J. Am. Chem. Soc.*, **120**, pp. 8963–8970.

32. Sun, P., Dang, Q., Li, B., Chen, T., Wang, Y., Lin, H., Jin, Q., Ding, D., and Shi, A.-C. (2005). Mobility, miscibility, and microdomain structure in nanostructured thermoset blends of epoxy resin and amphiphilic poly (ethylene oxide)-b lock-poly (propylene oxide)-b lock-poly (ethylene oxide) triblock copolymers characterized by solid-state NMR, *Macromolecules*, **38**, pp. 5654–5667.

33. Drobny, J. G. (2014). *Handbook of Thermoplastic Elastomers*, 2nd ed. (Elsevier, USA).

34. Bonart, R. (1979). Thermoplastic elastomers, *Polymer*, **20**, pp. 1389–1403.
35. Zuchowska, D. (1980). Polybutadiene modified by epoxidation. 1. Effect of polybutadiene microstructure on the reactivity of double bonds, *Polymer*, **21**, pp. 514–520.
36. Wang, Q., Zhang, X., Wang, L., and Mi, Z. (2009). Epoxidation of hydroxyl-terminated polybutadiene with hydrogen peroxide under phase-transfer catalysis, *J. Mol. Catal. A: Chem.*, **309**, pp. 89–94.
37. Wang, Q., Zhang, X., Wang, L., and Mi, Z. (2009). Kinetics of epoxidation of hydroxyl-terminated polybutadiene with hydrogen peroxide under phase transfer catalysis, *Ind. Eng. Chem. Res.*, **48**, pp. 1364–1371.
38. Nikje, M. M. A., Soleimani, M., and Mozaffari, Z. (2009). In situ generated dimethyl dioxirane (DMD)/copper (II)(2,2′-diamino-4,4′-bithiazole) complex as a new oxidant for polybutadiene and hydroxyl terminated polybutadiene (HTPB) epoxidation, *J. Elastom. Plast.*, **41**, pp. 41–64.
39. Mizuno, N., Yamaguchi, K., and Kamata, K. (2005). Epoxidation of olefins with hydrogen peroxide catalyzed by polyoxometalates, *Coord. Chem. Rev.*, **249**, pp. 1944–1956.
40. Kurusu, Y., Masuyama, Y., and Miyamoto, M. (1994). Reactivity and microstructure in the epoxidation of polybutadiene, *Polym. J.*, **26**, pp. 1163–1169.
41. Jian, X., and Hay, A. S. (1990). Epoxidation of unsaturated polymers with hydrogen peroxide, *J. Polym. Sci. Part C: Polym. Lett.*, 28, pp. 285–288.
42. Gregório, J. R., Engel Gerbase, A., Martinelli, M., Maldaner Jacobi, M. A., de Luca Freitas, L., Ambros von Holleben, M. L., and Dutra Marcico, P. (2000). Very efficient epoxidation of 1, 4-polybutadiene with the biphasic system methyltrioxorhenium (MTO)-CH_2Cl_2/H_2O_2, *Macromol. Rapid Commun.*, **21**, pp. 401–403.
43. Chen, Y., Zhuang, J., Liu, X., Gao, J., Han, X., Bao, X., Zhou, N., Gao, S., and Xi, Z. (2004). On the nature of reaction-controlled phase transfer catalysts for epoxidation of olefin: A 31P NMR investigation, *Catal. Lett.*, **93**, pp. 41–46.
44. Venturello, C., and D'Aloisio, R. (1988). Quaternary ammonium tetrakis (diperoxotungsto) phosphates (3-) as a new class of catalysts for efficient alkene epoxidation with hydrogen peroxide. *J. Org. Chem.*, **53**, pp. 1553–1557.
45. Jørgensen, K. A. (1989). Transition-metal-catalyzed epoxidations, *Chem. Rev.*, **89**, pp. 431–458.

46. Kaczmarczyk, E., Janus, E., and Milchert, E. (2005). Epoxidation of 1, 4-diallyloxybutane to 1-allyloxy-4-glycidyloxybutane by the method of phase transfer catalysis, *J. Mol. Catal. A: Chem.*, **235**, pp. 52–56.

47. Salles, L., Piquemal, J.-Y., Thouvenot, R., Minot, C., and Brégeault, J.-M. (1997). Catalytic epoxidation by heteropolyoxoperoxo complexes: From novel precursors or catalysts to a mechanistic approach, *J. Mol. Catal. A: Chem.*, **117**, pp. 375–387.

48. Nikje, M. A., Rafiee, A., and Haghshenas, M. (2006). Epoxidation of polybutadiene using in situ generated dimethyl dioxirane (DMD) in the presence of tetra-n-butyl ammonium bromide, *Des. Monom. Polym.*, **9**, pp. 293–303.

49. Nikje, M. M. A., Mozaffari, Z., and Rfiee, A. (2007). Synthesis of epoxidized polybutadienes (EPBs) using a novel dimethyl dioxirane (DMD)/MoO_3 complex and characterization of product using spectroscopy methods, *Des. Monom. Polym.*, **10**, pp. 119–130.

50. Rajabi, F., Nikje, M. A., and Taslimipour, T. (2010). Epoxidation of styrene–butadiene rubber (SBR) using in situ generated dimethyldioxirane (DMD): Characterization and kinetic study, *Des. Monom. Polym.*, **13**, pp. 535–546.

51. Nikje, M. M. A., and Mozaffari, Z. (2007). Chemoselective epoxidation of hydroxyl-terminated polybutadiene (HTPB) using in-situ generated dimethyl dioxirane (DMD), *Des. Monom. Polym.*, **10**, pp. 67–77.

52. Bonnaud, L., Pascault, J., Sautereau, H., Zhao, J., and Jia, D. (2004). Use of reactive polyetherimide to modify epoxy thermosets. I. Synthesis of an amino-grafted polyetherimide, *Eur. Polym. J.*, **40**, pp. 2637–2643.

53. Wu, J., and Xia, H.-G. (2005). Tertiary amines as highly efficient catalysts in the ring-opening reactions of epoxides with amines or thiols in H_2O: Expeditious approach to β-amino alcohols and β-aminothioethers, *Green Chem.*, **7**, pp. 708–710.

54. Kim, S., Nealey, P. F., and Bates, F. S. (2011). Decoupling bulk thermodynamics and wetting characteristics of block copolymer thin films, *ACS Macro Lett.*, **1**, pp. 11–14.

55. Bates, F. S., and Fredrickson, G. H. (1990). Block copolymer thermodynamics: Theory and experiment. *Ann. Rev. Phys. Chem.*, **41**, pp. 525–557.

56. Fredrickson, G. H., Liu, A. J., and Bates, F. S. (1994). Entropic corrections to the Flory-Huggins theory of polymer blends: Architectural and conformational effects, *Macromolecules*, **27**, pp. 2503–2511.

57. Greenberg, C., Foster, M., Turner, C., Corona-Galvan, S., Cloutet, E., Butler, P., Hammouda, B., and Quirk, R. (1999). Effective interaction

parameter between topologically distinct polymers, *Polymer*, **40**, pp. 4713–4716.

58. Dudowicz, J., and Freed, K. F. (1993). Relation of effective interaction parameters for binary blends and diblock copolymers: Lattice cluster theory predictions and comparisons with experiment, *Macromolecules*, **26**, pp. 213–220.

59. Semenov, A. (1985). Contribution to the theory of microphase layering in block copolymer melts. *Zh. Eksp. I Teor. Fiz.*, **88**, pp. 1242–1256.

60. Ceniceros, H. D., and Fredrickson, G. H. (2004). Numerical solution of polymer self-consistent field theory. *Mult. Model. Simul.*, **2**, pp. 452–474.

61. Ren, Y., Lodge, T. P., and Hillmyer, M. A. (2000). Synthesis, characterization, and interaction strengths of difluorocarbene-modified polystyrene-polyisoprene block copolymers, *Macromolecules*, **33**, pp. 866–876.

62. Widin, J. M., Schmitt, A. K., Schmitt, A. L., Im, K., and Mahanthappa, M. K. (2012). Unexpected consequences of block polydispersity on the self-assembly of ABA triblock copolymers, *J. Am. Chem. Soc.*, **134**, pp. 3834–3844.

63. Lynd, N. A., and Hillmyer, M. A. (2005). Influence of polydispersity on the self-assembly of diblock copolymers, *Macromolecules*, **38**, pp. 8803–8810.

64. Li, Y., Qian, H.-J., and Lu, Z.-Y. (2013). The influence of one block polydispersity on phase separation of diblock copolymers: The molecular mechanism for domain spacing expansion, *Polymer*, **54**, pp. 3716–3722.

65. Beardsley, T., and Matsen, M. (2008). Effects of polydispersity on the order-disorder transition of diblock copolymer melts, *Eur. Phys. J. E*, **27**, pp. 323–333.

66. Lynd, N. A., and Hillmyer, M. A. (2007). Effects of polydispersity on the order-disorder transition in block copolymer melts, *Macromolecules*, **40**, pp. 8050–8055.

67. Guo, Q., Figueiredo, P., Thomann, R., and Gronski, W. (2001). Phase behavior, morphology and interfacial structure in thermoset/thermoplastic elastomer blends of poly(propylene glycol)-type epoxy resin and polystyrene–b-polybutadiene, *Polymer*, **42**, pp. 10101–10110.

68. Meng, F., Zheng, S., Zhang, W., Li, H., and Liang, Q. (2006). Nanostructured thermosetting blends of epoxy resin and

amphiphilic poly(ε-caprolactone)-block-polybutadiene-block-poly(ε-caprolactone) triblock copolymer, *Macromolecules*, **39**, pp. 711–719.

69. Larranaga, M., Arruti, P., Serrano, E., De la Caba, K., Remiro, P., Riccardi, C., and Mondragon, I. (2006). Towards microphase separation in epoxy systems containing PEO/PPO/PEO block copolymers by controlling cure conditions and molar ratios between blocks. Part 2. Structural characterization, *Colloid Polym. Sci.*, **284**, pp. 1419–1430.
70. Jyotishkumar, P., Moldenaers, P., George, S. M., and Thomas, S. (2012). Viscoelastic effects in thermoplastic poly (styrene-acrylonitrile)-modified epoxy–DDM system during reaction induced phase separation, *Soft Matter*, **8**, pp. 7452–7462.
71. Jyotishkumar, P., Pionteck, J., Özdilek, C., Moldenaers, P., Cvelbar, U., Mozetic, M., and Thomas, S. (2011). Rheology and pressure–volume–temperature behavior of the thermoplastic poly (acrylonitrile-butadiene-styrene)-modified epoxy-DDS system during reaction induced phase separation, *Soft Matter*, **7**, pp. 7248–7256.
72. Gan, W., Yu, Y., Wang, M., Tao, Q., and Li, S. (2003). Viscoelastic effects on the phase separation in thermoplastics-modified epoxy resin, *Macromolecules*, **36**, pp. 7746–7751.
73. Choe, Y., Kim, M., and Kim, W. (2003). In situ detection of the onset of phase separation and gelation in epoxy/anhydride/thermoplastic blends, *Macromol. Res.*, **11**, pp. 267–272.
74. Cassagnau, P., and Fenouillot, F. (2004). Rheological study of mixing in molten polymers: 2-mixing of reactive systems, *Polymer*, **45**, pp. 8031–8040.
75. Yu, H., Zhu, J., and Jiang, W. (2008). Effect of binary block-selective solvents on self-assembly of ABA triblock copolymer in dilute solution, *J. Polym. Sci. Part B: Polym. Phys.*, **46**, pp. 1536–1545.
76. Schillén, K., Bryskhe, K., and Mel'Nikova, Y. S. (1999). Vesicles formed from a poly (ethylene oxide)-poly (propylene oxide)-poly (ethylene oxide) triblock copolymer in dilute aqueous solution, *Macromolecules*, **32**, pp. 6885–6888.
77. Liu, J. D., Sue, H.-J., Thompson, Z. J., Bates, F. S., Dettloff, M., Jacob, G., Verghese, N., and Pham, H. (2009). Strain rate effect on toughening of nano-sized PEP–PEO block copolymer modified epoxy, *Acta Mater.*, **57**, pp. 2691–2701.
78. Liu, J. D., Sue, H.-J., Thompson, Z. J., Bates, F. S., Dettloff, M., Jacob, G., Verghese, N., and Pham, H. (2009). Effect of crosslink density on fracture behavior of model epoxies containing block copolymer nanoparticles, *Polymer*, **50**, pp. 4683–4689.

79. Liu, J., Thompson, Z. J., Sue, H.-J., Bates, F. S., Hillmyer, M. A., Dettloff, M., Jacob, G., Verghese, N., and Pham, H., Toughening of epoxies with block copolymer micelles of wormlike morphology, *Macromolecules*, **43**, pp. 7238–7243.
80. Liu, J., Sue, H.-J., Thompson, Z. J., Bates, F. S., Dettloff, M., Jacob, G., Verghese, N., and Pham, H. (2008). Nanocavitation in self-assembled amphiphilic block copolymer-modified epoxy, *Macromolecules*, **41**, pp. 7616–7624.
81. Harrats, C., Thomas, S., Groeninckx, G. (2005). *Micro-and Nanostructured Multiphase Polymer Blend Systems: Phase Morphology and Interfaces* (CRC Press, USA).
82. Thompson, Z. J., Hillmyer, M. A., Liu, J., Sue, H.-J., Dettloff, M., and Bates, F. S. (2009). Block copolymer toughened epoxy: Role of cross-link density, *Macromolecules*, **42**, pp. 2333–2335.
83. Declet-Perez, C., Redline, E. M., Francis, L. F., and Bates, F. S. (2012). Role of localized network damage in block copolymer toughened epoxies, *ACS Macro Lett.*, **1**, pp. 338–342.
84. Declet-Perez, C., Francis, L. F., and Bates, F. S. (2013). Cavitation in block copolymer modified epoxy revealed by in situ small-angle x-ray scattering, *ACS Macro Lett.*, **2**, pp. 939–943.
85. Dean, J. M., Verghese, N. E., Pham, H. Q., and Bates, F. S. (2003). Nanostructure toughened epoxy resins, *Macromolecules*, **36**, pp. 9267–9270.
86. Dean, J., Lipic, P., Grubbs, R., Cook, R., and Bates, F. S. (2001). Micellar structure and mechanical properties of block copolymer-modified epoxies, *J. Polym. Sci. Part B: Polym. Phys.*, **39**, pp. 2996–3010.
87. Ruiz-Pérez, L., Royston, G. J., Fairclough, J. P. A., and Ryan, A. J. (2008). Toughening by nanostructure, *Polymer*, **49**, pp. 4475–4488.
88. Hermel-Davidock, T. J., Tang, H. S., Murray, D. J., and Hahn, S. F. (2007). Control of the block copolymer morphology in templated epoxy thermosets, *J. Polym. Sci. Part B: Polym. Phys.*, **45**, pp. 3338–3348.
89. Zhang, L., and Eisenberg, A. (1999). Thermodynamic vs kinetic aspects in the formation and morphological transitions of crew-cut aggregates produced by self-assembly of polystyrene-b-poly (acrylic acid) block copolymers in dilute solution, *Macromolecules*, **32**, pp. 2239–2249.
90. Epps, T. H., Chatterjee, J., and Bates, F. S. (2005). Phase transformations involving network phases in ISO triblock copolymer-homopolymer blends, *Macromolecules*, **38**, pp. 8775–8784.
91. Hashimoto, T., Yamasaki, K., Koizumi, S., and Hasegawa, H. (1993). Ordered structure in blends of block copolymers. 1. Miscibility

criterion for lamellar block copolymers, *Macromolecules*, **26**, pp. 2895–2904.

92. Hashimoto, T., Tanaka, H., and Hasegawa, H. (1990). Ordered structure in mixtures of a block copolymer and homopolymers. 2. Effects of molecular weights of homopolymers, *Macromolecules*, **23**, pp. 4378–4386.

93. Ganß, M., Staudinger, U., Thunga, M., Knoll, K., Schneider, K., Stamm, M., and Weidisch, R. (2012). Influence of S/B middle block composition on the morphology and the mechanical response of polystyrene-poly (styrene-co-butadiene)-polystyrene triblock copolymers, *Polymer*, **53**, pp. 2085–2098.

94. Cohen, Y., Albalak, R. J., Dair, B. J., Capel, M. S., and Thomas, E. L. (2000). Deformation of oriented lamellar block copolymer films, *Macromolecules*, **33**, pp. 6502–6516.

95. Honeker, C. C., and Thomas, E. L. (1996). Impact of morphological orientation in determining mechanical properties in triblock copolymer systems, *Chem. Mater.*, **8**, pp. 1702–1714.

96. Yamaoka, I., and Kimura, M. (1993). Effects of morphology on mechanical properties of a SBS triblock copolymer, *Polymer*, **34**, pp. 4399–4409.

97. Pakula, T., Saijo, K., Kawai, H., and Hashimoto, T. (1985). Deformation behavior of styrene-butadiene-styrene triblock copolymer with cylindrical morphology, *Macromolecules*, **18**, pp. 1294–1302.

98. Spontak, R. J., Fung, J. C., Braunfeld, M. B., Sedat, J. W., Agard, D. A., Kane, L., Smith, S. D., Satkowski, M. M., Ashraf, A., and Hajduk, D. A. (1996). Phase behavior of ordered diblock copolymer blends: Effect of compositional heterogeneity, *Macromolecules*, **29**, pp. 4494–4507.

99. Kloninger, C., and Rehahn, M. (2004). Bicontinuous gyroidic morphologies in ferrocenyldimethylsilane-b-methyl methacrylate diblock copolymer blends, *Macromolecules*, **37**, pp. 8319–8324.

100. Hajduk, D. A., Harper, P. E., Gruner, S. M., Honeker, C. C., Kim, G., Thomas, E. L., and Fetters, L. J. (1994). The gyroid: A new equilibrium morphology in weakly segregated diblock copolymers, *Macromolecules*, **27**, pp. 4063–4075.

101. Tureau, M. S., Kuan, W.-F., Rong, L., Hsiao, B. S., and Epps, T. H. (2012). Inducing order from disordered copolymers: On demand generation of triblock morphologies including networks, *Macromolecules*, **45**, pp. 4599–4605.

Chapter 7

Nanostructuration of Unsaturated Polyester Resins Using Block Copolymers

Daniel H. Builes
Research and Development, Andercol S.A., Autopista Norte, 95-84 Medellín, Colombia
daniel.builes@andercol.com.co

This chapter deals with the preparation and characterization of nanostructured thermosetting materials based on unsaturated polyester modified with different block copolymers. Immiscible block of block copolymers microphase separated from the unsaturated polyester matrices leading to nanostructured thermosetting materials with ability to act as template for inorganic and organic nanoentities. The effect of miscibility, crystallinity and morphology on the final properties such as optical and mechanical is also discussed.

7.1 Introduction

Unsaturated polyester (UP) resins are among the most commonly used resins for thermosets and composites, comprising more

Block Copolymer Nanocomposites
Edited by Galder Kortaberria and Agnieszka Tercjak

ISBN 978-981-4669-54-2 (Hardcover), 978-981-4669-55-9 (eBook)
www.panstanford.com

than 80% of the global market [1]. UP resins exhibit an excellent Young's modulus and thermal and chemical stability obtained mainly due to the high crosslinking density achieved after curing [2]. UP resins technology was born out of the patent of Carleton Ellis [3]. Then, the Scott Bader Company designed in 1941 the first polyester resin cured at room temperature [4, 5]. Thirteen years after, this company developed the first unsaturated polyester resin for marine applications, which was used to make the first large boat in the world manufactured using glass-reinforced plastic in Europe. UP resins are extensively used in applications such as construction [6], adhesives [7], coatings [8] and automotive industry [9], mainly due to properties as rapid strength gain, excellent resistance to abrasion, chemicals, corrosion, heat, impact and compressive stress [6].

After curing, the unmodified UP resins present high brittleness, bad surface aspect and suffer a macroscopic shrinkage, this last behavior being responsible for the variation of dimension, shape and surface aspects of the final UP resin-based pieces. This has currently encouraged the work of many researchers to modify UP resins trying to avoid these drawbacks using low molecular weight polymers. Considering structure as paramount parameter to tough [10–15], researchers have focused a great part of their efforts in this route to toughening thermosets (see previous chapters). To improve the low toughness of the UP thermosets, it is also usual to mix the UP thermosetting precursors with conventional fillers and reinforcements or non-reactive thermoplastic low profile additives (LPA) [2, 13, 14, 16].

In the past decade, researchers have focused on modifying thermosetting UP resins at nanoscale level as an effective route to improve their properties. Modification with nanofillers such as organoclay nanoparticles [17], carbon nanotubes [18] or carbon nanofibers [19, 20] and block copolymers [11, 21–25] has been used as an effective way to control the morphology of UP thermosetting matrices with improved mechanical properties. In the case of shrinkage, its decrease depends on the size of phase-separated domains and the number of formed microvoids. Thus, a route to avoid it is the adequate selection of a block copolymer and curing conditions which promote the achievement of an accurate control of the domain size at nanoscale [11].

In this chapter, some phenomena that govern the modification of UP resin in order to obtain nanostructured matrices using the amphiphilic capabilities of block copolymers (BCPs) are explained.

7.2 Curing Process Chemistry of Unsaturated Polyester Resins

The simplest commercial UP resin is a mixture of macromolecules with molecular weight between 1000–10000 g mol^{-1}, dissolved in a vinyl-like co-reactant such as styrene (St) or methyl methacrylate, which acts as diluent and unsaturated monomer or crosslinking agent [4, 26, 27]. The macromolecules are unsaturated polyester oligomers (UPol) obtained by polycondensation of glycols with unsaturated and saturated di-acids [28, 29]. For the majority of commercial UP resins, maleic anhydride provides the unsaturation, propylene glycol, ethylene glycol and diethylene glycol are the main glycols, and ortho-phthalic, tere-phthalic or iso-phthalic acids are the main saturated acids used. An example of an ortho-phthalic UPol chain is shown in Fig. 7.1.

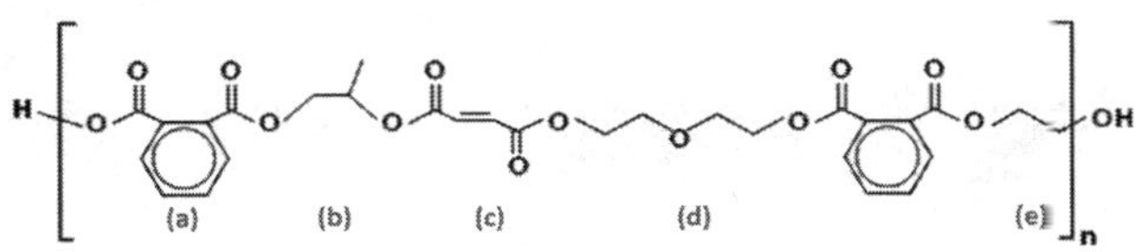

Figure 7.1 Scheme of an UPol made from (a) ortho-phthalic anhydride, (b) propylene glycol, (c) fumaric acid (or maleic anhydride), (d) diethylene glycol and (e) ethylene glycol.

A cured UP resin consists of a three-dimensional network formed by the homo- or co-polymerization carried out between the C=C double bond of UPol and vinyl-like co-reactant through a free-radical chain polymerization. Simulation of a UP curing process versus experimental data has been reported by Capehart et al. [30] and dell'Erba et al. [31]. To cure a UP resin free radicals are required that could be obtained from the dissociation of an initiator activated by heat or redox metal promoters [32]. This crosslinking process is exothermic [33] and inhomogeneous [34] due to three general types of reactions that could occur [16, 35–37]: UPol-St reaction, St-St homopolymerization and inter- and intra-molecular reactions between oligomers (UPol-UPol* and UPol-UPol)—the first promotes the gelation and the second retards it [38] (see Fig. 7.2).

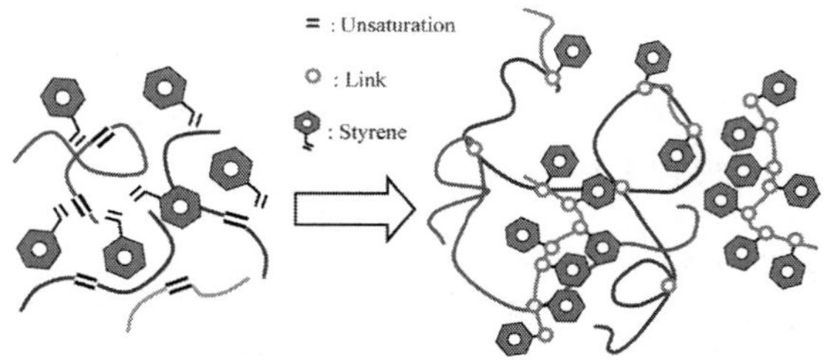

Figure 7.2 Schematic representation of several types of reactions during crosslinking process.

Furthermore, as explained in Table 7.1, during the global crosslinked network formation, almost four states can be distinguished since the non-reactive UP resin until the fully cured resin.

Table 7.1 Curing process steps of St-dissolved UP resin

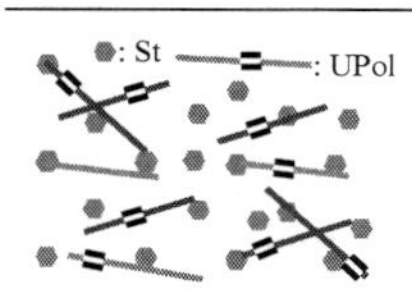	**UP thermosetting precursor or neat non-reactive UP resin.** The size of UPol coils depends on the chain length, chain stiffness, and compatibility with the St. It is the commercially available state of the resin.
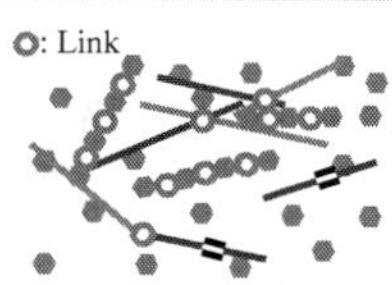	**Induction**. After mixing with an initiator, it dissociates forming free radicals that start the crosslinking reaction, thereby an increase of coil size occurs.
	**Microgelation**. Due to at the beginning intramolecular reactions are more frequent, and to St is poor solvent for UPol, cyclization occurs generating internally crosslinked compact molecules called microgels or crosslinked microparticles with diameter of ca. 10 nm. This step occurs at conversions below 5% and generates a UPol-rich and St-rich phases by polymerization-induced microphase separation [35, 37–43].

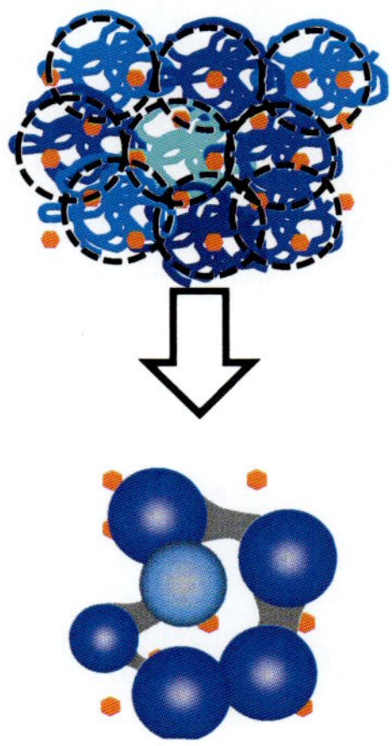

Percolation. Concentration of microgels increases continuously during a competition between intramolecular and intermicrogel crosslinking reactions (network formation), generating microgels percolation. This step occurs at conversions below 10% with consequences as shrinkage, macrogelation and a dramatically increase of viscosity of the resin [2]. The unsaturations located inside microgels produce a decrease of the final conversion, due to reduction of the chain mobility leads to a decrease of the network crosslinking density [36, 39]. The conversion increase and the "void" space that contain monomers is filled mainly by St homopolymerization, increasing the homogeneity of the matrix; however, the heterogeneous structure always persists.

Finally, according to the St and UPol reaction steps explained in Table 7.1, sphere-like morphology is achieved. Figure 7.3 clearly shows the microphase-separated structure formed by the microgels of a St-dissolved UP resin after curing.

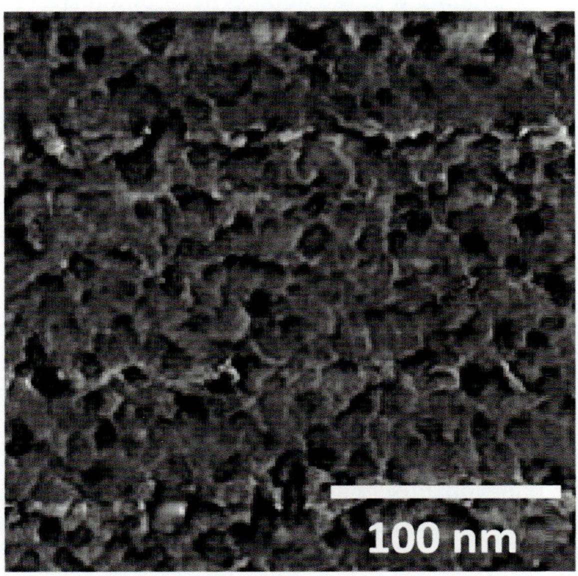

Figure 7.3 AFM phase image (250 nm × 250 nm) of cured UP resin.

Cheng et al. [39] and Sun et al. [40] investigated the curing process of an iso-phthalic UP resin, and reported that the size of microgels can change from 22 to 56 nm from the beginning of the reaction to near the macrogelation. They also proved that microgels can reduce the size during the curing process, which is justified mainly due to intramolecular crosslinking reactions that occur inside the UPol-rich phase.

7.3 General Considerations of Block Copolymer and Thermosetting Mixtures

To obtain a well-modified thermosetting material using homopolymers, nano-objects, BCPs, etc., a miscible or homogeneous non-reactive mixture with the thermosetting precursor is required. That is why the use of BCPs as a thermosetting modifier requires at least one miscible block with the thermosetting precursor. Thus, the miscible blocks should be capable of dispersing the non-miscible blocks homogenously through the continuous phase before curing. Once a submicrometric homogeneous mixture is achieved, the curing process could start. It is worth mentioning that the homogeneity of a mixture depends on the analyzed scale, e.g., a macrometric homogeneous mixture could be micrometrically or nanometrically heterogeneous.

Thermosetting precursors could act as a selective solvent [12] or as a solvent [44] for BCPs. In the first case, nanometric heterogeneities could be self-assembled forming several geometries (spherical, cylindrical, lamellar, etc.) in the liquid resin. These mesophases or micelles could be "frozen" during curing leading to a thermoset with a fixed morphology. In the second case, all blocks of the BCP are miscible in the thermosetting precursor, and during curing the molecular weight increases decreasing the conformational entropy of mixing changing the BPC-resin interactions. This phenomenon could trigger a phase separation of the previously miscible blocks in a process called polymerization-induced phase separation (PIPS). During PIPS, depending on miscibility of blocks, the BCP could remain homogeneous or could follow micro- or macrophase separation. Contrary to epoxy resins, UP resins curing process generates high compositional changes during short time intervals due to the high reaction rate, finishing the PIPS when gelation and vitrification is reached. Due to the low gel conversion the size of phase-separation is hindered [11].

Conversely to epoxy systems, only a few works have been reported related to the nanostructuration of UP resins employing BCPs. One of the main difficulties is based on the phase heterogeneity achieved during curing (see Table 7.1), which is due mainly to the extreme changes in miscibility at very low conversion leading to microphase separation of the St-crosslinked UPol before gelation. This behavior is also one of the main differences between epoxy and UP thermosets together with the conversion degree at which the gelation is reached—longer in the case of epoxies. Sinturel et al. [21] and Builes et al. [22] analyzed UP thermosetting matrices using commercially available PEO-composed BCPs and obtained self-assembled morphologies; lamellae [21] and sphere-like, worm-like and closed worm-like [22]. Serrano et al. [11] mixed synthesized several BCPs based on poly(butyl acrylate) and poly[(methyl methacrylate)-*co*-(*N*,*N*-dimethyl acrylamide)] with two different UP resins. Furthermore, Li et al. [45] used a solid state NMR method to analyze the interface between BCP and UP matrix modified with a PEO-b-PPO-b-PEO block copolymer. Some of these cases will be explained later.

7.4 Modification of an Unsaturated Polyester Matrix with PEO-b-PPO-b-PEO Block Copolymers

The amphiphilic character of several PEO-b-PPO-b-PEO block copolymers has been effectively proved for epoxy resins [15, 46], aqueous solutions [47, 48] and UP resins [22, 24, 25, 49, 50]. In some of these systems, self-assembled structures as a function of BCP content and temperature were achieved. Taking into consideration the miscibility of each block of the BCP with the UP resin, the PEO and PPO blocks could be miscible with the UP resin or the PPO central block phase could separate from the UP resin. Builes et al. [22, 49] obtained the nanostructuration of an UP matrix using two different PEO-b-PPO-b-PEO block copolymers with structures $E_{75}P_{34}E_{75}$ and $E_{20}P_{69}E_{20}$. In this nomenclature, E was used for ethylene oxide and P for propylene oxide. The subscript numbers indicate the quantity of repeated units.

Due to UP resin composition, miscibility of any UP resin modifier depends on the modifier interactions with the vinyl-like co-reactant, with the UPol, and on the relative contents of the specific tri-component system. In the following section, miscibility of the UP resin with PEO and PPO homopolymers will be explained.

7.4.1 Miscibility of Unsaturated Polyester Resin with PEO Homopolymers

Behavior of an UP resin modified with the E_{182} homopolymer was analyzed [24]. The UP resin used was St-dissolved and commercially available, composed of UPols made from maleic and ortho-phthalic anhydrides, with a C=C molar ratio between St and UPols of ca. 1. Non-reactive UP/E_{182} mixtures with E_{182} content between 5–15 wt% exhibited high transparency and homogeneity, indicating a lack of macrophase separation at room temperature. DSC thermal transitions of these mixtures depicted that the UP resin hindered the crystallization of PEO and showed a common behavior of a partially miscible crystalline/amorphous polymeric mixture [51]. This results are in a good agreement with the miscibility study of UPol/PEO and cured UP resin/PEO mixtures reported by Zheng et al. [52] and Zheng et al. [53]. In the first research work, they calculated the interaction parameter (χ) of these systems proving the strong miscibility of the systems in the molten state. In the second one, the effect of the hydrogen bonding interactions between hydroxyl groups of UPol and ether groups of PEO on the miscibility was investigated. This behavior can be thermodynamically explained taking into account that UP resin/PEO systems have exothermic enthalpy of mixing (lower critical solution temperature (LCST) behavior), thus macrophase separation could take place with increase of temperature [49, 50, 54]. Miscibility and morphology obtained after curing UP/E_{182} mixtures will be explained in the next section.

7.4.2 Miscibility of Unsaturated Polyester Resin with PPO Homopolymers

In order to better understand the effect of the PPO block on miscibility of the investigated non-reactive mixtures, Bailes et al. [24] studied the miscibility of a UP resin with PPO homopolymers with different molecular weights. Figure 7.4 shows the effect of PPO content and temperature and the content of PPO on the miscibility of non-reactive UP/P_{34} and UP/P_{69} systems.

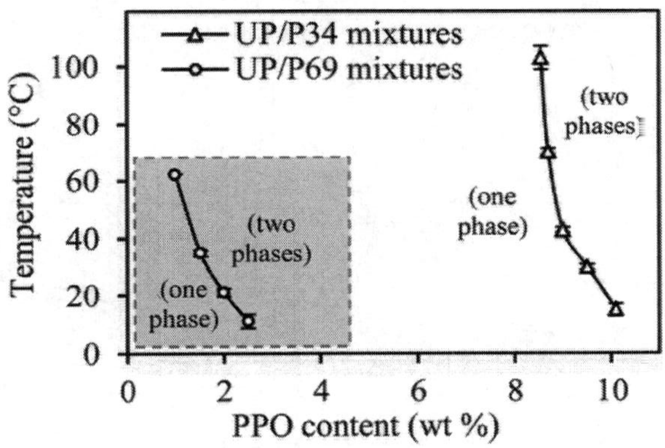

Figure 7.4 Temperature-PPO content phase diagram of non-reactive UP/PPO mixtures for: (-○-) UP/P69 and (-△-) UP/P34 systems. (one phase): transparent and homogeneous mixtures. (two phases): cloudy mixtures. Reproduced with permission of [24]. Copyright 2014, American Chemical Society.

Applying the Hoftyzer and Van Krevelen method [55], the calculated interaction parameters for UPol-St, UPol-PPO and PPO-St were $\chi_{UPol\text{-}St} = 2.85$, $\chi_{UPol\text{-}PPO} = 4.1$ and $\chi_{PPO\text{-}St} = 0.36$, respectively [22, 24]. Those values indicate high contribution of St and low contribution of UPol to the miscibility of PPO with UP resin. Furthermore, in order to verify the miscibility between UPol and PPO, several UPol/P_{34} and UPol/P_{69} mixtures were analyzed. The visual appearance of neat UPol, P_{34} and P_{69} were transparent, as well as the mixture of UPol with 4 wt% of P_{34}. On the contrary, mixtures with 7 and 15 wt% of P_{34}, or with 2 and 4 wt% of P_{69} were opaque at room temperature.

The thermal behavior of UPol/P_{34} and UPol/P_{69} mixtures is shown in Fig. 7.5. As can be seen, the modification of UPol with 4 wt% of P_{34} provoked a shift of T_g from −16 to −41°C (Fig. 7.5a). However, the increase of P_{34} content to 7 and 15 wt%, shifted the T_gs to values similar to T_g of UPol. In fact, mixture with 15 wt% of P_{34} exhibited a weak T_g at -65°C associated with a P_{34}-rich phase. These results corroborated the existence of a maximum miscibility threshold between UPol and P_{34}. In the case of UPol modified with P_{69}, the T_gs of the mixtures containing 4 and 7 wt% of P_{69} (−19 and −17°C, respectively) proved the strong immiscibility between the components UPol and P_{69} (Fig. 7.5b).

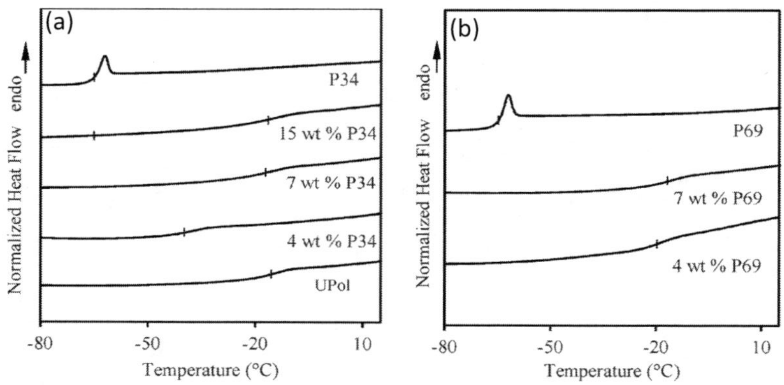

Figure 7.5 DSC thermograms of (a) UPol, P_{34} and UPol/P_{34} mixtures and (b) P_{69} and UPol/P_{69} mixtures. Reproduced with permission of [24]. Copyright 2014, American Chemical Society.

According to the phase diagram in Fig. 7.4, and DSC analysis, UP/PPO mixtures presented LCST behavior strongly influenced by the content and molecular weight of PPO. Moreover, the strong immiscibility of UP resin with P_{69} homopolymer was confirmed and the fact that miscibility of P_{34} with UP resin was related to the miscibility with St and partial miscibility with UPol. The latter proves that not only the group interactions, strongly related with enthalpy contribution, but molecular weight, strongly related with entropy contribution, are responsible for this miscibility.

7.4.3 Modification of an Unsaturated Polyester Matrix with $E_{75}P_{34}E_{75}$ Block Copolymer

Builes et al. [24] investigated the nanostructuration of UP resin mixed with the $E_{75}P_{34}E_{75}$ block copolymer (named here as EPE75). In that work was analyzed the role played by the PEO lateral blocks and PPO central block on final properties of these materials, and the influence of curing temperature and EPE75 content on miscibility, morphology, transparency and mechanical properties of designed mixtures. Similar to UP/E_{182}, UP/EPE75 mixtures presented high transparency and homogeneous appearance between 5–15 wt% of EPE75 content indicating a lack of macrophase separation at room temperature.

7.4.3.1 UP/$E_{75}P_{34}E_{75}$ and UP/E_{182} non-reactive mixtures

UP/EPE_{75} and UP/E_{182} systems were compared taking into account that E182 and EPE75 present similar molecular weight. Thermal transitions of non-reactive UP/EPE_{75} and UP/E_{182} mixtures were measured using DSC. Figure 7.6 shows the DSC thermograms of the non-reactive mixtures. The UP/modifier mixtures were named taking into account the modifier content (wt%) followed by the name of the modifier, e.g., the mixture named 15%EPE75 contained 15 wt% of EPE75 and 75 wt% of UP resin.

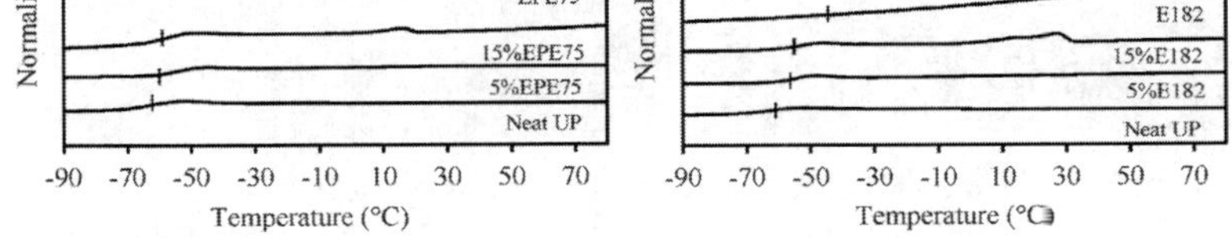

Figure 7.6 DSC thermograms for (a) UP/EPE75 and (b) UP/E_{182} non-reactive mixtures. The lines (|) denote the midpoints of the T_g transitions. Adapted and reproduced with permission of [24]. Copyright 2014, American Chemical Society.

As can be seen, the UP/EPE75 mixtures showed a single T_g between the glass transitions temperatures of the neat components, indicating partial miscibility between them. Due to the last, the UP/EPE75 system exhibited a thermal behavior similar to a no-micellar system [24]. On the other hand, taking into account that EPE75 and E_{182} had similar molecular weights, the phase diagram of Fig. 7.4, and that the PPO block content in 15%EPE75 mixture (ca. 3 wt%), it could be concluded that no phase separation of PPO central block took place in the analyzed non-reactive UP/EPE75 system.

7.4.3.2 Cured UP/$E_{75}P_{34}E_{75}$ and UP/E_{182} mixtures

After curing at 80°C, the UP/EPE75 and UP/E_{182} mixtures remained visually transparent, similar to the neat UP resin. The latter indicated the absence of macrophase separation. Morphology and miscibility analysis of studied cured mixtures clearly exhibited that final thermosets were composed of UPol-rich, polystyrene-rich and EPE75-rich or E_{182}-rich phases. A comparison between AFM images of cured UP matrix (Fig. 7.3), 15%EPE75 and 15%E_{182} mixtures (Fig. 7.7) allow to distinguish two different of micro-separated phases. The first one, related to the microphase separation of the UP matrix (UP-rich and polystyrene-rich phases), and the second one related of microphase separation of EPE75 or E_{182}.

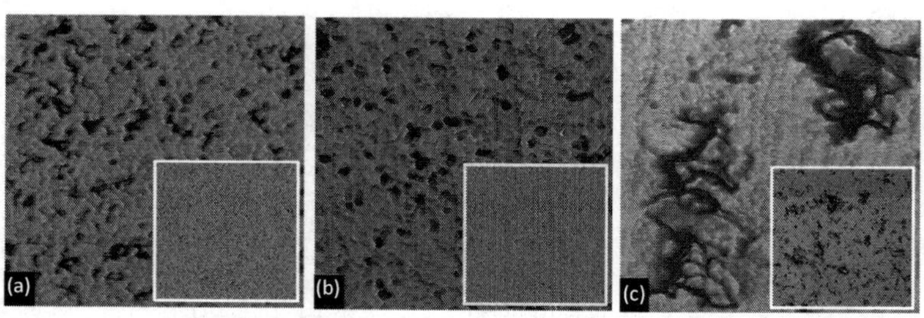

Figure 7.7 TM-AFM phase images (500 nm × 500 nm) of 15 wt% modified thermosets cured at 80°C for: (a) 15%EPE75, (b) 15%E_{182} and (c) 15%EPE20 mixtures. The insets correspond to 3 μm × 3 μm AFM images. These AFM images correspond to internal surfaces of the cured mixtures trimmed using an ultramicrotome. Reproduced with permission of [24]. Copyright 2014, American Chemical Society.

Since the non-reactive mixtures did not form self-assembled structures, during the formation of the crosslinked network, one could deduce that the PIPS process was followed to achieve these microphase-separated morphologies. This phase behavior was due to the decrease of PEO miscibility with UP resin during curing.

The morphology of the UP/EPE75 system evolved from sphere-like to worm-like domains with increasing EPE75 content from 5 to 15 wt%. However, morphology of UP/E_{182} mixtures resulted in well-defined sphere-like domains with the same change of modifier content [24]. In order to explain the effect of PEO and PPO blocks of the EPE75 on the morphology of cured UP/EPE75 mixtures, the morphology of 12% E_{75} and 3%P_{34} mixtures cured at 80°C were also analyzed (see Fig. 7.8). This two mixtures were prepared with PEO and PPO homopolymers with similar contents and molecular weights of the PEO and PPO blocks in 15%EPE75 mixture.

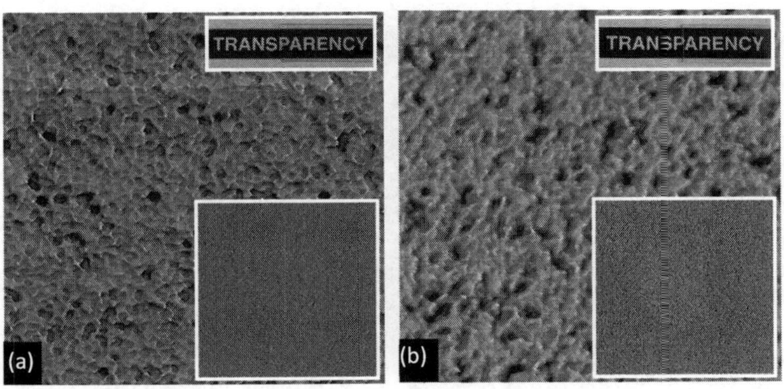

Figure 7.8 TM-AFM phase images (500 nm × 500 nm) of (a) 12%E_{75} and (b) 3%P_{34} mixtures cured at 80°C. The insets at the top of each image correspond to the digital image of transparency of a sheet of 1 mm of thickness. The insets at the bottom correspond to 3 μm × 3 μm AFM images. These AFM images correspond to internal surfaces of the cured mixtures trimmed using an ultramicrotome. Reproduced with permission of [24]. Copyright 2014, American Chemical Society.

As can be noted, the 12%E_{75} mixture (Fig. 7.8a) exhibited a sphere-like morphology whereas the 3%P_{34} mixture (Fig. 7.8b) showed a tendency to form worm-like phase separated domains. Subsequently, comparing the morphologies of 15%EPE75- and 15%E_{182}-cured mixtures (see Fig. 7.7a,b, respectively) one could conclude that the worm-like morphology of 15%EPE75 was associated with the presence of PPO central block. Relationships between morphology and thermoset optical and mechanical properties were reported elsewhere [24]. By means of the Fox equation [2], it can be found that if E_{182} homopolymer is completely miscible with the UP matrix, the glass transition temperature of UPol-rich matrix, $T'_{g\text{-UP}}$, would shift from 165°C for neat UP resin to 110°C for the cured 15%E_{182} mixture. However, a dynamic mechanical analysis of the cured UP/EPE75 and UP/E182 mixtures (see Fig. 7.9) shows that $T'_{g\text{-UP}}$, (ascribed as the maximum of tan δ peaks) had almost no reduction after modification with the EPE75 or E_{182} thermoplastics. Consequently, it can be concluded that EPE75 and E_{182} were weakly miscible with the UP-cured matrix, which is in agreement with the morphology analysis (see Fig. 7.7).

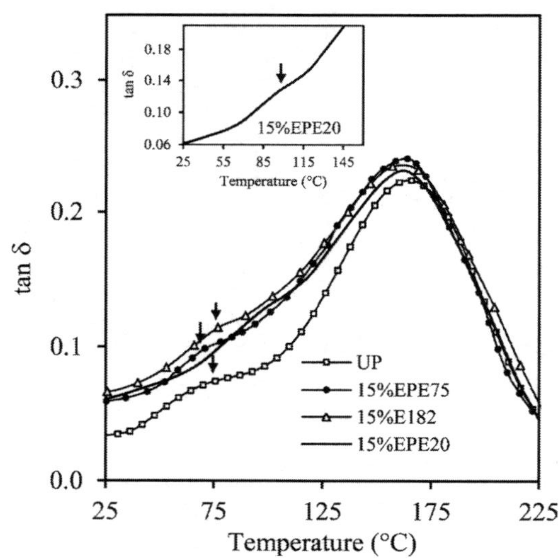

Figure 7.9 Loss factor (tan δ) curves as a function of temperature for: (-□-) UP, (-●-) 15%EPE75, (-Δ-) 15%E_{182} and (-) 15%EPE20 mixtures cured at 80°C. The inset details the shoulder (signed by an arrow) in tan δ curve for 15%EPE20 thermoset. Reproduced with permission of [24]. Copyright 2014, American Chemical Society.

Additionally, the shoulder in the tan δ peak between 40 and 100°C is related to the polystyrene-rich phase [56] verifying the heterogeneity of the neat matrix. The relative changes observed between the $T'_{g\text{-UP}}$ and shoulder temperatures indicate that EPE75 can modify selectively the polystyrene-rich or UPol-rich phases. This phenomenon was verified by means of DSC analysis [24] and agrees with the miscibility parameter of PPO and PEO analyzed above. Phase separation of the EPE75 was also detected by means of the T_gs measured using a DSC at low temperature values, $T_{g\text{-cold}}$, (see Fig. 7.10). These results also agree with AFM analysis, and showed that despite the miscibility of PEO with UP resin, the increase of molecular weight during network formation changed the miscibility of the system. An explanation of these results using solubility parameters was reported elsewhere [24].

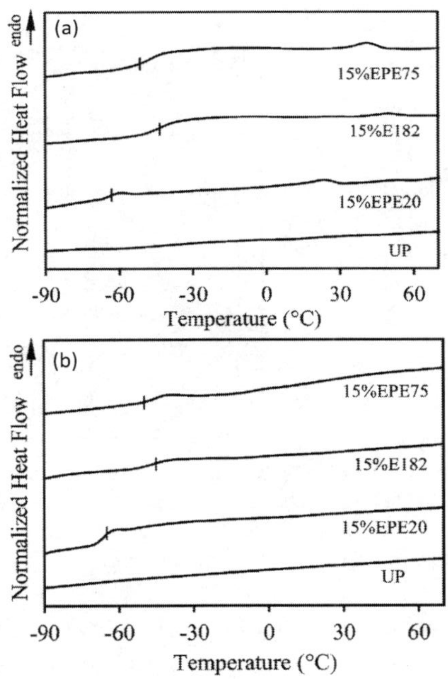

Figure 7.10 DSC thermograms of UP and 15%EPE75, 15%E_{182} and 15%EPE20 mixtures cured at (a) 80°C and (b) 60°C. The lines (|) denote the midpoints of the glass transition temperatures. Reproduced with permission of [24]. Copyright 2014, American Chemical Society.

DSC results shown in Fig. 7.10 also indicate that crystallization of PEO in cured mixtures could be hindered depending on the curing temperature. Consequently, higher polymer–polymer interactions between PEO and the UP network can be expected at lower curing temperatures. This effect with curing temperature was also observed in final morphology of cured UP/EPE75 and E_{182} mixtures [24].

15%EPE75 mixture cured at 80°C shown a $T_{g\text{-cold}}$ of –50°C, which is a value between the T_gs of the neat components, viz. –58 and 165°C for EPE75 and the neat UP-cured resin, respectively. This result indicates that the separated phase of cured 15%EPE75 mixture was a mixture between EPE75 and UP matrix.

Figure 7.11 shows a schematic explanation of the morphology obtained and phase location in cured mixtures as a function of curing temperature.

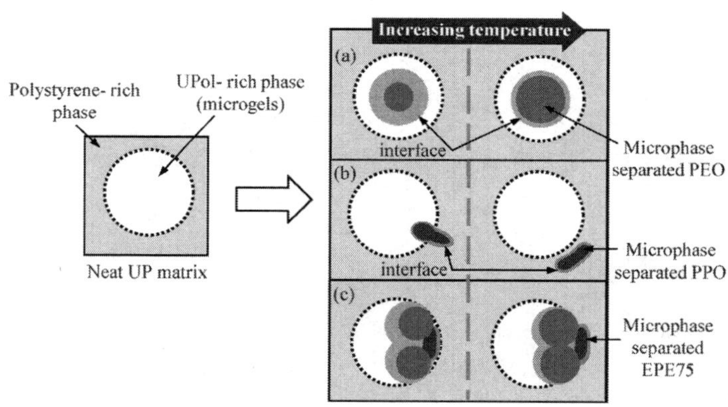

Figure 7.11 Schematic representation of morphology achieved after modifying the UP matrix with (a) E_{182}, (b) P_{34} and (c) EPE75. Right side corresponds to mixtures cured at higher temperature than the left side.

Since PEO miscibility with UPol is higher than miscibility with St, it is expected PEO remains inside UPol-rich microgels (white circles) as sphere-like domains (dark gray circles) surrounded by an interface (light gray), which becomes thinner at higher curing temperatures (see Fig. 7.11a, right). Taking into account that P_{34} exhibited total miscibility with St, and partial miscibility with UPol, P_{34} nanodomains (black ovals in Fig. 7.11b) could be located partially into microgels or into polystyrene-rich phase depending on P_{34} content and curing temperature. Therefore, considering the composition of EPE75 block copolymer, it is expected that phase location and morphology of cured UP/EPE75 mixtures show a combination of the effects observed in Figs. 7.11a,b (see Fig. 7.11c).

7.4.4 Modification of an Unsaturated Polyester Matrix with $E_{20}P_{69}E_{20}$ Block Copolymer

The effect of the relative amount and molecular weight of PEO and PPO blocks of PEO-b-PPO-b-PEO block copolymers on properties such as miscibility, morphology and mechanical properties was investigated by researches comparing UP/EPE75 and UP/$E_{20}P_{69}E_{20}$ systems ($E_{20}P_{69}E_{20}$ was named as EPE20). This work allowed a better understanding of the role of PPO central block in the non-reactive and cured mixtures. Despite P_{69} homopolymer presented strong immiscibility with the UP resin (Fig. 7.4), experimental data shows that UP/EPE20 non-reactive system was miscible until 60 wt% of EPE20 content (see Fig. 7.12).

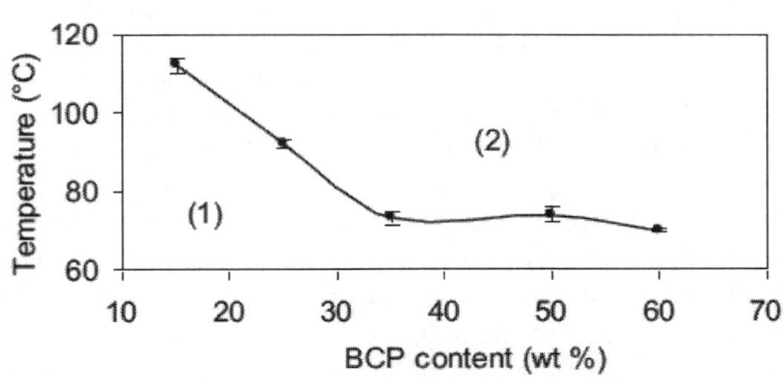

Figure 7.12 Temperature–BCP content phase diagram of UP/EPE20 non-reactive mixtures. (1) Homogeneous phase region at visible light scale. (2) Macrophase separated region. Reproduced with permission of [49]. Copyright 2012, Elsevier Ltd.

Obtained results of the last phase diagram suggest that hydrogen bonding interactions between PEO blocks and UP resin for UP/EPE20 non-reactive systems were enough to disperse the PPO block homogenously through the UP resin, and also that these interactions are thermally labile (LCST behavior).

Figure 7.13 shows the effect of curing temperature on phase separation of the 25%EPE20 mixture.

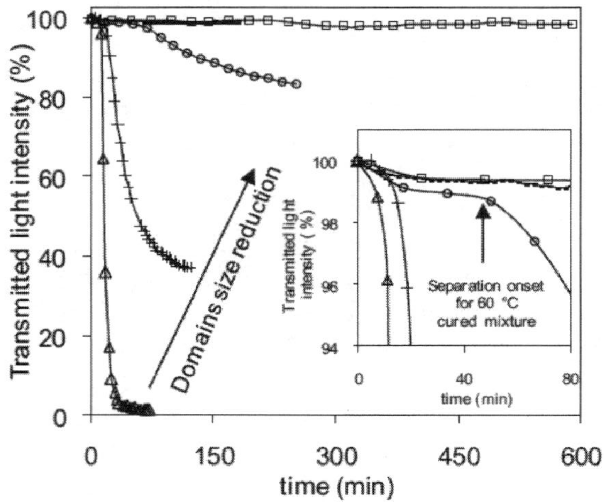

Figure 7.13 Transmitted light intensity as a function of curing time for (–=–) neat UP resin cured at 80°C and for 25%EPE20 mixture cured at: (–□–) 35°C, (–O–) 60°C, (–+–) 70°C and (–Δ–) 80°C. The inset shows the reduction in transmitted light intensity vs. time curves at low transmitted light intensity range. Reproduced with permission of [49]. Copyright 2012, Elsevier Ltd.

As can be seen, optical transparency (transmitted light intensity) decreases when curing temperature increases. That means that there were a continuous increase of phase separated domains until macrophase separation was achieved. Thus, since the transparency changes of UP/EPE20 mixture cured at 35°C were similar to that of the neat UP resin, one could conclude that no macrophase separation was achieved at this curing temperature.

7.4.4.1 Thermal behavior of UP/$E_{20}P_{69}E_{20}$ non-reactive mixtures

Similarly UP/EPE75 mixtures, the UP/EPE20 non-reactive mixtures were visually homogeneous and transparent at room temperature. However, markedly different phenomena were detected between both systems. One of these differences was observed by means of DSC thermal behavior of UP/EPE20 non-reactive mixtures (see Fig. 7.14).

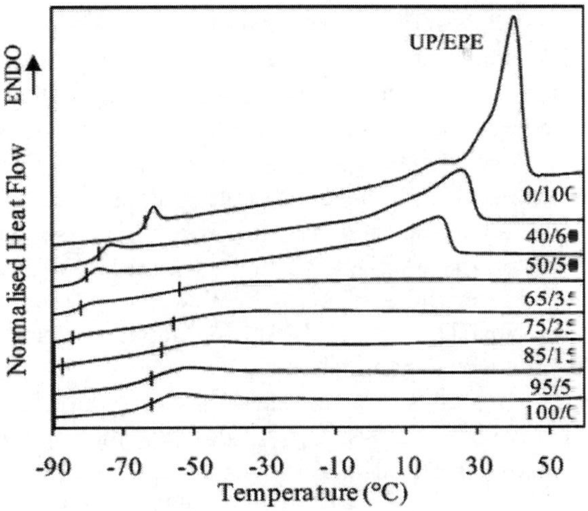

Figure 7.14 DSC thermograms of non-reactive neat UP resin, pure EPE20 (named as EPE in figure legend) and UP/EPE20 non-reactive mixtures. The lines (|) denote the midpoints of the transitions. Reproduced with permission of [22]. Copyright 2013, American Chemical Society.

Figure 7.14 shows that pure EPE20 exhibits a T_g = −63°C, and melting point, T_m = 39°C. A shoulder and lower peak were also observed. UP/EPE20 mixtures between 35 and 15 wt% of EPE20 content depicted two T_gs, one associated with the UP resin ($T_{g\text{-UP}}$) and the other with EPE20 ($T_{g\text{-EPE20}}$). Changes of EPE20 content leads to shifts in $T_{g\text{-UP}}$ and $T_{g\text{-EPE20}}$. Here, one can observe two uncommon phenomena of these partially miscible polymer mixtures. First, the increase of EPE20 content shifted the $T_{g\text{-UP}}$ of mixtures to higher values if compared with the $T_{g\text{-UP}}$ of neat UP resin, *viz.*, the $T_{g\text{-UP}}$ of mixtures did not tend to be between the T_{gs} of pure EPE20 and the T_g of neat UP resin. Second, the decrease of EPE20 content shifted of $T_{g\text{-EPE20}}$ of mixtures to lower temperatures if compared with T_g of pure EPE20. These two phenomena can be explained considering the interaction parameters of the UPol/St/EPE20 ternary system analyzed above, which showed high compatibility between PPO and St and a very low one between UPol and PPO. Due to that, $T_{g\text{-UP}}$ is associated with a UPol-rich phase composed of UPol, St and PEO, whereas $T_{g\text{-EPE20}}$ is associated with a PPO-rich separated phase composed of PPO and St. Consequently, the changes in both $T_{g\text{-UP}}$ and $T_{g\text{-EPE20}}$ were related to a distribution of St between UPol-rich and PPO-rich phase acting as a dissolvent of this phase. These DSC results were in agreement with dynamic light scattering measures, where self-assembled micelles at length-scales below the wavelength of visual light, were detected in UP/EPE20 non-reactive mixtures [22]. These structures consist of a PPO-rich separated phase surrounded by an UPol-rich phase. The ability of EPE20 to self-assembly in aqueous solutions [57] forming structures such as sphere-like, wormlike, very long wormlike micelles [58], vesicles [59] and multi-lamellar vesicles [60], viz. "onions" has been well established.

A miscibility analysis of full cured mixtures performed using dynamic mechanical analysis [49] showed that conversely to UP/EPE20 mixtures cured at 80°C (see Fig. 7.9) T_g's of mixtures upon EPE20 amount from 25 to 50 wt%, cured at 35°C and postcured at 170°C, shifted to lower temperatures with increased EPE20 content (from 164 to 130°C for neat UP matrix and 50%EPE20 mixture, respectively). Shifting of T_g's of matrix to lower values indicates partial miscibility of PEO blocks with UPol-rich phase. On the other hand, these mixtures exhibited a T_g at low temperatures almost with similar value of the neat EPE20 due to its phase separation (no shown here).

7.4.4.2 Morphology analysis of cured UP/$E_{20}P_{69}E_{20}$ mixtures

Morphology of UP/EPE20 mixtures cured at 35°C was analyzed elsewhere [49]. Furthermore, Builes et al. [22] found that mixtures cured at 25°C nanostructured with smaller microphase-separated EPE20 domains if compared to the mixtures cured at 35°C, which agrees with transparency changes observed during curing as a function of curing temperature and time (Fig. 7.13).

Figure 7.15 shows the morphology of the fully cured mixtures performed by means of TM-AFM. As can be seen, all investigated mixtures reached nanostructuration of the UP matrix. For the 5%EPE20 mixture, the sphere-like nanostructure due to the microphase-separated EPE20 was clearly distinguished (see Fig. 7.15b). Progress was also observed in the morphology from sphere-like to worm-like nanostructures with a progressive growth of length domains generated by the increase of EPE20 content. Similar evolution in morphology has been observed in other systems [44, 61, 62] and explained via the combination of many mechanisms such as the coalescence and elongation-distortion of the spherical micelles.

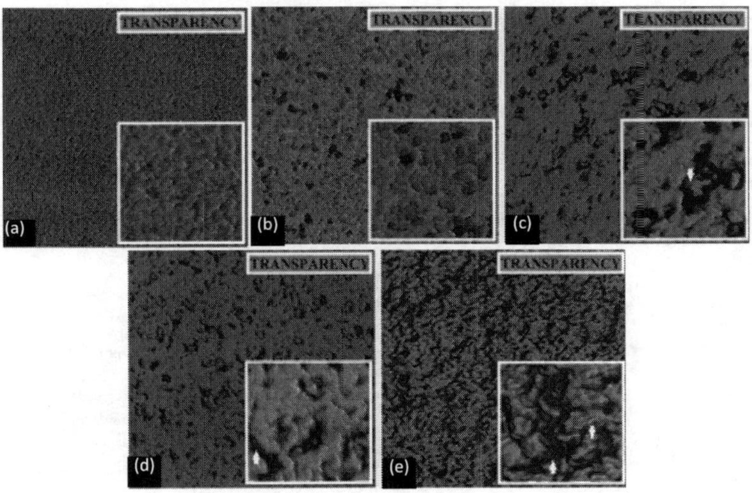

Figure 7.15 TM-AFM phase images (1 μm × 1 μm) of cured: (a) neat UP; (b) 5%EPE20; (c) 15%EPE20, (d) 25%EPE20 and (e) 50%EPE20. The insets at the top of each image correspond to the digital image of transparency of a sheet of 1 mm of thickness. The insets at the bottom correspond to 200 nm × 200 nm AFM images. Reproduced with permission of [22]. Copyright 2013, American Chemical Society.

As can be seen in Figs. 7.15c–e, the mixtures with EPE20 content higher than 15 wt% revealed the coexistence of worm-like structures with rather more complex structures encasing another phase surrounded by phase separated domains which were harder than the phases of EPE20 blocks (see the arrows in the insets at the bottom of Figs. 7.15c–e). The two types of these complex structures were better visualized in the 50%EPE20 mixture. As can be clearly observed inside of the worm-like domains (see the inset at the bottom of Fig. 7.15e indicated by the arrow pointing upward) long, narrowed and almost parallel structures were clearly distinguished. These structures could be generated by the fraction of St that remained inside the PPO-rich phase. Thus, homopolymerization of that fraction of St phase separates from the PPO-rich phase by PIPS. In addition, as can be also easily distinguished in the inset at the bottom of Fig. 7.15e, the arrow pointing downward shows a closed interconnected wormlike structure, which encases a crosslinked UPol-rich network within. The observed interconnected worm-like structures resembled the interconnected worm-like structures reported in the literature for other thermosetting systems [12, 61, 63].

After 15%EPE20 mixture was cured at 80°C, macrophase separation was obtained. Morphology achieved is shown in Fig. 7.7c. A schematic explanation of the morphology obtained and phase location in cured UP/EPE20 mixtures as a function of curing temperature is shown in Fig. 7.16.

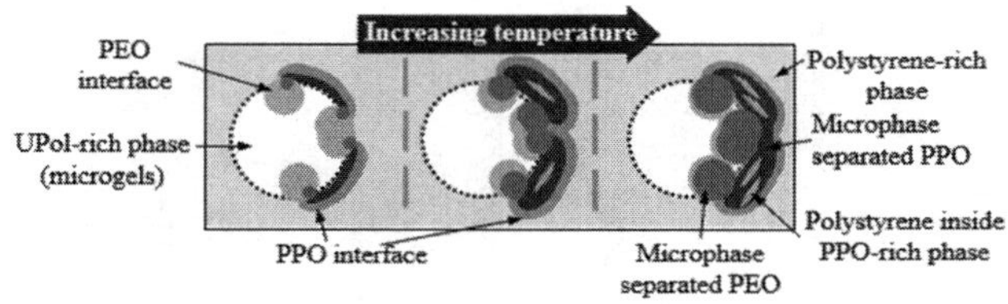

Figure 7.16 Schematic representation of morphology achieved after modifying the UP matrix with EPE20. The right side corresponds to mixtures cured at higher temperature than the left side.

7.4.5 Nanostructuration Mechanism of Cured UP/$E_{20}P_{69}E_{20}$ System

The instability of the structures of these BCPs in aqueous solutions could be also a property of UP/EPE20 non-reactive mixtures, which would be one of the reasons for the very irregular microstructures achieved in UP/EPE20-cured mixtures. The anomalous micellization due to polydispersity, fractions of PPO and PEO homopolymers, and PEO-b-PPO diblock copolymer content in commercial PEO-b-PPO-b-PEO triblock copolymers [64–66] could also provoke irregularities in UP/EPE20 mixtures. Furthermore, in the case of epoxy mixtures, PEO blocks lose their miscibility from the matrix during curing [67] which has been proposed elsewhere [62] as one of the reasons for the deformation from sphere-like to worm-like domains in epoxy mixtures modified with a PEO-b-PEP block copolymer. This loss of miscibility also occurs in UP systems mixed with BCPs composed of PEO blocks [45], and similarly could have an effect on the deformation of nanodomains. Another explanation of the irregular nanostructures was presented elsewhere [49]. Figure 7.17 summarizes the process of the self-assembled structure formation of UP/EPE20 system by means of a schematic representation.

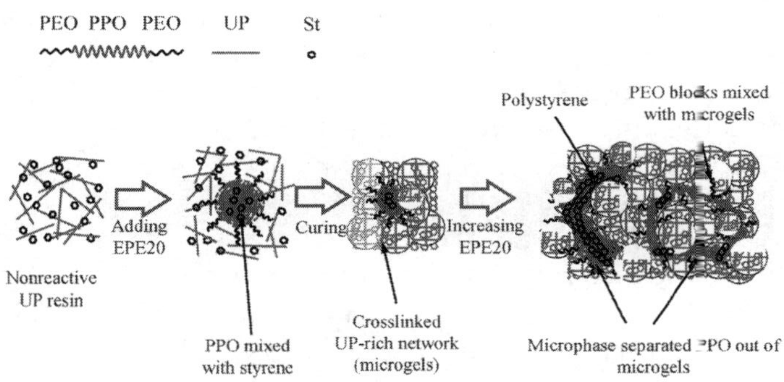

Figure 7.17 Schematic description of the structures achieved in UP/EPE20 mixtures: (a) Non-reactive system before and after adding EPE20, (b) Changes of morphology of cured mixtures upon increase of EPE20 content. Reproduced with permission of [22]. Copyright 2013, American Chemical Society.

Figure 7.17a shows the distribution of the components after the self-assembling process in non-reactive mixtures, and Figure 7.17b the evolution of domains morphology after curing upon increasing EPE20 content.

The effect of the molecular weight of PEO block lateral blocks on dispersion of PPO central block can be analyzed by comparing UP/$E_5P_{69}E_5$ and UP/EPE20 non-reactive mixtures. In the case of UP/$E_5P_{69}E_5$ with 5 wt% of BCP, strong immiscibility was observed with UP resin at room temperature, which is attributed to the low molecular weight of the PEO lateral blocks of $E_5P_{69}E_5$ (5 repeated units for $E_5P_{69}E_5$ and 20 for $E_{20}P_{69}E_{20}$). As explained elsewhere [68], lateral miscible blocks of a linear triblock copolymer could act as compatibilizers of the immiscible central block as a direct consequence of an extending effect. Thus, it is indeed expected that EPE5 block copolymer has a lateral PEO blocks with no enough molecular weight (directly related to amount of hydrogen bonds) to achieve a homogeneous mixture with the UP resin. This effect resembles the case of epoxidation degree of poly(styrene-b-butadiene) diblock copolymer in a mixture with an epoxy resin [12].

7.5 Modification of an Unsaturated Polyester Matrix with a PE-b-PEO Diblock Copolymer

Behavior of a St-dissolved UP resin composed of UPol made from neopentyl glycol, propylene glycol and maleic anhydride was analyzed by Sinturel et al. [21]. This resin was modified with poly[(ethylene)-*block*-(ethylene oxide)] diblock copolymer (PE-b-PEO), with an average M_n of 1400 g mol^{-1} and composed of 50:50 wt% ratio of PE/PEO blocks.

Although PE is completely immiscible with UP resin mainly due to the non-polar and linear hydrocarbon chain, UP/PE-b-PEO non-reactive mixtures were liquid, homogeneous and transparent at room temperature. However, at temperatures higher than 75°C, macroscopic phase separation was observed, exhibiting LCST behavior.

As judged by SAXS patterns of several UP/PE-b-PEO non-reactive mixtures (Fig. 7.18), self-assembled nanostructures at room temperature were detected (see the maximum at the low *q* values pointed out by arrows).

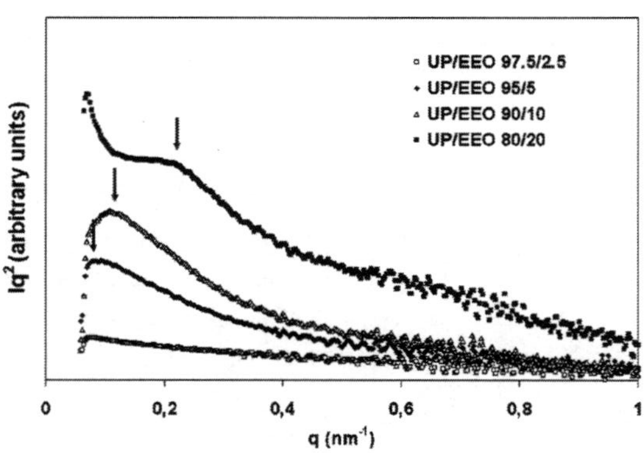

Figure 7.18 SAXS patterns at room temperature of UP/PE-b-PEO non-reactive mixtures composed of 2.5, 5, 10 and 20 wt% of PE-b-PEO (named as EEO in the legend) diblock copolymer content. Reproduced with permission of [21]. Copyright 2007, American Chemical Society.

It should be pointed out that the pure PE-b-PEO block copolymer exhibits a SAXS pattern with the typical hallmarks of a lamellar structure [69]. Comparing these results with UP/PE-b-PEO SAXS patterns, it can be concluded that the lamellar structure of the pure BCP was lost after mixing with UP resin and replaced by some correlated microphase-separated domains.

A thermal analysis carried out using a DSC to the non-reactive (Fig. 7.19) and cured mixtures (Fig. 7.20), showed that contrary to PE block, PEO block crystallinity was suppressed in both non-reactive and cured systems.

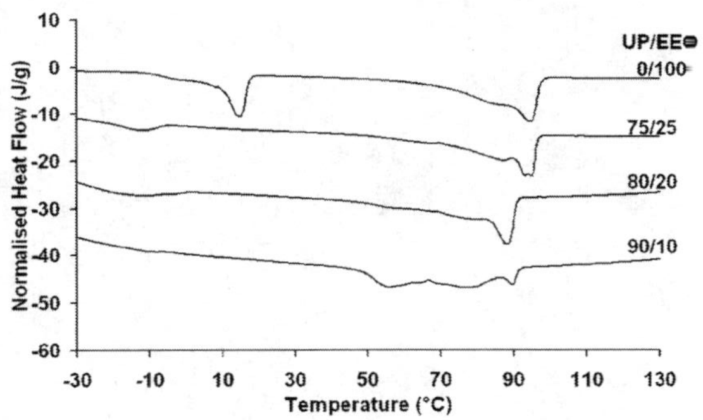

Figure 7.19 DSC thermograms obtained during cooling for non-reactive mixtures of UP resin mixed with PE-b-PEO block copolymer (named here as EEO). Reproduced with permission of [21]. Copyright 2007, American Chemical Society.

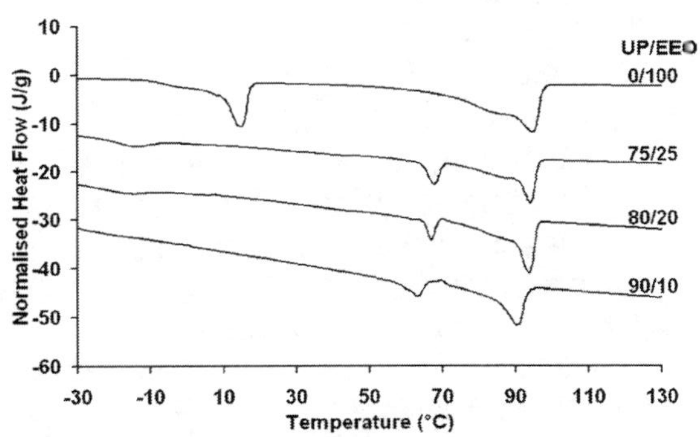

Figure 7.20 DSC thermograms obtained during cooling for cured mixtures of UP resin modify with PE-b-PEO block copolymer (named here as EEO). Reproduced with permission of [21]. Copyright 2007, American Chemical Society.

These results agree with the miscibility analysis of PE and PEO blocks with a UP resin. Additionally, the fact that PE melting peak was almost unchanged during non-reactive and cured mixtures if compared with the pure PE-b-PEO block copolymer shows that PE block phase separation and self-assembling phenomenon were carried out in this system.

TEM images of a cured mixture shown in Fig. 7.21 revealed the morphology of a UP resin modified with 5 wt% of PE-b-PEO block copolymer. As can be seen, spread nanometric lamellae structures in a continuous matrix were observed.

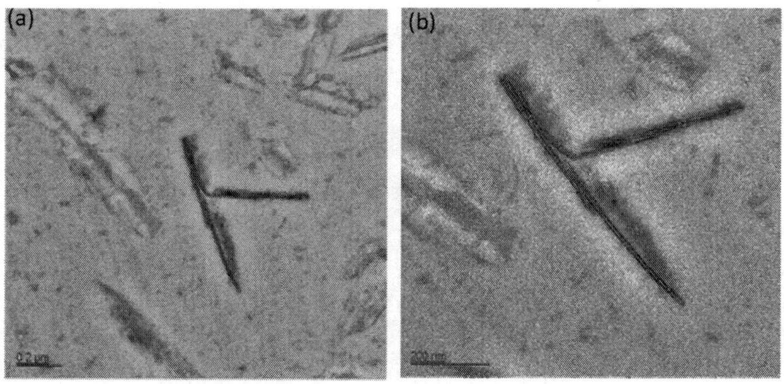

Figure 7.21 TEM micrographs of UP resin mixed with 5 wt% of PE-b-PEO diblock copolymer, cured at 70°C and postcured at 135°C, at various magnifications: (a) ×11500 and (b) ×20000. Reproduced with permission of [21]. Copyright 2007, American Chemical Society.

Researchers found that homogeneous dispersion of isotropic lamellae orientation changed with the increase of PE-b-PEO content obtaining domains with preferential orientation. Increasing of BCP content depicted a behavior of nanometric domains that is in agreement with the high aspect ratio of a lamellae structure inducing steric hindrance and leading to a quasi-parallel orientation along a specific axis. Nevertheless, isotropy of the mixtures was maintained since variations of the orientation axis throughout all the mixture. Several types of lamellae cross section were observed with structures consisting of BCP platelets with a high aspect ratio and thickness of ca. 6 nm, whereas the length of the lamellae was between 0.5 and 1 μm. In some cases, the lamellae normal was within the plane of the cut. Lamellae appeared with a white cross section of PE in the middle surrounded by dark area of PEO in contact with a white area of UP matrix.

Figure 7.22 shows a schematic representation of PE-b-PEO diblock copolymer conformation in cured UP matrix. The scheme shows the immiscible and dispersed PE blocks in an organized structure surrounded and stabilized by the UP-miscible PEO block.

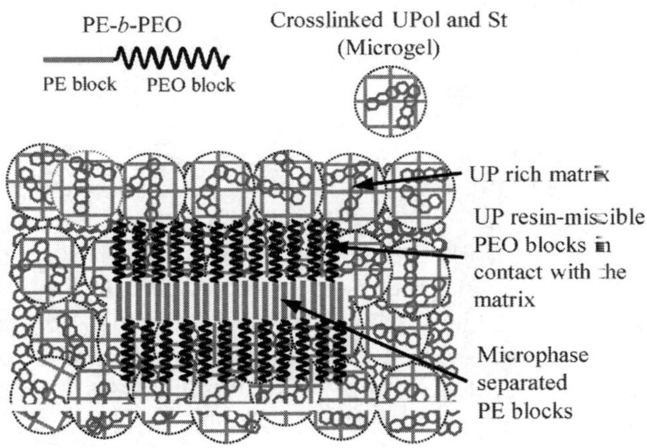

Figure 7.22 Schematic description of the meso-structure configuration of PE-b-PEO diblock copolymer mixed with a UP resin.

7.6 Modification of Two Unsaturated Polyester Matrices with Several P[(MMA-*co*-DMA)-b-BA-b-(MMA-*co*-DMA)] Triblock Copolymers

Another interesting system of a nanostructured UP-cured resin was reported by Serrano et al. [11]. They evaluated two different St-dissolved UP resins composed of propylene glycol, maleate and ortho-phthalic/maleate UPol modified with amphiphilic poly[(methyl methacrylate)-*co*-(*N*,*N*-dimethyl acrylamide)-*block*-(butyl acrylate)-*block*-(methyl methacrylate)-*co*-(*N*,*N*-dimethyl acrylamide)] triblock copolymers named as P[BA-b-(MMA-*co*-DMA)$_2$]. These modifications allowed to achieve transparent cured mixtures with improved mechanical properties when compared with the neat UP resins.

Several P[BA-b-(MMA-*co*-DMA)$_2$] triblock copolymers were designed according to a miscibility analysis of each block with the two designed UP resins at several St contents. The synthesis of both P(MMA-*co*-DMA) random copolymers and P[BA-b-(MMA-*co*-DMA)$_2$] triblock copolymers has been reported in literature [70]. Miscibility of UP resins with PBA and the random copolymer will be explained later.

7.6.1 Miscibility of PBA Homopolymer

The elastomeric PBA homopolymer is St-soluble in the overall content range; however, macrophase separation occurs during St polymerization. This behavior agrees with solubility parameters (δ) of St, PBA and polystyrene, 19.0, 19.2 and 21.6 J$^{1/2}$ cm$^{-3/2}$, respectively [11, 40]. On the other hand, macrophase separation was observed after adding UPol (δ_{UPol} between 20.2 and 20.7 J$^{1/2}$ cm$^{-3/2}$) to the PBA/St non-reactive liquid system until a ternary mixture composed of 35:65 wt% ratio of St/UPol was obtained (i.e., PBA is immiscible with UP resin in both non-reactive and cured systems). Although miscibility of PBA homopolymer was higher with St than with UPol, and the increase of St content to 50:50 wt% ratio of St/UPol, immiscible systems were still obtained for ortho-phthalic-composed UPol. Conversely, PBA was miscible when maleate-composed UPol was used. However, macrophase separation took place during curing in both systems. Consequently, taking into account that PBA phase separation could occurs before or during curing, phase separation mechanism of PBA-composed BCPs when mixed with UP resins could be modified depending on UPol composition and St content. Due to the last, the miscible random copolymer would be designed in order to obtain homogeneous mixtures of these UP resins with PBA-composed BCPs.

7.6.2 Miscibility of P(MMA-*co*-DMA) Random Copolymer

Poly(*N*,*N*-dimethylacrylamide) (PDMA) random copolymer (δ_{PDMA} = 22.7 J$^{1/2}$ cm$^{-3/2}$) was miscible with UP resins before and after curing, whatever the chemical composition of UPol, and St and PDMA contents. However, miscibility increases with the increase of St content. It was found that irrespective of the St content and the two UPol chemical composition analyzed here, a P(MMA-*co*-DMA) with DMA amount higher than 12 mol% can be used as a miscible block of a BCP in a UP resin before and after curing. It is worth mentioning that PMMA (δ_{PMMA} = 18.7 $J^{1/2}$ $cm^{-3/2}$) was immiscible with a UP resin composed of 35 wt% of St. Moreover, different behaviors depending on the DMA content in the random copolymer when mixed with non-reactive UP resin was also observed. That is the case of a UCST behavior observed at DMA contents lower than 12 ml% in mixtures with maleate-composed UPol. In the case of ortho-phthalic-composed UPol, at any St content and even at room temperature, full miscibility in non-reactive mixtures at DMA content higher than 4 mol% in the random copolymer was observed.

7.6.3 Design of P[BA-b-(MMA-*co*-DMA)$_2$] Triblock Copolymers to Modify the UP Resins

According to the aforementioned miscibility of PBA and P(MMA-*co*-DMA) with UP resins, P(MMA-*co*-DMA) can be used as miscible block to disperse the UP-immiscible PBA block. Due to that, the DMA units required to obtain miscibility of individual P(MMA-*co*-DMA) with the UP resins must be considerably increased. On the other hand, in order to decrease the UP/P[BA-b-(MMA-*co*-DMA)$_2$] mixtures viscosity, it was required to increase the St content of the neat UP resin to 50 wt%. Analyzing the mixtures with these diluted UP resins, it was found that the minimum amounts of DMA units in the random block to obtain transparent cured mixtures were 33 and 40 mol% for maleate and ortho-phthalic-composed UP resin, respectively. These minimum amounts of DMA were lower for the non-reactive mixtures.

As depicted by the TEM images in Fig. 7.23, micro- and macrophase separations were obtained in the maleate-composed UP resin modified with 10 wt% P[BA-b-(MMA-*co*-DMA)$_2$] triblock copolymers depending on the DMA content in the BCP.

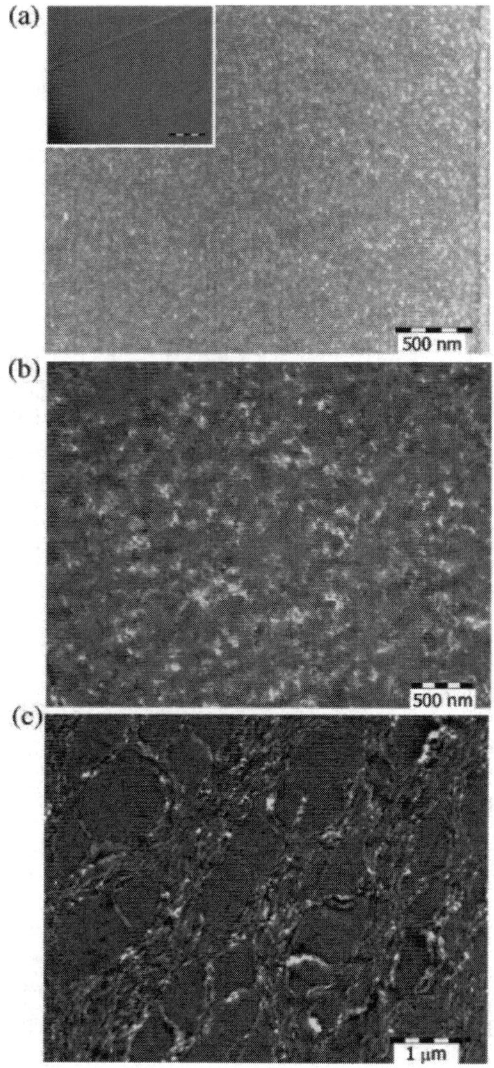

Figure 7.23 TEM micrographs of cured systems composed of maleate-composed UP resin mixed with 10 wt% of P[BA-b-(MMA-*co*-DMA)] with DMA units content of (a) 40 mol%, (b) 33 mol% and (c) 25 mol%. The inset in Fig. 7.24a shows the corresponding TEM image for the neat maleate-composed UP resin at the same magnification. Reproduced with permission of [11]. Copyright 2008, Wiley-VCH Verlag GmbH & Co. KGaA.

When P[BA-b-(MMA-*co*-DMA)$_2$] triblock copolymer is composed of DMA amount higher than 33 mol%, miscible systems before and after curing lead to a nanostructure matrix (see Fig. 7.23a,b, respectively). As can be seen in Fig. 7.23a, small micelles of around 20 nm were obtained with the P[BA-b-(MMA-*co*-DMA)$_2$] 40 mol%-based mixture, which aggregate in higher domains when the DMA content in random block decreases to 33 mol% (Fig. 7.23b). Conversely, the P[BA-b-(MMA-*co*-DMA)$_2$] 25 mol%-based mixture was opaque and led to a flocculated macrophase-separated structure (Fig. 7.23c).

In the case of ortho-phthalic-composed UP resin, the modification with 10 wt% PBA-b-P(MMA-*co*-DMA) triblock copolymer with 40 and 33 mol% of DMA in random block, micro- and macrophase separation was achieved, respectively, after curing process. Morphology of these final cured mixtures was observed by means of TEM analysis (see Fig. 7.24).

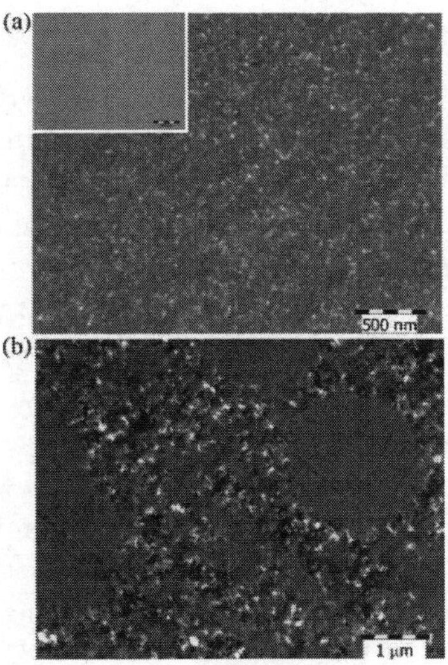

Figure 7.24 TEM micrographs of ortho-phthalic-composed UP resin-cured mixtures modified with 10 wt% of block copolymers P[BA-b-(MMA-*co*-DMA)] with DMA units content of (a) 40 mol% and (b) 33 mol%. The inset in Fig. 7.25a shows the corresponding TEM image for the neat UP-cured resin at the same magnification. Reproduced with permission of [11]. Copyright 2008, Wiley-VCH Verlag GmbH & Co. KGaA.

The larger DMA content required to keep the miscibility of the random block of the BCP with ortho-phthalic-composed UP resin than with the maleate-composed one is explained considering that contrary to the first UP resin, the PBA block was miscible in the non-reactive mixtures with the second UP resin. It should be pointed out that the PBA/-*co*-DMA mol ratio in BCPs required to obtain transparent cured mixtures increases from 0.8 to 1.6 for ortho-phthalic and for maleate-composed UP resin-cured mixtures, respectively. As can be clearly seen in Fig. 7.24a, for the ortho-phthalic-composed UP matrix modified with the BCP with 40 mol% of DMA units in random block, a micellar nanostructure was achieved. On the contrary, for the UP matrix modified with the BCP with 33 mol% of DMA units, macrophase separation with flocculated domains was observed (Fig. 7.24b).

7.6.4 Nanostructuration Mechanism

Taking into account the previous section, one can conclude that P(MMA-*co*-DMA) random copolymers require considerably lower molar fraction of DMA units to be miscible when are used as individual random copolymers than when are used as compatibilizer-block of PBA-composed BCPs. This fact can be explained considering the competition between the PBA elastomeric-immiscible block and the P(MMA-*co*-DMA) glassy-miscible one during the phase separation process.

Considering the differences of miscibility between the PBA homopolymer and the two investigated UP resins, two different nanostructuration processes can be suggested for cured UP/P [BA-b-(MMA-*co*-DMA)] triblock copolymers mixtures shown in Figs. 7.23a and 7.24a. As explained above, maleate-composed UP/PBA non-reactive mixtures were miscible; thus the nanostructuring could take place through PIPS of the PBA block. On the other hand, since the ortho-phthalic-composed UP/PBA non-reactive mixtures were immiscible, one could deduce that the final structure of cured mixtures self-assembled since the non-reactive mixtures until cured mixtures.

Toughening performance of UP resin modified with 5 wt% of P[BA-b-(MMA-*co*-DMA)$_2$] triblock copolymers is reported elsewhere [11].

References

1. Lucintel (2010). Global Unsaturated Polyester Resin Market 2010–2015.
2. Pascault, J. P., Sautereau, H., Verdu, J., Williams, R. J. J. (2002). *Thermosetting Polymers* (Marcel Dekker: New York).
3. Ellis, C. (1933). Catalyzed Urea Resin. U.S. Patent 1'905'999, April 25.
4. Scott Bader Home Page. (2005). http://www.scottbader.com/ (accessed May 10, 2013). Crystic Composite Handbook. Wollaston, Wellingborough, Northamptonshire NN29 7RL.
5. Plastics Historical Society Home Page. http://www.plastiquarian.com/index.php?id=84 (accessed May 8, 2013).
6. Irfan, M. H. (1998.). *Chemistry and Technology of Thermosetting Polymers in Construction Applications* (Kluwer Academic Publishers: Dordrecht).
7. Chenier, P. J. (2002). *Survey of Industrial Chemistry*. 3rd ed. (Kluwer Academic/Plenum Publishers: New York).
8. Cohen, E. D., Gutoff, E. B., Coating Methods, Survey, in: *Encyclopedia of Polymer Science and Technology* [Online], vol. 1, John Wiley & Sons, Posted Oct 22, 2001. http://onlinelibrary.wiley.com/doi/10.1002/0471440264.pst061/abstract (accessed Sep 10, 2005).
9. Boyard, N., Serre, C., Vayer, M. (2007). A physical approach to define a class A surface in polymer thermosetting composite materials. *J. Appl. Polym. Sci.,* **103**, pp. 451–461.
10. Wegst, U. G. K., Ashby, M. F. (2004). The mechanical efficiency of natural materials. *Philos. Mag.,* **84**, 2167–2186.
11. Serrano, E., Gerard, P., Lortie, F., Pascault, J. P., Portinha, D. (2008). Nanostructuration of unsaturated polyester by all-acrylic block copolymers, 1-use of high-molecular-weight block copolymers. *Macromol. Mater. Eng.,* **293**, pp. 820–827.
12. Ocando, C., Tercjak, A., Martín, M. D., Ramos, J. A., Campo, M., Mondragon, I. (2009). Morphology development in thermosetting mixtures through the variation on chemical functionalization degree of poly(styrene-b-butadiene) diblock copolymer modifiers. Thermomechanical properties. *Macromolecules,* **42**, pp. 6215–6224.
13. De la Caba, K., Guerrero, P., Gavalda, J., Mondragon, I. (1999). Fracture behavior-morphology relationships in an unsaturated polyester resin modified with a liquid oligomer. *J. Polym. Sci. Part B: Polym. Phys.,* **37**, pp. 1677–1686.

14. Bakar, M., Djaider, F. (2007). Effect of plasticizers content on the mechanical properties of unsaturated polyester resin. *J. Thermoplast Compos. Mater.*, **20**, 53–64.

15. Larrañaga, M., Serrano, E., Martín, M. D., Tercjak, A., Kortaberria, G., de la Caba, K., Riccardi, C., Mondragon, I. (2007). Mechanical properties-morphology relationships in nano-/microstructured epoxy matrices modified with PEO-PPO-PEO block copolymers. *Polym. Int.*, **56**, pp. 1392–1403.

16. De la Caba, K. Influencia de la Temperatura de Curado y del Contenido de Modificador en la Conducta de Curado y en las Propiedades de Resinas Poliéster Insaturadas con Diversas Composiciones. Ph.D. Thesis, University of the Basque Country, December 1997.

17. You, C., Xu, J., Zeng, Y., Li, Y., Jia, D., Nishi, T. (2010). Morphology and performance of unsaturated polyester nanocomposites modified with organoclay and thermoplastic polyurethane. *Chin. J. Polym. Sci.*, **28**, pp. 913–922.

18. Kayatin, M., Davis, V. A. (2009). Viscoelasticity and shear stability of single–walled carbon nanotube/unsaturated polyester resin dispersions. *Macromolecules*, **42**, pp. 6624–6632.

19. Monti, M., Puglia, D., Natali, M., Torre, L., Kenny, J. M. (2011). Effect of carbon nanofibers on the cure kinetics of unsaturated polyester resin: Thermal and chemorheological modelling. *Compos. Sci. Technol.*, **71**, pp. 1507–1516.

20. Monti, M., Natali, M., Petrucci, R., Puglia, D., Terenzi, A., Valentine, L., Kenny, J. M. (2011). Composites based on nanocomposite matrices, in: *Wiley Encyclopedia of Composites* [Online], 2nd ed. (Nicolais, L., Borzacchiello, A., eds.), John Wiley & Sons, Posted Sep 15, 2011. http://onlinelibrary.wiley.com/doi/10.1002/9781118097298.weoc025/pdf (accessed May 9, 2013).

21. Sinturel, C., Vayer, M., Erre, R., Amenitsch, H. (2007). Nanostructured polymers obtained from polyethylene-block-poly (ethylene oxide) block copolymer in unsaturated polyester. *Macromolecules,* **40**, pp. 2532–2538.

22. Builes, D. H., Hernández, H., Mondragon, I., Tercjak, A. (2013). Relationship between the morphology of nanostructured unsaturated polyesters modified with PEO-b-PPO-b-PEO triblock copolymer and their optical and mechanical properties. *J. Phys. Chem. C*, **117**, 3563–3571.

23. Builes, D. H., Labidi, J., Eceiza, A., Mondragon, I., Tercjak, A. (2013). Unsaturated polyester nanocomposites modified with fibrillated cellulose and PEO-b-PPO-b-PEO block copolymer. *Compos. Sci. Technol.*, **89**, pp. 120–126.
24. Builes, D., Hernández-Ortiz, J. P., Corcuera, M. A., Mondragon, I., Tercjak, A. (2014). Effect of poly(ethylene oxide) homopolymer and two different poly(ethylene oxide-b-poly(propylene oxide)-b-poly(ethylene oxide) triblock copolymers on morphological, optical, and mechanical properties of nanostructured unsaturated polyester. *ACS Appl. Mater. Interfaces*, **6**, pp. 1073–1081.
25. Builes, D. Study of the effect of modification of nano/microstructured thermosetting systems with block copolymers on morphology and mechanical properties. Ph.D. Thesis, University of the Basque Country, July 2013.
26. Yang, Y. S., Pascault, J. P. (1997). Modeling of unsaturated polyester prepolymer structures. I. Chain branches and overall chain end numbers. *J. Appl. Polym. Sci.*, **64**, pp. 133–145.
27. Yang, Y. S., Pascault, J. P. (1997). Modeling of unsaturated polyester prepolymer structures. II. Hydroxyl and carboxyl functionalities. *J. Appl. Polym. Sci.*, **64**, pp. 147–156.
28. Nova, H. (2004). Polyesters, unsaturated, in: *Encyclopedia of Polymer Science and Technology* [Online], vol. 11, John Wiley & Sons, Posted Mar 15, http://onlinelibrary.wiley.com/doi/10.1002/0471440264.pst061/abstract (accessed Aug 25, 2005).
29. Odian, G. (2004). *Principles of Polymerization*, 4th ed. (John Wiley & Sons: Canada).
30. Capehart, T. W., Kia, H. G., Abujoudeh, T. (2007). Cure simulation of thermoset composite panels. *J. Compos. Mater.*, **41**, pp. 1339–1360.
31. Dell'Erba, R. (2000). Unsaturated polyester resins: Numerical solution of a system of equations able to describe both styrene and unsaturation consumption. *J. Appl. Polym. Sci.*, **78**, pp. 124–132.
32. Penczek, P., Czub, P., Pielichowski, J. (2005). Unsaturated polyester resins: Chemistry and technology. *Adv. Polym. Sci.*, **184**, pp. 1–95.
33. Atta, A. M., Elsaeed, A. M., Farag, R. K., El-Saeed, S. M. (2007). Synthesis of unsaturated polyester resins based on rosin acrylic acid adduct for coating applications. *React. Funct. Polym.*, **67**, pp. 549–563.
34. Pascault, J. P., Williams, R. J. J. (1990). Glass transition temperature versus conversion relationships for thermosetting polymers. *J. Polym. Sci. Part B: Polym. Phys.*, **28**, pp. 85–95.

35. Nebioglu, A., Soucek, M. D. (2006). Reaction kinetics and microgel particle size characterization of ultraviolet-curing unsaturated polyester acrylates. *J. Polym. Sci. Part A: Polym. Chem.*, **44**, pp. 6544–6557.

36. Hild, G., Okasha, R. (1985). Kinetic investigation of the free radical crosslinking copolymerization in the pre-gel state, 1. Styrene/m-and p-divinylbenzene systems. *Macromol. Chem.*, **186**, pp. 93–110.

37. Pucic, I., Ranogajec, F. (2003). Phase separation during radiation crosslinking of unsaturated polyester resin. *Radiat. Phys. Chem.*, **67**, 415–419.

38. Hsu, C. P., Lee, L. (1993). Free-radical crosslinking copolymerization of styrene/unsaturated polyester resins: 1. Phase separation and microgel formation. *J. Polym.*, **34**, pp. 4496–4505.

39. Chen, J. S., Yu, T. L. (1998). Microgelation of unsaturated polyester resins by static and dynamic light scattering. *J. Appl. Polym. Sci.*, **69**, pp. 871–878.

40. Sun, M. C., Chang, Y. F., Leon Yu, T. (2001). Microgelation of unsaturated polyester resins in the presence of poly(vinyl acetate) by static and dynamic light scattering. *J. Appl. Polym. Sci.*, **79**, pp. 1439–1449.

41. Lin, H. L., Yu, T. L., Cheng, C. H. (1999). Dynamic light scattering from sol–gel transition of unsaturated polyester resins. *Macromolecules*, **32**, pp. 690–696.

42. Nebioglu, A., Soucek, M. D. (2008). Microgel formation and thermo-mechanical properties of UV-curing unsaturated polyester acrylates. *J. Appl. Polym. Sci.*, **107**, pp. 2364–2374.

43. Van Assche, G., Verdonck, E., Van Mele, B. (2000). TMDSC and dynamic rheometry, gelation, vitrification and autoacceleration in the cure of an unsaturated polyester resin. *J. Therm. Anal. Calorim.*, **59**, pp. 305–318.

44. Meng, F., Xu, Z., Zheng, S. (2008). Microphase separation in thermosetting blends of epoxy resin and poly(ε-caprolactone)-block-polystyrene block copolymers. *Macromolecules*, **41**, pp. 1411–1420.

45. Li, X., Fu, W., Wang, Y., Chen, T., Liu, X., Lin, H., Sun, P., Jin, Q., Ding, D. (2008). Solid–state NMR characterization of unsaturated polyester thermoset blends containing PEO-PPO-PEO block copolymers. *Polymer*, **49**, 2886–2897.

46. Zheng, S. (2010). In: *Epoxy Polymers. New Materials and Innovations.* (Pascault, J. P., Williams, J. J., eds.), Wiley-VCH: Weinheim.

47. Guo, C., Wang, J., Liu, H.-Z., Chen, J.-Y. (1999). Hydration and conformation of temperature–dependent micellization of PEO-PPO-

PEO block copolymers in aqueous solutions by FT-Raman. *Langmuir,* **15**, pp. 2703–2708.

48. Pruitt, J., Husseini, G., Rapoport, N., Pitt, W. (2000). Stabilization of pluronic P-105 micelles with an interpenetrating network of *N,N*-diethylacrylamide. *Macromolecules,* **33**, pp. 9306–9309.

49. Builes, D. H., Tercjak, A., Mondragon, I. (2012). Nanostructured unsaturated polyester modified with poly[(ethylene oxide)-b-(propylene oxide)-b-(ethylene oxide)] triblock copolymer. *Polymer,* **53**, pp. 3669–3676.

50. Boyard, N., Sinturel, C., Vayer, M., Erre, R. (2006). Morphology and cure kinetics of unsaturated polyester resin/block copolymer blends. *J. Appl. Polym. Sci.,* **102**, pp. 149–165.

51. Guo, Q., Thomann, R., Gronski, W. (2003). Nanostructures, semicrytalline morphology, and nanoscale confinement effect on the crystallization kinetics in self-organized block copolymer/thermoset blends. *Macromolecules,* **36**, pp. 3635–3645.

52. Zheng, H., Zheng, S., Guo, Q. (1997). Thermosetting polymer blends of unsaturated polyester resin and poly (ethylene oxide). I. Miscibility and thermal properties. *J. Polym. Sci. Part A: Polym. Chem.,* **35**, pp. 3161–3168.

53. Zheng, H., Zheng, S., Guo, Q. (1997). Thermosetting polymer blends of unsaturated polyester resin and poly (ethylene oxide). II. Hydrogen–bonding interaction, crystallization kinetics, and morphology. *J. Polym. Sci. Part A: Polym. Chem.,* **35**, pp. 3169–3179.

54. Sinturel, C., Vayer, M., Erre, R., Amenitsch, H. (2009). Nanostructured polymers obtained from polyethylene-block-poly(ethylene oxide) block copolymer in unsaturated polyester. *Eur. Polym. J.,* **45**, pp. 2505–2538.

55. Van Krevelen, D. W., Te Nijenhuis, K. (2009). *Properties of Polymers,* 4th ed. (Elsevier: Amsterdam).

56. Calvo, S., Escribano, J., Prolongo, M. G., Masegosa, R. M., Salom, C. (2011). Thermomechanical properties of cured isophthalic polyester resin modified with poly(ε-caprolactone). Journal of thermal analysis and calorimetry. *J. Therm. Anal. Calorim.,* **103**, pp. 195–203.

57. Jansson, T., Schillén, K., Olofsson, G., Cardoso da Silva, R., Loh, W. (2004). The interaction between PEO-PPO-PEO triblock copolymers and ionic surfactants in aqueous solution studied using light scattering and calorimetry. *J. Phys. Chem. B,* **108**, pp. 82–92.

58. Chaibundit, C., Ricardo, N., Ricardo, N., Costa, F., Wong, M., Hermida-Merino, D., Rodriguez-Perez, D., Hamley, I. W., Yeates, S. G., Booth, C. (2008). Effect of ethanol on the micellization and gelation of pluronic P123. *Langmuir*, **24**, pp. 12260–12266.
59. Chen, S., Yang, B., Guo, S., Ma, J. H., Yang, L. R., Liang, X., Hua, C., Liu, H. Z. (2008). Spontaneous vesicle formation of poly(ethylene oxide)-poly(propylene oxide)-poly(ethylene oxide) triblock copolymer. *J. Phys. Chem. B*, **112**, pp. 15659–15665.
60. Zipfel, J., Lindner, P., Tsianou, M., Alexandridis, P., Richtering, W. (1999). Shear–induced formation of multilamellar vesicles ("onions") in block copolymers. *Langmuir*, **15**, pp. 2599–2602.
61. Ramos, J. A., Espósito, L., Fernández, R., Zalakain, I., Goyanes, S., Avgeropoulos, A., Zafeiropoulos, N. E., Kortaberria, G., Mondragon, I. (2012). Block copolymer concentration gradient and solvent effects on nanostructuring of thin epoxy coatings modified with epoxidized styrene-butadiene-styrene block copolymers. *Macromolecules*, **45**, pp. 1483–1491.
62. Hermel-Davidock, T., Sean Tang, H. S., Murray, D. J., Hahn, S. F. (2007). Control of the block copolymer morphology in templated epoxy thermosets. *J. Polym. Sci. Pol. Phys.*, **45**, pp. 3338–3348.
63. Fan, W., Wang, L., Zheng, S. (2009). Nanostructures in thermosetting blends of epoxy resin with polydimethylsiloxane-block-poly(ε-caprolactone)-block-polystyrene ABC triblock copolymer. *Macromolecules*, **42**, pp. 327–336.
64. Nakashima, K., Bahadur, P. (2006). Aggregation of water-soluble block copolymers in aqueous solutions: Recent trends. *Adv. Colloid. Interface Sci.*, **123**, pp. 75–96.
65. Lodge, T. P., Bang, J., Hanley, K. J., Krocak, J., Dahlquist, S., Sujan, B., Ott, J. (2003). Origins of anomalous micellization in diblock copolymer solutions. *Langmuir*, **19**, pp. 2103–2109.
66. Linse, P. (1994). Micellization of poly(ethylene oxide)-poly(propylene oxide) block copolymer in aqueous solution: Effect of polymer impurities. *Macromolecules*, **27**, pp. 2685–2693.
67. Sun, P., Dang, Q., Li, B., Chen, T., Wang, Y., Lin, H., Jin, Q., Ding, D. (2005). Mobility, miscibility, and microdomain structure in nanostructured thermoset blends of epoxy resin and amphiphilic poly(ethyleneoxide)-block-poly(propyleneoxide)-block-poly(ethylene oxide) triblock copolymers characterized by solid-state NMR. *Macromolecules*, **38**, pp. 5654–5667.

68. Hadjichristidis N., Pispas S., Floudas G. (2003). *Block Copolymers: Synthetic Strategies, Physical Properties and Applications*. (New Jersey: John Wiley & Sons).

69. Sun, L., Liu, Y., Zhu, L., Hsiao, B. S., Avila-Orta, C. A. (2004). Self-assembly and crystallization behavior of a double-crystalline polyethylene-block-poly(ethylene oxide) diblock copolymer. *Polymer*, **45**, pp. 8181–8193.

70. Phan, T. N. P., Maiez-Tribut, S., Pascault, J. P., Bonnet, A., Gerard, P., Guerret, O., Bertin, D. (2007). Synthesis and characterizations of block copolymer of poly(*n*-butyl acrylate) and gradient poly(methyl methacrylate-*co*-*N*,*N*-dimethyl acrylamide) made via nitroxide-mediated controlled radical polymerization. *Macromolecules*, **40**, pp. 4516–4523.

Chapter 8

Block Copolymers as Template for the Design of Advanced Multifunctional Hybrid Nanostructured Thermosetting Materials

Junkal Gutierrez,[a] Connie Ocando,[b] and Agnieszka Tercjak[a]

[a]*Materials + Technologies Group (GMT),*
Department of Chemical and Environmental Engineering,
Polytechnic School, University of the Basque Country (UPV/EHU),
Plaza Europa 1, 20018 Donostia-San Sebastián, Spain
[b]*Instituto de Tecnología en Polímeros y Nanotecnología (ITPN),*
Facultad de Ingeniería, Universidad de Buenos Aires,
Las Heras 2214, Buenos Aires, 1127AAR/Capital Federal, Argentina

juncal.gutierrez@ehu.eus, agnieszka.tercjaks@ehu.eus, connie_ocando@hotmail.com

This chapter describes a few research works done mostly by Prof. Iñaki Mondragon and their co-workers on the fabrication and characterization of advanced multifunctional materials based on nanostructured thermosetting materials, described in detail in Chapters 5, 6, and 7, as template for dispersion and localization of different nano-objects such as nanoparticles, carbon nanotubes (CNTs), microfibrillated cellulose and others. The chapter

Block Copolymer Nanocomposites
Edited by Galder Kortaberria and Agnieszka Tercjak

ISBN 978-981-4669-54-2 (Hardcover), 978-981-4669-55-9 (eBook)
www.panstanford.com

focuses on novel strategies to achieve simple, flexible and highly versatile approaches to design advanced multifunctional hybrid nanostructured thermosetting materials.

8.1 Introduction

During the past decade, multicomponent thermosetting systems have been widely investigated in order to reach new requirements in the field of nanotechnology. Thermosetting materials based on epoxy resins [1], especially thermoset systems modified with block copolymers, have been studied [1–18], because block copolymers offer the possibility to control self-assembling at the nanoscale in designed materials and optimize their properties as described in detail in Chapters 5, 6 and 7.

On the other hand, extensive work has been focused on the synthesis and characterization of thermosetting systems modified with inorganic nanoparticles, carbon nanotubes, and nanofibers, among others. The interest to develop this kind of systems relies upon the unique opportunity to combine mechanic, optic, magnetic, and optoelectronic properties [19–22]. These properties can be tuned varying the concentration, size, or type of nanoentity.

As it is well known, the major problem in the preparation of this kind of nanocomposites is related to achieving homogeneous dispersion of these nano-objects in the epoxy matrix. Nowadays, preparations of both nanostructured thermosetting systems modified with block copolymers and thermosetting systems modified with inorganic nanoparticles, carbon nanotubes, nanofibers and others have merged with the aim to develop novel advanced multifunctional hybrid nanostructured thermosetting materials [23–33].

Nanostructured thermosetting systems designed using block copolymers (BCPs) are challenging to obtain multifunctional materials, since block copolymers can act as a template for both, the incorporation and selective location of nano-objects [34, 35].

In the case of these multifunctional thermosetting systems modified with nanoparticles, the main objective is the control of the distribution and size of synthesized nanoparticles, as these parameters limit possible applications of the designed materials

[34, 35]. Additionally, many researchers are currently looking for the selective confinement of nanoparticles in one of the microphase-separated domains playing with the capability of block copolymers to being nanostructure agent for the epoxy resins.

This kind of materials combine the properties of the nanostructured thermosetting matrix and the unique properties of the added inorganic nanoparticles, which leads to the development of interesting advanced materials with tunable properties, which can find application in many field of nanotechnology. Additionally, taking into account that this subject has not been still explored enough, the development, design and characterization of advanced multifunctional hybrid nanostructured thermosetting materials can open new challenges in the field of nanomaterials.

8.2 Hybrid Epoxy Systems Based on Block Copolymer and Commercial Inorganic Nanoparticles

Chemically functionalized poly(styrene-b-butadine-b-styrene) (SBSep) block copolymer has been used as nanostructuring and dispersing agent for the development of tough hybrid epoxy systems. As discussed in Chapter 6, it is well established that the employ of epoxidized SB and SBS block copolymers as modifier of brittle epoxy system permits the obtaining of tough thermosetting materials by the self-assembly and/or reaction-induced microphase separation of nanodomains arranged in the epoxy matrix. However, one disadvantage of this kind of organic modifiers is the possibility of sacrificing the stiffness of the epoxy system [36–38]. In this sense, and to avoid this plasticization effect on the modulus, a widely used alternative has been the addition of well-dispersed rigid inorganic nanoparticles, such as SiO_2, Al_2O_3, ZrO_2 and TiO_2, into the epoxy matrix [23, 24, 27, 39–59]. Furthermore, the addition of inorganic nanoparticles as modifier of the epoxy system not only acts as stiffening but also could act as toughening additive as well. Johnsen et al. [44] evaluated the toughening mechanisms of inorganic nanoparticle modified epoxy systems by means the observation of the fracture surfaces using SEM and AFM techniques and found that the nanoparticles were surrounded by voids, which provided evidence of toughening mechanisms, such as debonding of the nanoparticles and subsequent plastic void growth. An analytical model of plastic void growth was used to confirm whether this mechanism was responsible for the increased toughness in the nanocomposites. On the other hand, Zhao et al. [45] determined that crack deflection, debonding and plastic deformation of the debonded matrix (plastic void growth) were the possible toughening mechanism for nanoparticles with a relatively weak interface. On the contrary, it was also observed that stronger interface led to enhanced crack deflection and microcracking, and, as a result, a further improvement in properties.

Ocando et al. [23] investigated the effect of combining the benefits of epoxidized SBS and commercially Al_2O_3 nanoparticles into an epoxy system on morphological and mechanical properties. For this purpose, the solution mixing method combined with ultrasonic dispersion was used to fabricate the ternary hybrid inorganic/organic thermosetting systems [49, 54]. The dispersion and morphology of composites were studied by means of TM-AFM.

Figure 8.1 shows the AFM images for the epoxy system containing 1 wt% (epoxy-1) and 3 wt% (epoxy-3) Al_2O_3 nanoparticles and Fig. 8.2 shows the AFM images for the nanostructured epoxy systems modified with 30 wt% SBSep48 containing 1 wt% Al_2O_3 nanoparticles (30SBSep48-1) and 30 wt% SBSep44 containing 3 wt% Al_2O_3 nanoparticles (30C500ep44-3).

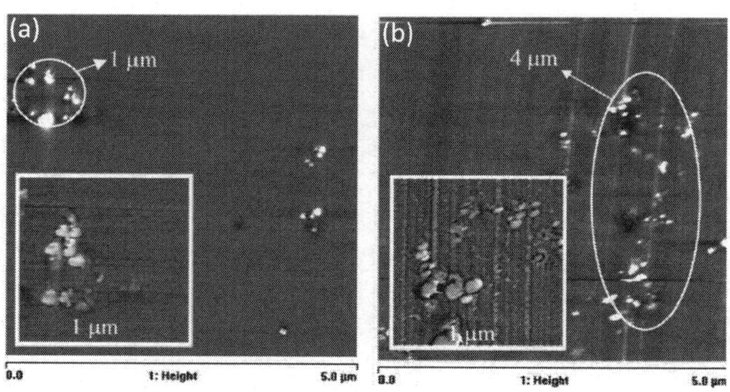

Figure 8.1 AFM height image of (a) epoxy-1 and (b) epoxy-3. The inset shows the phase image. Reproduced with permission of [23]. Copyright Elsevier 2010.

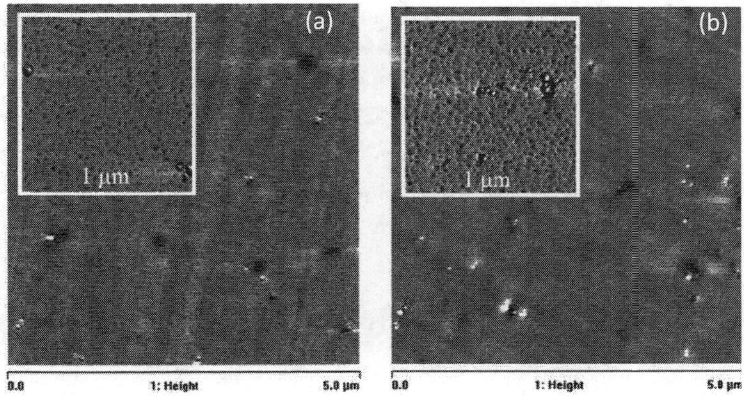

Figure 8.2 AFM height image of (a) 30SBSep48-1 and (b) 30SBSep44-3. The inset shows the phase image. Reproduced with permission of [23]. Copyright Elsevier 2010.

It was noticed that the nanostructured system modified with block copolymer presented a better dispersion of the Al_2O_3 nanoparticles in the epoxy matrix. This improvement in the dispersion of Al_2O_3 nanoparticles in 30SBSep48-1 and 30SBSep44-3 in contrast with epoxy-1 and was explained taking into account that the block copolymer can act as surfactant for the inorganic nanoparticles before curing reaction. In addition, it was also proposed that a second factor could be related to the fact that nanostructured epoxy systems lead to better dispersion of Al_2O_3 nanoparticles throughout the matrix because the microphase separation of PS block domains prevents their agglomeration [23].

It is worth highlighting that the K_{Ic} value increased from 0.8 MPa $m^{1/2}$ for neat epoxy matrix to 1.26 MPa $m^{1/2}$ for the epoxy system modified with 30 wt% SBSep44 containing 3 wt% Al_2O_3 nanoparticles without sacrificing the stiffness of the epoxy system (Fig. 8.3). These facts demonstrated the fascinating synergistic effect of having ternary hybrid thermosetting systems based in inorganic nanoparticles and block copolymers on mechanical properties [23]. Similar synergistic effect on mechanical properties by combining three type modifiers for epoxy matrix was obtained by Mirmohseni and Zavareh [55, 56]; they evaluated the development of an toughened epoxy nanocomposite-based on poly(acrylonitrile-co-butadiene-co-styrene) (ABS), clay (layered nanofiller) and TiO_2 (particulate nanofiller). For ternary nanocomposite, they established that the addition of both ABS and nanofiller was more effective for the improvement of mechanical properties than adding each of modifiers alone.

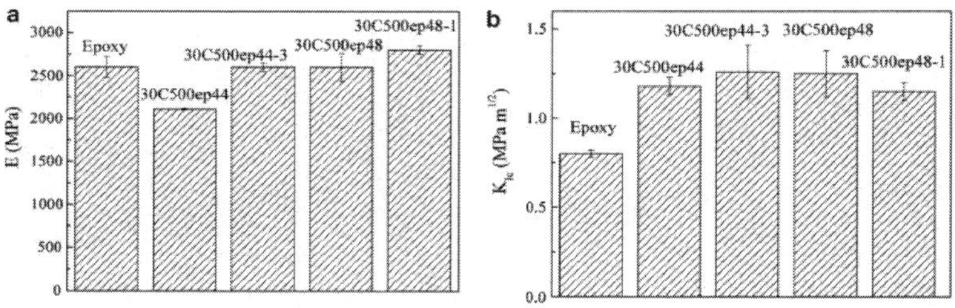

Figure 8.3 (a) Flexural modulus and (b) critical stress intensity factor of mixtures and composites: neat epoxy matrix, 30SBSep44, 30SBSep44-3, 30SBSep48 and 30SBSep48-1. Reproduced with permission of [23]. Copyright Elsevier 2010.

It is well known that low-molecular-weight liquid crystals can act as dispersing agent for the arrangement of diverse types of semiconductive nanoparticles leading materials with strong electro-optic response [57–59]. Interestingly, Tercjak et al. [60] successfully employed nematic liquid crystals (4′-(hexyloxy)-4-biphenyl carbonitrile) (HOBC) as surfactant and template agent of commercially TiO_2 nanoparticles for the selective location of this nanoparticles in the self-assembled PS phase of the poly(styrene-b-ethylene oxide) (PSEO) block copolymer matrix.

Taking into consideration this last approach, in a subsequent work, Tercjak et al. [24] designed and developed a quaternary hybrid inorganic/organic thermosetting systems based on commercial TiO_2 nanoparticles and PSEO block copolymer using 4′-(hexyl)-4-biphenyl-carbonitrile (HBC) liquid crystals as a dispersing agent for the semiconductive nanoparticles in order to obtain thermoresponsive epoxy materials.

The morphological study by means of atomic force microscopy (AFM) and transmission electron microscopy (TEM) and conductive behavior using electrostatic force microscopy (EFM) of 1 wt% TiO_2-15 wt% PSEO-40 wt% HBC-(DGEBA/MXDA) epoxy system suggested that nanoparticles were confined within HBC/PS-nanodomains that were microphase separated from the PEO-block/HBC/epoxy-rich phase (Fig. 8.4). In addition, from TEM analyses (Fig. 8.4aII), it was noticed that the microphase-separated domains consisted of two different phases, corresponding to the PS and the HBC phases.

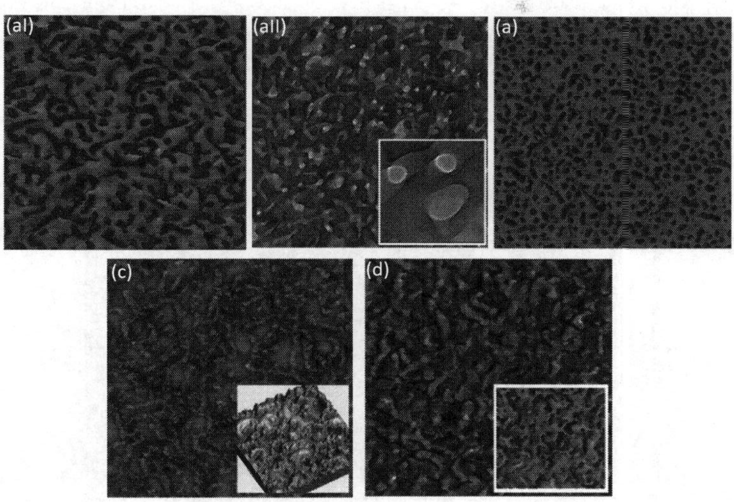

Figure 8.4 (aI) AFM phase image (2 × 2 μm^2) of 1 wt% TiO_2-15 wt% PSEO-40 wt% HBC-(DGEBA/MXDA) epoxy system. (aII) TEM image (2 × 2 μm^2) of 1 wt% TiO_2-15 wt% PSEO-40 wt% HBC-(DGEBA/MXDA) epoxy system; inset: a higher magnification detail. AFM phase images (2 × 2 μm^2) of (b) 15 wt% PSEO-40 wt% HBC-(DGEBA/MXDA) epoxy system; (c) 1 wt% TiO_2-15 wt% PSEO-40 wt% HBC-(DGEBA/MXDA) epoxy system after 3 h of UV light exposure; inset: 3D image (d) 1 wt% TiO_2-15 wt% PSEO-40 wt% HBC-(DGEBA/MXDA) epoxy system after 3 h of applying 12 V by using EFM tip; inset: negative image. Reproduced with permission of [24]. Copyright ACS 2009.

Taking into account that in the ternary system of TiO_2/HBC/PSEO, the inorganic nanoparticles were surrounded by the liquid crystal [60], it was proposed that the TiO_2 nanoparticles were segregated with the HBC liquid crystal inside the PSEO/HBC rich phase. This last fact was corroborated by EFM, where the brightest spherical nanodomains can be clearly recognized as TiO_2 nanoparticles, which are spontaneously arranged as nanodots mainly located in the interface with epoxy rich and the HBC/PS-block phase.

It is worth highlighting that the quaternary hybrid inorganic/organic thermosetting systems maintained the properties of the nematic liquid crystals, because they were able to switch from the opaque to the transparent state by applying thermal gradients (Fig. 8.5). In addition, the employ of TiO_2 nanoparticles led to materials with high contrast between opaque and transparent during the nematic-isotropic transition [24].

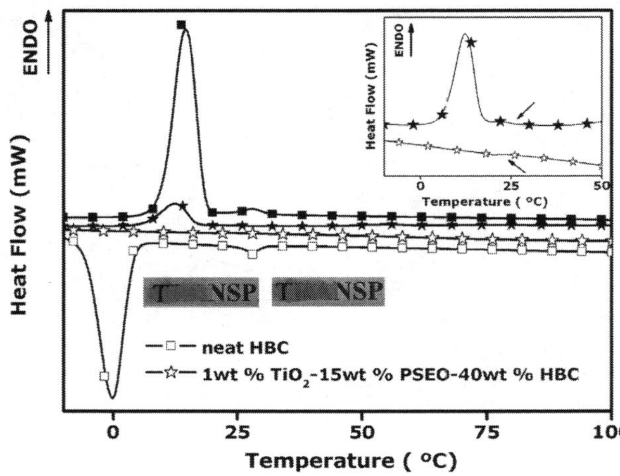

Figure 8.5 DSC thermograms of (-9-/-0- heating/cooling) neat HBC liquid crystal and (-f-/-g- heating/cooling) 1 wt% TiO_2-15 wt% PSEO-40 wt% HBC-(DGEBA/MXDA) epoxy system; inset: details of DSC thermograms of the 1 wt% TiO_2-15 wt% PSEO-40 wt% HBC-(DGEBA/MXDA) epoxy system. Transparency of this blend at 16°C (digital image on the left) and at room temperature (digital image on the right). Reproduced with permission of [24]. Copyright ACS 2009.

8.3 Multifunctional Nanostructured Composites Based on Sol-Gel Synthesized TiO_2 Nanoparticles

Nowadays in order to reach new requirements in nanotechnology, ternary thermosetting systems modified with inorganic nanoparticles and block copolymers have attracted attention.

The use of amphiphilic poly(styrene-b-ethylene oxide) (SEO) block copolymer as template for the development of hybrid nanostructured materials has been extensively reported. On the one hand, as mentioned in Chapter 4, numerous studies from different research groups have attempted to design hybrid inorganic/organic materials (HIOM) based on this block copolymer in combination with sol–gel chemistry [61–64]. On the other hand, several authors employ amphiphilic SEO block copolymer as nanostructuring agent to fabricate self-assembled thermosetting materials [4, 26, 27, 65] (see Chapter 5).

Taking this into account, Prof. Iñaki Mondragon and his co-workers of Group "Materials + Technologies" (GMT) proposed to merge both pathways with the aim to design advanced multifunctional hybrid nanostructured thermosetting materials. They carried out different research works using amphiphilic SEO block copolymer as nanostructuring agent and simultaneously as template for the dispersion of sol-gel synthesized nanoparticles.

This idea was based on the miscibility of components. Thus, as it is well known from literature, PEO-block of the SEO block copolymer is epoxyphilic block which is miscible with the epoxy resin [66–68] and simultaneously miscible with a titanium sol-gel network (SG) [69].

The aim of the investigation work described in different publications of this research group [26–30] was the preparation and characterization of different ternary inorganic/organic nanostructured thermosetting composites based on two different SEO block copolymers as template agents and sol-gel synthesized TiO_2 nanoparticles.

Playing with both block copolymer and TiO_2 nanoparticle contents, transparent multifunctional advanced thermosetting materials with tunable optical and conductive properties were successfully designed [26–30].

To fabricate ternary inorganic/organic thermosetting composites a diglicydylether of bisphenol A epoxy monomer (DGEBA) epoxy resin was cured with a stoichiometric amount of 4,4′-methylene-bis(3-chloro 2,6-diethylaniline) (MCDEA) amine hardener. This study also discussed the influence of the ratio between blocks of the SEO block copolymer on the final morphology of the thermosetting systems using two different molecular weight SEO block copolymers with high (HSEO) and low (LSEO) PS block contents. Titanium isopropoxide (TTIP) was employed to generate TiO_2 nanoparticles via sol-gel synthesis [26, 27, 30].

This research, based on the preparation of novel inorganic/organic nanostructured thermosetting systems, discusses the influence of both SEO block copolymer and TiO_2 nanoparticles content on the network curing formation and morphological variations in the final composites. Playing with the ratio between SEO block copolymer and TiO_2 nanoparticles novel transparent nanostructured composites with well-dispersed inorganic nanoparticles were developed.

First, to better understand morphological results of ternary systems it is necessary to take in mind the microphase separation mechanism of binary SEO block copolymer/epoxy systems (see Chapter 5). As mentioned above and in previous work developed in Group "Materials + Technologies" [66, 70], PEO block of the SEO block copolymer is miscible with the epoxy matrix, which provokes microphase separation of the PS-block domains during network formation.

With the objective to reach good dispersion of the TiO_2 nanoparticles, 20 wt% (HSEO:SG)-(DGEBA/MCDEA) system and 40 wt% (LSEO:SG)-(DGEBA/MCDEA) system were chosen to produce different ternary inorganic/organic thermosetting systems. In both cases, different thermosetting systems were designed varying the ratio between block copolymer (HSEO or LSEO) and SG.

Morphological analysis was carried out by means of AFM and TEM and focused on the study of the influence of the relation between modifiers on the final morphology of designed thermosetting systems.

Figure 8.6 shows AFM phase images of binary 20 wt% HSEO-(DGEBA/MCDEA) system and different 20 wt% (HSEO:SG)-(DGEBA/MCDEA) ternary systems.

For 20 wt% HSEO-(DGEBA/MCDEA) system, the morphology was long interconnected wormlike micelles (also presenting a bilayered structure) well dispersed in the continuous PEO-block/epoxy-rich phase. In this case, the bilayered structure means layers of the epoxy-rich phase surrounded by PS-block-rich layer. The generated morphology was very regular and repeatable being the wide of the long wormlike micelles around 60–80 nm. In the case of ternary 20 wt% (HSEO:SG)-(DGEBA/MCDEA) thermosetting systems TiO_2 nanoparticles are the hardest component of these systems [26] and therefore, in the AFM images, their appeared as spherical bright domains.

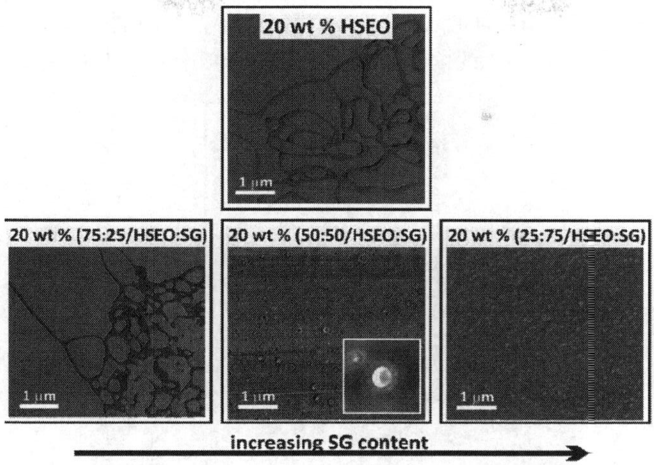

Figure 8.6 AFM phase images (5 μm × 5 μm) of binary 20 wt% HSEO-(DGEBA/MCDEA) system and ternary 40 wt% (HSEO:SG)-(DGEBA/MCDEA) systems.

For all designed thermosetting systems, neither TiO_2 nor their aggregates were detected in the epoxy continuous matrix. For 20 wt% (75:25/HSEO:SG)-(DGEBA/MCDEA) system, synthesized TiO_2 nanoparticles were located in some of the separated PEO-block/epoxy-rich domains and confined between microphase-separated long PS-block wormlike micelles. Here, it should be noted that the generated nanostructure was similar to the binary 20 wt% HSEO. The increase of SG content up to a ratio 50:50 between HSEO and SG provoke significant changes in the generated morphology [26, 30].

In this case of 20 wt% (50:50/HSEO:SG)-(DGEBA/MCDEA) system TiO_2 nanoparticles were located into the epoxy-rich phase confined between the bilayered PS-block-rich phase formed vesicle structure. This morphology can be clearly observed in the high magnification AFM phase image (white square inset on the corresponding image of Fig. 8.6). For 20 wt% (25:75/HSEO:SG)-(DGEBA/MCDEA) system—the inorganic content being higher than the block copolymer content—a spherical morphology was observed.

Different morphological behavior was observed for ternary systems based on LSEO block copolymer due to the high amount of epoxyphilic PEO block [27, 30]. Figure 8.7 shows AFM phase images of binary 40 wt% LSEO-(DGEBA/MCDEA) system and different ternary 40 wt% (SEO:SG) thermosetting systems.

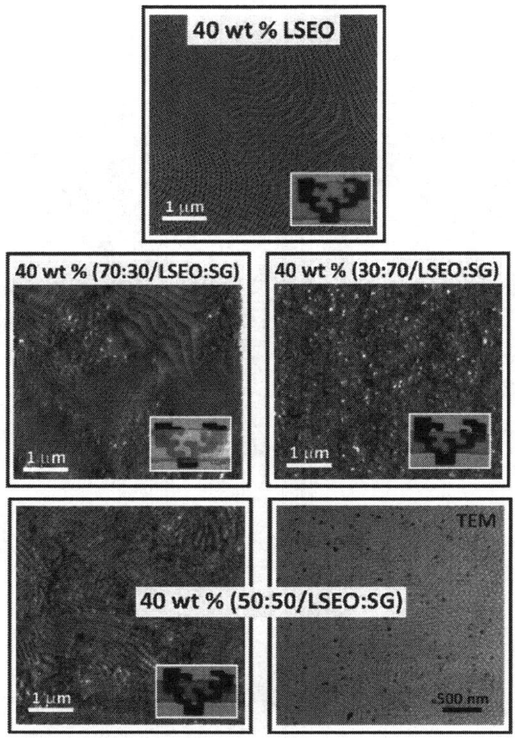

Figure 8.7 AFM phase images (5 μm × 5 μm) of binary 40 wt% LSEO-(DGEBA/MCDEA) system and different ternary 40 wt% (LSEO:SG)-(DGEBA/MCDEA) systems. TEM image of 40 wt% (50:50/LSEO:SG)-(DGEBA/MCDEA) system.

The addition of 40 wt% of LSEO block copolymer to the epoxy matrix led to a hexagonally ordered structure where PS block cylinders are parallel and perpendicularly oriented to the cut surface.

Ternary inorganic/organic thermosetting systems showed good dispersion of synthesized TiO_2 nanoparticles and simultaneously maintained nanostructure of SEO block copolymer.

Here, it should be pointed out that the size of the TiO_2 nanoparticles in all ternary inorganic/organic thermosetting systems was almost the same, with a diameter around ~25 nm.

40 wt% (70:30/LSEO:SG)-(DGEBA/MCDEA) system showed a hexagonally packed morphology, similar to the system modified only with 40 wt% LSEO. In this system, PS-block cylinders (dark areas) were located parallel and perpendicular to the cut surface. The increase of SG content up to a ratio 50:50 between LSEO block copolymer and SG did not have a significant effect on the microphase separation of PS-block phase in epoxy matrix. Nevertheless, in the corresponding AFM images higher amount of bright domains (TiO_2 nanoparticles) embedded in the epoxy matrix can be clearly distinguished. Additionally, TEM micrograph of this system confirmed the good dispersion of TiO_2 nanoparticles. Thus, black domains corresponding to TiO_2 nanoparticles were preferentially located near the interface between the epoxy-rich matrix and microphase-separated PS-block domain. Moreover, the increase of SG content led to detection of higher quantity of bright TiO_2 nanoparticles in the corresponding AFM phase image. The 40 wt% (30:70/LSEO:SG)-(DGEBA/MCDEA) system showed spherical morphology due to the low block copolymer content. Uniformly dispersed spherical PS-block domains with an average diameter of ~50 nm in the continuous epoxy matrix with well-dispersed spherical TiO_2 nanoparticles were easily detected.

Obtained results confirmed that systems based on both block copolymers, HSEO and LSEO, TiO_2 nanoparticles were preferentially confined at epoxy-rich matrix, near the interface between epoxy rich matrix and microphase-separated PS-block domains regardless of the ratio between SEO and SG. The good distribution of TiO_2 nanoparticles clearly indicated the strong interaction between PEO-block/epoxy-rich phase and sol-gel synthesized TiO_2 nanoparticles.

Taking into account the optical properties of TiO_2 nanoparticles and the actual interest to develop transparent and UV filtering composites, UV-vis spectroscopy measurements were also performed to investigate optical properties of designed materials. Both UV light shielding property of TiO_2 nanoparticles in the thermosetting systems and the transparency level were analyzed. Performed measurements confirmed that synthesized ternary inorganic/organic thermosetting systems were transparent in the visible range and exhibited almost perfect UV light shielding efficiency [26–28].

With the aim of study the conductive properties of TiO_2 nanoparticles embedded in nanostructured thermosetting composites electrostatic force microscopy (EFM) was employed [28, 29]. Figure 8.8 shows EFM phase images of binary and ternary nanostructured thermosetting systems with different voltages applied to the EFM tip. In the case of epoxy system modified with 40 wt% SEO block copolymer EFM measurements demonstrate that no charge domains were detected on the surface regardless of both the value and the sign of the applied voltage.

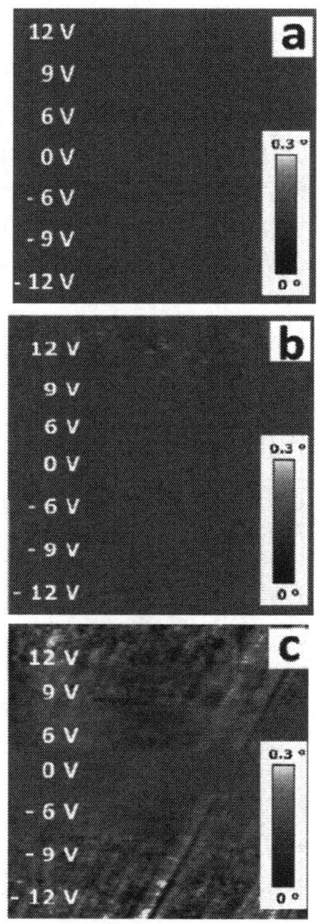

Figure 8.8 EFM phase images (5 × 5 μm) of (a) 40 wt% LSEO-(DGEBA/MCDEA), (b) 40 wt% (70:30/LSEO:SG)-(DGEBA-MCDEA) and (c) 40 wt% (30:70/LSEO:SG)-(DGEBA/MCDEA) systems applying different negative and positive voltages.

On the contrary, for hybrid inorganic/organic ternary thermosetting systems well dispersed small bright domains corresponding to the TiO_2 nanoparticles were detected. Here, it should be pointed out that a higher quantity of bright nanoparticles appeared with the increasing of sol-gel content. Additionally, one can easily note that ternary systems were able to respond, regardless of the sign of the applied voltage [28, 29].

Finally, one can concluded that designed thermosetting systems modified with both TiO_2 nanoparticles and PS-b-PEO block copolymer, can open a new strategy for preparation of well-defined transparent multifunctional thermosetting materials which have a large number of applications in variety of fields taking into account their optical and conductive properties.

8.4 Nanostructured Thermosetting Materials as Template for Carbon Nanotubes

González-Domínguez et al. [71] developed the epoxy resin-based thermosetting materials modified with the single-walled carbon nanotubes (SWNTs) covered with poly(ethylene oxide)-b-poly(propylene oxide)-b-poly(ethylene oxide) PEO-b-PEO-PEO block copolymer (Pluronic F68). The authors used a triglycidyl p-aminophenol (TGAP) as epoxy precursor and a 4,4′-diaminodiphenyl sulfone (DDS) as a hardener. They prepared and analyzed different systems using as modifier pristine SWNTs and SWNTs coated with 0.1, 0.25, 0.5, 1 and 2 wt% PEO-b-PEO-PEO block copolymer. They confirmed using optical microscopy that all investigated systems modified with PEO-b-PPO-b-PEO-coated SWNTs was homogenous with no visible agglomeration. On the contrary, TGAP/DDS-cured system modified with pristine SWNTs showed carbon nanotubes agglomeration at the macroscopic scale. In order to confirm good dispersion of PEO-b-PPO-b-PEO-coated SWNTs in epoxy matrix authors performed TEM and SEM measurement. As visualized in Fig. 8.9 the systems without PEO-b-PPO-b-PEO coating possessed large inhomogeneous agglomeration of SWNTs indicating poor adhesion between carbon nanotubes and epoxy-based thermosetting matrix. On the contrary, thermosetting system modified with PEO-b-PPO-b-PEO-coated SWNTs showed considerable increase of dispersion of the carbon nanotubes in the matrix. Individual SWNTs can be observed in both TEM and SEM images (marked with arrows in Figs. 8.9b,d).

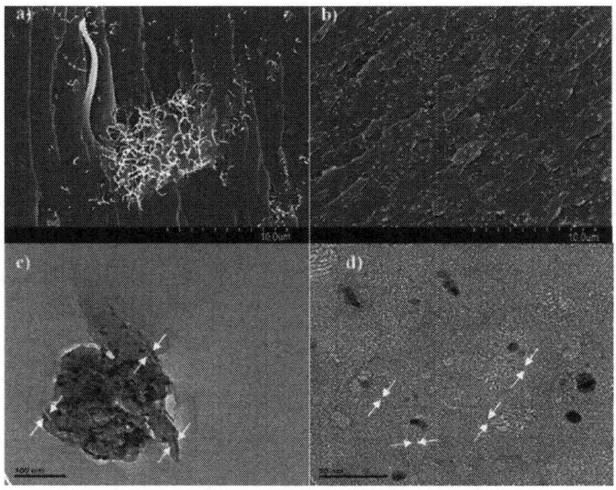

Figure 8.9 SEM (top) and TEM (bottom) images for thermosetting systems modified with 0.5 wt% of pristine SWNTs (a, c) and PEO-b-PPO-b-PEO-coated SWNTs (b, d). Arrows indicate the presence of individual SWNTs. Reproduced with the permission of [71]. Copyright 2011, American Chemical Society.

Additionally, no agglomeration of SWNTs was detected indicating that PEO-b-PPO-PEO coating is effective pathway to design thermosetting materials with very good dispersion of SWNTs.

González-Domínguez et al. [71] investigated also mechanical properties of the TGAP/DDS-cured systems modified with pristine SWNTs and PEO-b-PPO-b-PEO-coated SWNTs. For comparison, they investigated also neat TGAP/DDS-cured system and this system modified only with PEO-b-PPO-b-PEO block copolymer. Results obtained by authors using tensile test machine are shown in Fig. 8.10.

As expected the neat TGAP/DDS-cured system possessed high Young's modulus value (4.3GPa) but very low toughness. The addition of the pristine SWNTs into thermosetting system led to materials with higher Young's modulus value (4.7%) since SWNTs acted as reinforcement for epoxy matrix and also increased the toughness (34%) of this thermosetting system.

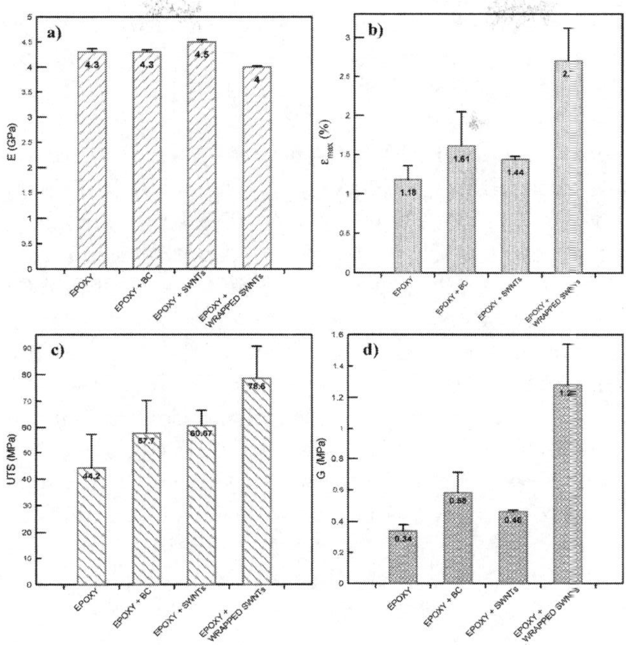

Figure 8.10 Mechanical parameter extracted from stress-strain curves: (a) Young's modulus, (b) maximum strain, (c) ultimate tensile strength, and (d) toughness. Reproduced with the permission of [71]. Copyright 2011, American Chemical Society.

As mentioned by authors the very interesting point of their work is that the Young's modulus value of the thermosetting system modified with PEO-b-PPO-b-PEO-coated SWNTs decrease slightly if compare with the neat TGAP/DDS-cured system; however, simultaneously the toughness of modified thermosetting system increase 276%. This phenomenon as pointed out by authors is due to a synergistic toughening effect between SWNTs and PEO-b-PPO-b-PEO block copolymer.

In the same research field, Espósito et al. [31] used nanostructured thermosetting systems modified with poly(isoprene-b-methyl methacrylate) (PI-b-PMMA) block copolymer to dispersed polyisoprene-grafted carbon nanotubes (PI-g-CNT). Authors used diglycidyl ether of bisphenol-A (DGEBA) as epoxy precursor and 4,4′-methylenebis(3-chloro-2,6-diethylaniline) (MCDEA) as hardener. Similarly to previous work, authors investigated the difference in dispersion and mechanical properties between the thermosetting systems modified with PI-g-CNT and the thermosetting systems modified with both PI-g-CNT and PI-b-PMMA.

Authors employed AFM to study the dispersion of the PI-g-CNT in the neat DGEBA/MCDEA-cured system and DGEBA/MCDEA-cured system nanostructured with PI-b-PMMA. They proved that presence of PI-b-PMMA block copolymer improved the dispersion of the PI-g-CNT compared with DGEBA/MCDEA-cured system modified only with PI-g-CNT. Moreover, Espósito et al. [31] proved also that the addition of the PI-g-CNT into PI-b-PMMA-(DGEBA/MCDEA)-cured systems do not affected on the microseparation of the PI block rich phase. Consequently, authors concluded that the functionalization of MWCNT with grafted polyisoprene was not enough to totally disperse them into the epoxy matrix; however, an improvement of the dispersion of carbon nanotubes was achieved since PI-b-PMMA block copolymer acted not only as nanostructuring agent for epoxy matrix but also as dispersion agent for PI-g-CNT.

Authors studied also thermomechanical properties of investigated systems noticed slight increase in these properties with increase of PI-g-CNT content in nanostructured thermosetting systems. For the systems modified only with PI-g-CNT the thermo-mechanical properties decrees for high content of functionalized MWCNT probably due to the present of agglomerates.

In the second research work developed by Espósito et al. [32], authors used epoxidized polystyrene-b-polybutadiene-b-polystyrene block copolymer (epSBS) for nanostructuration of thermosetting materials based on epoxy. In this case, diglycidyl ether of bisphenol A (DGEBA) was used as epoxy precursor and 1-(2-aminoethyl)piperazine:1,3-bis(aminomethyl)benzene (APAB) was used as hardener. The nanostructured epSBS-(DGEBA/APAB) thermosetting systems were modified with different wt% of multi-walled carbon nanotubes (a-MWCNT) functionalized with carboxylic acid groups. The functionalization was done in order to reach higher dispersion of the functionalized a-MWCNT. The thermosetting materials modified with both SBS and functionalized a-MWCNT was studied from the point of view of their potential application as industrial coatings. Thermosetting materials modified with epSBS block copolymer were previously described in Chapter 6. The authors showed that these nanostructured thermosetting materials acted as effective templates for dispersion of the functionalized a-MWCNT leading to coatings with interesting surface properties.

As shown in Fig. 8.11, employment of AFM techniques allowed authors to visualize the higher dispersion of functionalized a-MWCNT in the coatings based on SBS-(DGEBA/APAB)-cured systems.

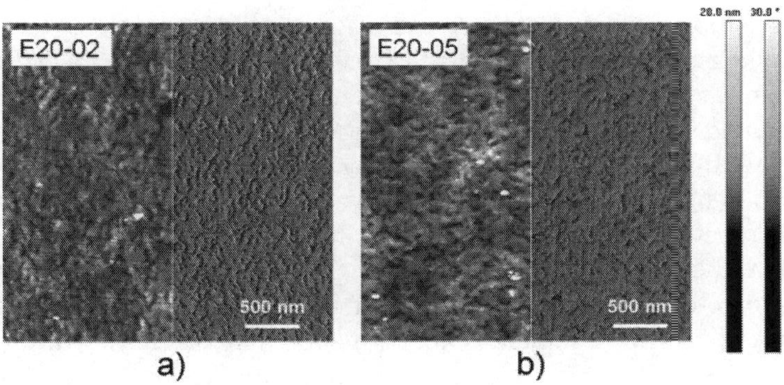

Figure 8.11 Cross-sectional AFM height (left) and phase (right) images (3 μm × 3 μm) of epoxy systems containing 20 wt% ep52SBS with: (a) 0.2 wt% and (b) 0.5 wt% a-MWCNT. Reproduced with the permission of [32]. Copyright 2014, Elsevier.

Additionally, similarly to the thermosetting systems described before, the addition of different wt% of functionalized a-MWCNT do not interrupted microphase separation of PS block rich phase allowed to design ternary nanostructured systems with good dispersion of carbon nanotubes. As pointed out by authors, AFM images showed cross-section of the coatings and consequently nanotubes appeared as weight dots.

8.5 Nanostructured Thermosetting Materials as Template for Microfibrillated Cellulose

Builes et al. [33] used nanostructured thermosetting materials, in their particular case, unsaturated polyester (UP) (described in detail in Chapter 7) modified with PEO-b-PPO-b-PEO block copolymer [33] for dispersion of sisal microfibrillated cellulose (MFC).

To reach the microfibrils, they employed chemical–mechanical treatment. In the first step of the treatment they used Heavy-Duty Cutting Mill Retsch SM 2000 with a 2–4 mm sieve in order to reduce the original size of sisal fibers. Next, they used a solvent extraction process according to TAPPI T204 standard. After this process, authors performed the treatment in 7.5 wt% aqueous solution of sodium hydroxide during 90 min in reflux. Finally, the obtained pulp was washed repeated times with 10 wt% acetic acid solution and distillate water until it neutralization. Chemically treated fibers was then dispersed in water and pumped throughout a Panda 2 K Niro Soavi homogenizer.

Authors proved using optical microscopy that 100 passes in homogenizer reduced effectively the size of the sisal fibers. As shown in Fig. 8.12, they verified, using AFM, that after chemical–mechanical treatment the sisal fibers had micro- and nanometric size. AFM results confirmed that sisal fibers were broken down to smaller fibers, which formed a network of fibrils. Fibers with a diameter of ~160 nm (indicated by the arrow pointing to the right) were fibrillated in fibrils with diameter of ~10 nm (indicated by the arrow pointing as can be clearly observed from AFM phase image in Fig. 8.12. The detailed analysis of AFM results indicated that each fiber consisted of 5–10 nm diameter microfibrils. Thus, the fibrillated structure of the fibers confirmed the effectiveness of using a high-pressure homogenizer.

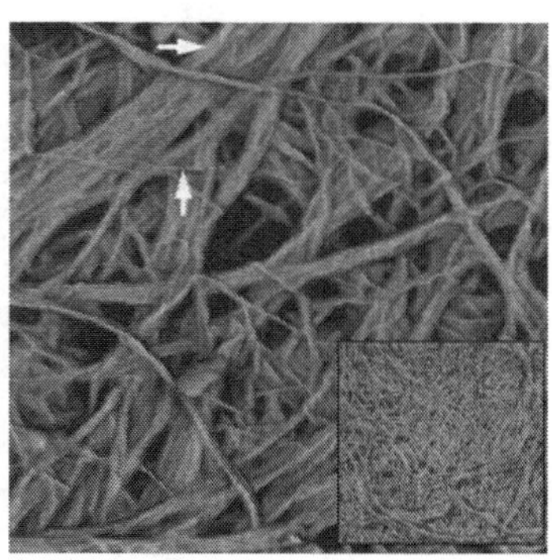

Figure 8.12 AFM image (1 μm × 1 μm) of fibrillated sisal fibers. The inset corresponds to 5 μm × 5 μm AFM image. Reproduced with permission of [33]. Copyright 2013, Elsevier.

Nanostructured thermosetting systems were prepared in a parallelepipedic mould of 1 ± 0.1 mm thickness; consequently obtained results are related to the bulk behavior. The authors confirmed that in thermosetting systems modifies with sisal MFC PEO-b-PPO-b-PEO block copolymer act not only as nanostructuring agent for thermosetting matrix but also as dispersing agent for MFC. As visualized by optical microscopy in Fig. 8.13 the addition 1 wt% MFC into UP lead to macroscopic agglomeration of sisal fibers. On the contrary, the ternary system modified with 1 wt% MFC and 5 wt% PEO-b-PPO-b-PEO did not show any macroscopic agglomeration indicating good dispersion of MFC on the nanometric scale.

Figure 8.13 Optical micrographs of (a) 5%PEO-b-PPO-b-PEO/UP, (b) 1%MFC/UP and (c) 1%MFC/5%PEO-b-PPO-b-PEO/UP-cured systems. The inset of each image corresponds to a digital image of 1 mm sheet. Reproduced with permission of [33]. Copyright 2013, Elsevier.

In order to confirm the good dispersion of MFC in 1%MFC/5%PEO-b-PPO-b-PEO/UP-cured system, the authors studied the morphology of investigated systems by means of AFM. Figure 8.14 shows the morphology for UP-cured system and these system-modified PEO-b-PPO-b-PEO block copolymer and without and with MFC.

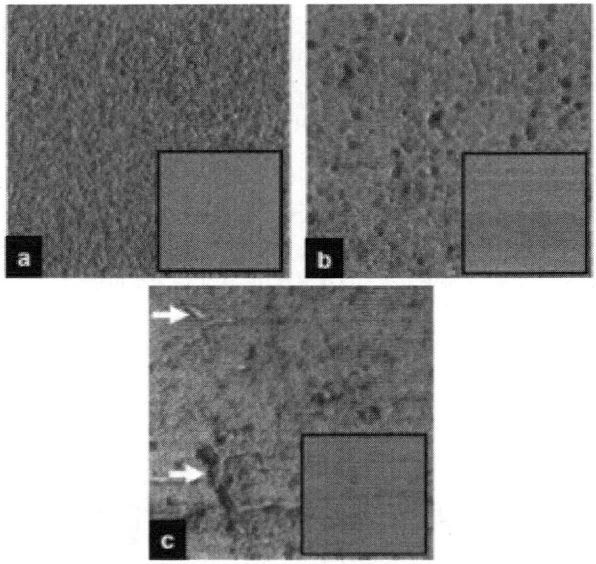

Figure 8.14 AFM images (0.75 μm × 0.75 μm) of (a) neat UP, (b) 5%PEO-b-PPO-b-PEO/UP and (c) 1%MFC/5% PEO-b-PPO-b-PEO/UP-cured systems. The inset in each image corresponds to 5 μm × 5 μm AFM phase image. Reproduced with permission of [33]. Copyright 2013, Elsevier.

The authors proved that the 5% PEO-b-PPO-b-PEO/UP-cured system showed spherical morphology related to the fact that PEO block of PEO-b-PPO-b-PEO block copolymer is miscible with UP matrix and PPO block microseparated forming spherical domains with the size ~10 nm. AFM phase image of 1%MFC/5%EPE20/UP-cured mixture confirmed good dispersion of the MFC without losing the nanostructuration of the UP matrix. As reported by the authors, this phenomenon could be related to the adsorption ability of MFC, which lead to location of the part of the PEO-b-PPO-b-PEO block copolymer on the surface of the fibrils. Thus, block copolymer acted as a dispersing agent instead of as a nanostructuring agent. Consequently, the microphase-separated PPO-block rich domains became smaller that for the system modified only with PEO-b-PPO-b-PEO block copolymer (see Figs. 8.14b,c).

Builes et al. [33] studied also mechanical properties of 1%MFC/5% PEO-b-PPO-b-PEO/UP-cured system if compare with the mechanical properties of neat UP and cured systems. As shown in Fig. 8.15, addition of 1 wt% MCF into 5% PEO-b-PPO-b-PEO/UP-cured system not only improved the toughness of the UP matrix but also improved the flexural modulus, *E*, of the cured system compared with the system modified only with 5 wt% PEO-b-PPO-b-PEO. Here, it should be pointed out that due to the plasticization effect of 5% PEO-b-PPO-b-PEO block copolymer, the *E* of 5% PEO-b-PPO-b-PEO/UP-cured system decreased from 4.1 ± 0.05 GPa to 3.5 ± 0.02 GPa. Simultaneously, as described in detail in Chapter 7, the toughness of 5% PEO-b-PPO-b-PEO/UP-cured system significantly increased. Moreover, in the 1%MFC/5% PEO-b-PPO-b-PEO/UP-cured system the MFC acted as reinforcement and made the flexural module almost equal to the UP-cured system.

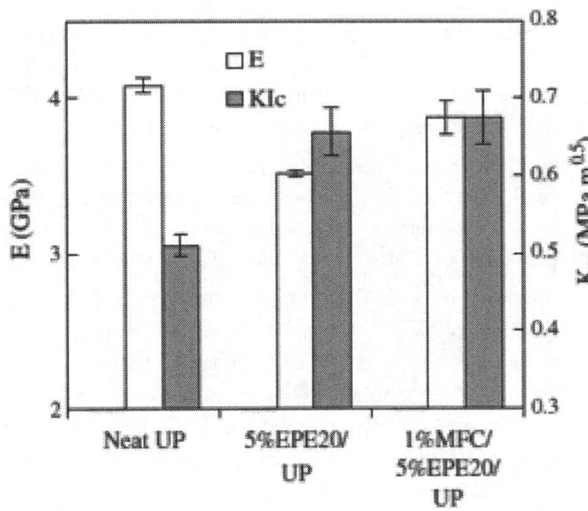

Figure 8.15 Flexural modulus, *E*, and the critical stress intensity factor, K_{Ic}, of neat UP, 5%PEO-b-PPO-b-PEO/UP and 1%MFC/5%PEO-b-PPO-b-PEO/UP-cured systems. Reproduced with permission of [33]. Copyright 2013, Elsevier.

References

1. Pascault, J. P., and Williams, R. J. J. (2010). *Epoxy Polymers. New Materials and Innovations*, Wiley-VCH, Weinheim, Germany. Libro Pascault.

2. Hillmyer, M. A., Lipic, P. M., Hadjuk, D. A., Almdal, K., and Bates, F. S. (1997). Self assembly and polymerisation of epoxy resin-amphiphilic block copolymer nanocomposites, *J. Am. Chem. Soc.*, **119**, pp. 2749–2750.
3. Maiez-Tribut, S., Pascault, J. P., Soule, E. R., Borrajo, J., and Williams, R. J. J. (2007). Nanostructured epoxies based on the self-assembly of block copolymers: A new miscible block that can be tailored to different epoxy formulations, *Macromolecules,* **40**, pp. 1268–1273.
4. Meng, F., Zheng, S., Li, H., Liang, Q., and Liu, T. (2006). Formation of ordered nanostructures in epoxy thermosets: A mechanism of reaction-induced microphase separation, *Macromolecules,* **39**, pp. 5072–5080.
5. Meng, F., Zheng, S., and Liu, T. (2006). Epoxy resin containing poly(ethylene oxide)-block-poly(ε-caprolactone) diblock copolymer: Effect of curing agents on nanostructures, *Polymer*, **47**, pp. 7590–7600.
6. Meng, F., Zheng, S., Zhang, W., Li, H., and Liang, Q. (2006). Nanostructured thermosetting blends of epoxy resin and amphiphilic poly(ε-caprolactone)-block-polybutadiene-block-poly(ε-caprolactone) triblock copolymer, *Macromolecules,* **39**, pp. 711–719.
7. Ocando, C., Serrano, E., Tercjak, A., Peña, C., Kortaberria, G., Calberg, C., Jerome, R., Carrasco, P. M., Mecerreyes, D., and Mondragon, I. (2007). Structure and properties of a semifluorinated diblock copolymer modified epoxy blend, *Macromolecules*, **40**, pp. 4068–4074.
8. Ritzenthaler, S., Court, F., David, L., Girard-Reydet, E., Leibler, L., and Pascault, J. P. (2002). ABC triblock copolymers/epoxy–diamine blends. 1. keys to achieve nanostructured thermosets, *Macromolecules*, **35**, pp. 6245–6254.
9. Ritzenthaler, S., Court, F., David, L., Girard-Reydet, E., Leibler, L., and Pascault, J. P. (2003). ABC triblock copolymers/epoxy–diamine blends. 2. parameters controlling the morphologies and properties, *Macromolecules*, **36**, pp. 118–126.
10. Serrano, E., Martin, M. D., Tercjak, A., Pomposo, J. A., Mecerreyes, D., and Mondragon, I. (2005). Nanostructured thermosetting systems from epoxidized styrene butadiene block copolymers, *Macromol. Rapid Commun.*, **26**, pp. 982–985.
11. Serrano, E., Tercjak, A., Kortaberria, G., Pomposo, J. A., Mecerreyes, D., Zafeiropoulos, N. E., Stamm, M., and Mondragon, I. (2006). Nanostructured thermosetting systems by modification with epoxidized styrene–butadiene star block copolymers. effect of epoxidation degree, *Macromolecules*, **39**, pp. 2254–2261.

12. Tercjak, A., and Mondragon, I. (2008). Relationships between the morphology and thermoresponsive behavior in micro/nanostructured thermosetting matrixes containing a 4′-(hexyloxy)-4-biphenylcarbonitrile liquid crystal, *Langmuir*, **24**, pp. 11216–11224.
13. Sinturel, C., Vayer, M., Erre, R., Amenitsch, H. (2007). Nanostructured polymers obtained from polyethylene-block-poly (ethylene oxide) block copolymer in unsaturated polyester, *Macromolecules,* **40**, pp. 2532–2538.
14. Sinturel, C., Vayer, M., Erre, R., Amenitsch, H. (2009). Nanostructured polymers obtained from polyethylene-block-poly(ethylene oxide) block copolymer in unsaturated polyester, *Eur. Polym. J.*, **45,** pp. 2505–2538.
15. Serrano, E., Gerard, P., Lortie, F., Pascault, J. P., Portinha, D. (2008). Nanostructuration of unsaturated polyester by all-acrylic block copolymers, 1-Use of high-molecular-weight block copolymers, *Macromol. Mater. Eng.*, **293**, pp. 820–827.
16. Builes, D. H., Hernández, H., Mondragon, I., Tercjak, A. (2013). Relationship between the morphology of nanostructured unsaturated polyesters modified with PEO-*b*-PPO-*b*-PEO triblock copolymer and their optical and mechanical properties, *J. Phys. Chem. C,* **117**, 3563–3571.
17. Builes, D. H., Tercjak, A., Mondragon, I. (2012). Nanostructured unsaturated polyester modified with poly[(ethylene oxide)-*b*-(propylene oxide)-*b*-(ethylene oxide)] triblock copolymer, *Polymer,* **53**, pp. 3669–3676.
18. Builes, D., Hernández-Ortiz, J. P., Corcuera, M. A., Mondragon, I., Tercjak, A. (2014). Effect of poly(ethylene oxide) homopolymer and two different poly(ethylene oxide-b-poly(propylene oxide)-b-poly(ethylene oxide)) triblock copolymers on morphological, optical, and mechanical properties of nanostructured unsaturated polyester. *ACS Appl. Mater. Interfaces,* **6**, pp. 1073–1081.
19. Amerio, E., Sangermano, M., Malucelli, G., Priola, A., and Voit, B. (2005). Preparation and characterization of hybrid nanocomposite coatings by photopolymerization and sol-gel process, *Polymer,* **46**, pp. 11241–11246.
20. Chau, J. L. H., Tung, Ch. T., Lin, Y. M., and Li, A. K. (2008). Preparation and optical properties of titania/epoxy nanocomposite coatings, *Mater. Lett.*, **62**, pp. 3416–3418.
21. Chiang, Ch. L., and Ma, Ch. M. (2002). Synthesis, characterization and thermal properties of novel epoxy containing silicon and

phosphorus nanocomposites by sol-gel method, *Eur. Polym. J.*, **38**, pp. 2219–2224.

22. Sangermano, M., Malucelli, G., Amerio, E., Bongiovanni, R., Priola, A., Di Gianni, A., Voit, B., and Rizza, G. (2006). Preparation and characterization of nanostructured TiO_2 epoxy polymeric films, *Macromol. Mater. Eng.*, **291**, pp. 517–523.
23. Ocando, C., Tercjak, A., and Mondragon, I. (2010). Nanostructured systems based on SBS epoxidized triblock copolymers and well-dispersed alumina/epoxy matrix composites, *Composites Sci. Technol.*, **70**, pp. 1106–1112.
24. Tercjak, A., Gutierrez, J., Peponi, L., Rueda, L., and Mondragon, I. (2009). Arrangement of conductive TiO_2 nanoparticles in hybrid inorganic/organic thermosetting materials using liquid crystal, *Macromolecules,* **42**, pp. 3386–3390.
25. Tercjak, A., Gutierrez, J., Ocando, C., and Mondragon, I. (2010). Conductive properties of switchable photoluminescence thermosetting systems based on liquid crystals, *Langmuir,* **26**, pp. 4296–4302.
26. Tercjak, A., Gutierrez, J., Martin, M. D., and Mondragon, I. (2012). Transparent inorganic/organic block copolymer-modified epoxy-based systems in the long scale microphase separation threshold, *Europ. Polym. J.,* **48**, pp. 16–25.
27. Gutierrez, J., Mondragon, I., and Tercjak, A. (2011). Morphological and optical behavior of thermoset matrix composites varying both polystyrene-block-poly(ethylene oxide) and TiO_2 nanoparticle content, *Polymer,* **52**, pp. 5699–5707.
28. Gutierrez, J., Tercjak, A., and Mondragon, I. (2010). Transparent nanostructured thermoset composites containing well-dispersed TiO_2 nanoparticles, *J. Phys. Chem. C,* **114**, pp. 22424–22430.
29. Gutierrez, J., Mondragon, I., and Tercjak, A. (2012). Nanostructured thermoset composites containing conductive TiO_2 nanoparticles, *Mater. Sci. Forum.,* **714**, pp. 147–152.
30. Gutierrez, J., Mondragon, I., and Tercjak, A. (2012). Multifunctional nanostructured composites based on TiO_2 nanoparticles, *Macromol. Symp.*, **321**, pp. 99–104.
31. Espósito, L. H., Ramos, J. A., Mondragon, I., Kortaberria, G. (2014). Nanostructured thermosetting epoxy systems modified with poly(isoprene-b-methyl metacrylate) diblock copolymer and polyisoprene-grafted carbon nanotubes, *J. Appl. Polym. Sci.,* **129**, pp. 1060–1067.

32. Espósito, L. H., Ramos, J. A., Kortaberria, G. (2014). Dispersion of carbon nanotubes in nanostructured epoxy systems for coating application, *Prog. Org. Coatings,* **77**, pp. 1452–1458.

33. Builes, D. H., Labidi, J., Eceiza, A., Mondragon, I., Tercjak, A. (2014). Unsaturated polyester nanocomposites modified with fibrillated cellulose and PEO-b-PPO-b-PEO block copolymer, *Comp. Sci. Technol.*, **89**, pp. 120–126.

34. Li, M., and Ober, C. K. (2006). Block copolymer patterns and templates, *Mater. Today,* **9**, pp. 30–39.

35. Hamley, I. W. (2009). Ordering in thin films of block copolymers: Fundamentals to potential application, *Prog. Polym. Sci.,* **34**, pp. 1161–1210.

36. Serrano, E., Tercjak, A., Ocando, C., Larrañaga, M., Parellada, M. D., Corona-Galván, S., Mecerreyes, D., Zafeiropoulos, N. E., Stamm, M., and Mondragon, I. (2007). Curing behavior and final properties of nanostructured thermosetting systems modified with epoxidized styrene-butadiene linear diblock copolymers, *Macromol. Chem. Phys.*, **208**, pp. 2281–2292.

37. Ocando, C., Tercjak, A., Serrano, E., Ramos, J. A., Corona-Galván, S., Parellada, M. D., Fernández-Berridi, M. J., and Mondragon, I. (2008). Micro-and macrophase separation of thermosetting systems modified with epoxidized styrene-block-butadiene-block-styrene linear triblock copolymers and their influence on final mechanical properties, *Polym. Int.*, **57**, pp. 1333–1342.

38. Ocando, C., Tercjak, A., Martín, M. D., Ramos, J. A., Campo, M., and Mondragon, I. (2009). Morphology development in thermosetting mixtures through the variation on chemical functionalization degree of poly (styrene-b-butadiene) diblock copolymer modifiers. Thermomechanical properties, *Macromolecules,* **42**, pp. 6215–6224.

39. Wetzel, B., Haupert, F., and Zhang, M. Q. (2003). Epoxy nanocomposites with high mechanical and tribological performance, *Composites Sci. Tech.*, **63**, pp. 2055–2067.

40. Wang, Z. Z., Gu, P., Zhang, Z., Gu, L., and Xu, Y. Z. (2011). Mechanical and tribological behavior of epoxy/silica nanocomposites at the micro/nano scale, *Tribol. Lett.*, **42**, pp. 185–191.

41. Uddin, M., and Sun, C. (2010). Improved dispersion and mechanical properties of hybrid nanocomposites, *Composites Sci. Tech.*, **70**, pp. 223–230.

42. Rajabi, L., Bostanipour, M., Rostamabadi, A., and Derakhshan, A. (2012). Cure behavior, thermal stability and mechanical properties of epoxy/alumina nanocomposites, *JANN*, **2**, pp. 259–267.

43. Lim, S., Zeng, K., and He, C. (2010). Morphology, tensile and fracture characteristics of epoxy-alumina nanocomposites, *Mater. Sci. Eng. A*, **527**, pp. 5670–5676.

44. Johnsen, B., Frømyr, T., Thorvaldsen, T., and Olsen, T. (2013). Preparation and characterisation of epoxy/alumina polymer nanocomposites, *Compos. Interfaces*, **20**, pp. 721–740.

45. Zhang, H., Zhang, Z., Friedrich, K., and Eger, C. (2006). Property improvements of in situ epoxy nanocomposites with reduced interparticle distance at high nanosilica content, *Acta Mater.*, **54**, pp. 1833–1842.

46. Zhao, S., Schadler, L. S., Hillborg, H., and Auletta, T. (2008). Improvements and mechanisms of fracture and fatigue properties of well-dispersed alumina/epoxy nanocomposites, *Composites Sci. Tech.*, **68**, pp. 2976–2982.

47. Zhao, S., Schadler, L. S., Duncan, R., Hillborg, H., and Auletta, T. (2008). Mechanisms leading to improved mechanical performance in nanoscale alumina filled epoxy, *Composites Sci. Tech.*, **68**, pp. 2965–2975.

48. Omrani, A., Simon, L. C., and Rostami, A. A. (2009). The effects of alumina nanoparticle on the properties of an epoxy resin system, *Mater. Chem. Phys.*, **114**, pp. 145–150.

49. West, R. D., and Malhotra, V. M. (2006). Rupture of nanoparticle agglomerates and formulation of Al_2O_3-epoxy nanocomposites using ultrasonic cavitation approach: Effects on the structural and mechanical properties, *Polym. Eng. Sci.*, **46**, pp. 426–430.

50. Chatterjee, A., and Islam, M. S. (2008). Fabrication and characterization of TiO_2–epoxy nanocomposite, *Mater. Sci. Eng. A*, **487**, pp. 574–585.

51. Medina, R., Haupert, F., and Schlarb, A. K. (2008). Improvement of tensile properties and toughness of an epoxy resin by nanozirconium-dioxide reinforcement, *J. Mater. Sci.*, **43**, pp. 3245–3252.

52. Johnsen, B., Kinloch, A., Mohammed, R., Taylor, A., and Sprenger, S. (2007). Toughening mechanisms of nanoparticle-modified epoxy polymers, *Polymer*, **48**, pp. 530–541.

53. Ma, J., Mo, M.-S., Du, X.-S., Rosso, P., Friedrich, K., and Kuan, H.-C. (2008). Effect of inorganic nanoparticles on mechanical property, fracture toughness and toughening mechanism of two epoxy systems, *Polymer*, **49**, pp. 3510–3523.

54. Lam, C.-K., Lau, K.-T., Cheung, H.-Y., and Ling, H.-Y. (2005). Effect of ultrasound sonication in nanoclay clusters of nanoclay/epoxy composites, *Mater. Lett.*, **59**, pp. 1369–1372.

55. Mirmohseni, A., and Zavareh, S. (2010). Preparation and characterization of an epoxy nanocomposite toughened by a combination of thermoplastic, layered and particulate nano-fillers, *Mater. Des.*, **31**, pp. 2699–2706.

56. Mirmohseni, A., and Zavareh, S. (2010). Epoxy/acrylonitrile-butadiene-styrene copolymer/clay ternary nanocomposite as impact toughened epoxy, *J. Polym. Res.*, **17**, pp. 191–201.

57. Kaczmarek, M., Buchnev, O., and Nandhakumar, I. (2008). Ferroelectric nanoparticles in low refractive index liquid crystals for strong electro-optic response, *Appl. Phys. Lett.*, **92**, pp. 1033071–10330713.

58. Jeng, S.-C., Kuo, C.-W., Wang, H.-L., and Liao, C.-C. (2007). Nanoparticles-induced vertical alignment in liquid crystal cell, *Appl. Phys. Lett.*, **91**, 0611121–0611123.

59. Da Cruz, C., Sandre, O., and Cabuil, V. (2005). Phase behavior of nanoparticles in a thermotropic liquid crystal, *J. Phys. Chem. B*, **109**, pp. 14292–14299.

60. Tercjak, A., Gutierrez, J., Ocando, C., Peponi, L., and Mondragon, I. (2009). Thermoresponsive inorganic/organic hybrids based on conductive TiO_2 nanoparticles embedded in poly (styrene-b-ethylene oxide) block copolymer dispersed liquid crystals, *Acta Mater.*, **57**, pp. 4624–4631.

61. Gutierrez, J., Tercjak, A., and Mondragon, I. (2010). Conductive behavior of high TiO_2 nanoparticles content of inorganic/organic nanostructured composites, *J. Am. Chem. Soc.*, **132**, pp. 873–878.

62. Kim, D. H., Sun, Z. C., Russell, T. P., Knoll, W., and Gutmann, J. S. (2005). Organic–inorganic nanohybridization by block copolymer thin film, *Adv. Funct. Mater.*, **15**, pp. 1160–1164.

63. Kannaiyan, D., Kim, E., Won, N., Kim, K. W., Jang, Y. H., Cha, M. A., Ryu, D. Y., Kim, S., and Kim, D. H. (2010). On the synergistic coupling properties of composite CdS/TiO_2 nanoparticle arrays confined in nanopatterned hybrid thin films, *J. Mater. Chem.*, **20**, pp. 677–682.

64. Nisticò, R., Scalarone, D., and Magnacca, G. (2014). Preparation and physico-chemical characterization of large-mesopore silica thin films template by block copolymers for membrane technology, *Microporous Mesoporous Mater.*, **190**, pp. 208–214.

65. Hu, D., and Zheng, S. (2009). Reaction-induced microphase separation in epoxy resin containing polystyrene-block-poly(ethylene oxide) alternating multiblock copolymer, *Eur. Polym. J.*, **45**, pp. 3326–3338.

66. Tercjak, A., Serrano, E., and Mondragon, I. (2007). Multifunctional Thermally reversible nanostructured thermosetting materials based

on block copolymers dispersed liquid crystal, *Macromol. Rapid Commun.*, **28**, pp. 937–941.

67. Guo, Q., Wang, K., Chen, L., Zheng, S., and Halley, P. J. (2006). Phase behavior, crystallization, and nanostructures in thermoset blends of epoxy resin and amphiphilic star-shaped block copolymers, *J. Polym. Sci. B Polym. Phys.*, **44**, pp. 975–985.

68. Larrañaga, M., Gabilondo, N., Kortaberria, G., Serrano, E., Remiro, P., Riccardi, C. C., and Mondragon, I. (2005). Micro- or nanoseparated phases in thermoset blends of an epoxy resin and PEO–PPO–PEO triblock copolymer, *Polymer*, **46**, pp. 7082–7093.

69. Gutierrez, J., Tercjak, A., Garcia, I., Peponi. L., and Mondragon, I. (2008). Hybryd titanium dioxide/PS-b-PEO block copolymer nanocomposites based on sol-gel synthesis, *Nanotechnology*, **19**, pp. 155607/1–155607/8.

70. Tercjak, A., Serrano, E., Martin, M. D., Marieta, C., and Mondragon, I. (2006). Influence of PS-*b*-PEO diblock copolymers on the compatibility of syndiotactic polystyrene modified epoxy blends, *J. Appl. Polym. Sci.*, **102**, pp. 479–488.

71. González-Domínguez, J. M., Ansón-Casaos, A., Díez-Pascual, A. M., Ashrafi, B., Naffakh, M., Backman, D., Stadler, H., Johnston, A., Marian Gomez, Martínez, M. T. (2011). Solvent-free preparation of high-toughness epoxy-SWNT composite materials, *ACS Appl. Mater. Interfaces*, **3**, pp. 1441–1450.

Chapter 9

Reversible Photoinduced Birefringence in Epoxy Polymers, Block Copolymers and Nanostructured Thermosetting Systems Containing Azobenzene Groups

Raquel Fernández

Materials + Technologies Group (GMT),
Department of Chemical and Environmental Engineering,
Polytechnic School, University of the Basque Country (UPV/EHU),
Plaza Europa 1, 20018 Donostia-San Sebastián, Spain

raquel.fernandez@ehu.eus

Azobenzene-containing polymers are fascinating materials with optical properties that can find many applications in diverse fields. In this chapter, the development of different materials modified with azobenzene chromophores is reported. On the one hand, various epoxy-based polymers containing azobenzene groups, both thermoplastics and thermosets, were synthesized and characterized. On the other hand, different block copolymers were modified with azobenzene moieties and, finally, azobenzene-containing nanostructured thermosetting epoxy systems using functionalized block copolymers to promote the compatibilization of one block with the thermosetting system were also investigated.

Block Copolymer Nanocomposites
Edited by Galder Kortaberria and Agnieszka Tercjak

ISBN 978-981-4669-54-2 (Hardcover), 978-981-4669-55-9 (eBook)
www.panstanford.com

9.1 Introduction

Azobenzene chromophores have been widely used as dyes and pH indicators for years, but in the past decades azobenzene-containing polymers or azopolymers have attracted considerable attention in material science and technology for their fascinating stimuli-responsive properties. Key to some of their most interesting applications is the reversible photoinduced isomerization between *trans* and *cis* isomeric states that azobenzene chromophores undergo upon exposure to light. The formation of *cis* isomers leads to remarkable changes of size, dipole moment, and geometry of azobenzene molecules. In addition, azobenzene compounds have the ability to photoorientate through optical induced *trans-cis-trans* isomerization cycles, which gives rise to birefringence and dichroism in azopolymers [1–3]. Figure 9.1 shows a scheme of the different light-induced motions in an azobenzene-containing polymer.

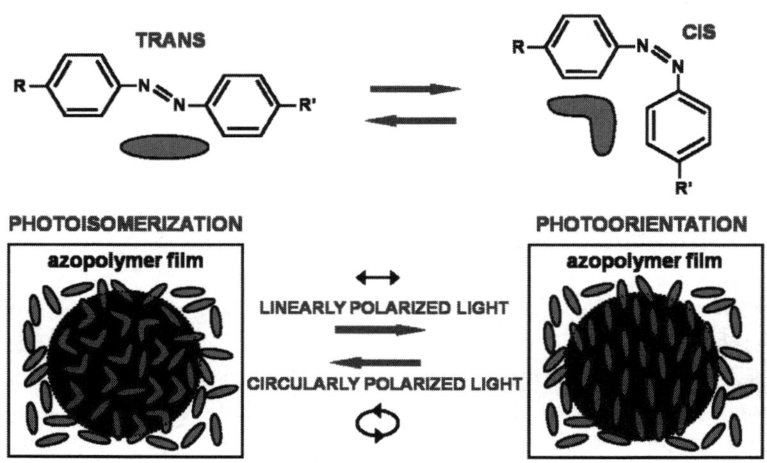

Figure 9.1 Schematic illustration of the azobenzene molecules photoinduced isomerization and orientation processes.

The main applications of azopolymers include reversible optical storage, which relies on the possibility of altering their refractive index by means of light. A single domain could encode one bit by either being isotropic or birefringent, a difference that is easily probed optically [4].

Another intriguing application is the photoinduced bending of azobenzene-containing polymer films. Yu et al. [5] showed that a film of a liquid crystal network containing an azobenzene chromophore could be repeatedly and precisely bent along any chosen direction by using linearly polarized light (Fig. 9.2). This striking photomechanical effect results from a photoselective volume contraction and may be useful in the development of high-speed actuators for microscale or nanoscale applications, for example in microrobots in medicine or optical microtweezers.

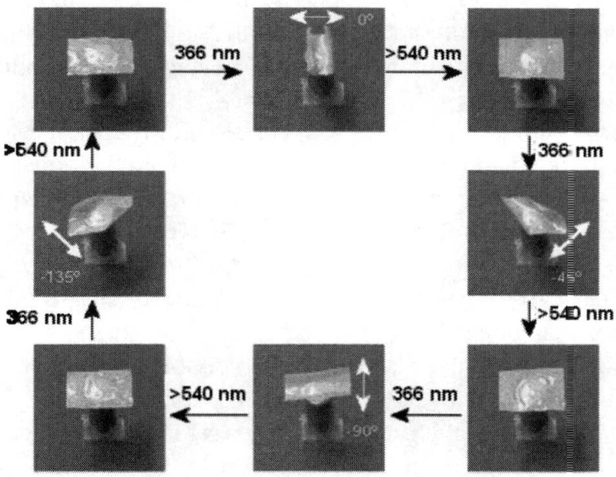

Figure 9.2 Photographic frames of the film bending in different directions in response to irradiation by linearly polarized light of different angles of polarization at 366 nm, and being flattened again by visible light longer than 540 nm. The flat film (4.5 mm × 3 mm × 7 μm) lay on a copper stick fixed to a copper plate; a stage under the plate was set at 85°C to control the temperature of the film, which was covered by a piece of blue paper. The bending time for the four different bending directions was within 10 s, when the light intensity of 366 nm linearly polarized light was 3.5 mW·cm^{-2} after exposure to visible light longer than 540 nm (547 nm, 24.2 mW·cm^{-2}; 577 nm, 26.8 mW·cm^{-2}), the bent film reverted to the flat state in about 10 s. (Reproduced with permission of [5]. Copyright 2003, Nature Publishing Group).

In drug delivery applications using azobenzene compounds, light can be used to control the loading-unloading of small molecule drugs. Moreover, the properties of surfaces modified with azobenzene groups can be also optically manipulated. Ichimura et al. [6] reported the photoinduced macroscopic motion of liquids on a flat solid surface covered by a photoisomerizable monolayer containing azobenzene units. When a liquid droplet of several millimeters in diameter was placed on a substrate surface modified with a calix[4]resorcinarene derivative having photochromic azobenzene units, asymmetrical photoirradiation caused a gradient in the surface free energy due to the photoisomerization of the azobenzenes groups, leading to the directional motion of the droplet. The direction and velocity of the droplet motion can be tuned by varying the direction and steepness of the gradient in light intensity.

Furthermore, Lim et al. [7, 8] developed fluorinated azobenzene-modified nanoporous substrates that were photoreversibly converted between superhydrophobicity and superhydrophilicity as a result of photoirradiation, and Goulet-Hanssens et al. [9] developed a photoreversible cell culture substrate containing azobenzene groups, whose capacity to support cell growth could be altered using light.

In general, materials that are suitable for such optical applications have to display a high thermal and mechanical stability as well as photostability, in order to retain the photoinduced orientation of the azobenzene groups. Polymers play an important role in the development of materials for photonics. They are relatively inexpensive materials that can be functionalized to achieve the required properties. Epoxy resins represent a family of thermosetting polymers known from many decades ago. The wide diffusion they have found on the market is a consequence of their extreme versatility and their excellent attributes, such as abrasion resistance, thermal stability, chemical resistance and mechanical properties. Those features make them primary candidates for a wide range of applications among which epoxy thermosets stand out as high performance matrices in advanced composites, coatings, adhesives, and laminates. Thus, researchers have focused mainly on the study of their mechanical and thermal properties [10–12]. However, epoxy networks are also transparent and, therefore, suitable for optical applications, even though these have not been explored so much as the mechanical ones.

On the other hand, recently, the development of block copolymers (BCP) containing azobenzene units has gained importance [13–29]. It is well known that BCP can form microphase-separated nanostructures with cylindrical, lamellar, spherical or bicontinuous morphologies [30–36]. The confinement of photoresponsive units in nanosized block copolymer domains gives these materials unique properties, because of the possibility of altering their self-assembly behavior, while allowing the incorporation of molecules with optical features. Additionally, the reorientation of nanosized domains themselves, induced by the irradiation of this type of polymers, has also been reported. In particular, Ikeda and coworkers [37, 38] demonstrated a molecular cooperative motion between azobenzene moieties and other photoinert groups in azo-containing BCP with specifically designed structures. The azochromophores became aligned, triggered by the irradiation with a polarized laser beam at 488 nm, and the photoinert groups were oriented together with the azobenzenes by supramolecular cooperative motions, although they did not absorb the actinic light (Fig. 9.3).

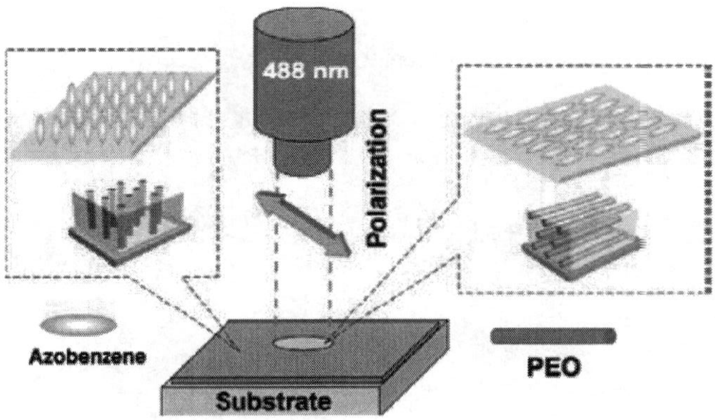

Figure 9.3 Schematic illustration of azobenzene units orientation and poly(ethylene oxide) (PEO) microphase-separated structures in the irradiated and unirradiated area of the block copolymer films. Reproduced with permission of [38]. Copyright 2006, American Chemical Society.

In addition, block copolymers, owing to their capability to form nanoscale structures, are widely used as templates for generating nanostructured thermosets with long-range order

[39–42]. This type of thermosets is a relatively new class of advanced materials which are able to improve the mechanical properties of conventional thermosetting polymers retaining their optical transparency [43–45]. On this basis and taking into account the need for new materials with specific functionalities appropriate for emerging technological applications, preparation of nanostructured thermosets based on epoxy resins for optical applications can be of interest. One way to achieve this goal is through reaction-induced microphase separation (RIMS) of one of the blocks during curing, both being initially miscible with the epoxy resin. In the case of BCP with both blocks immiscible with the epoxy system after curing, one feasible approach to promote the compatibilization of one of the blocks with the epoxy matrix can be its chemical modification.

In this chapter, different strategies for the preparation and characterization of azo-containing epoxy thermoplastics and thermosets, as well as different types of nanostructured polymers are discussed. Figure 9.4 shows a schematic of the different kinds of materials modified with azobenzene groups investigated.

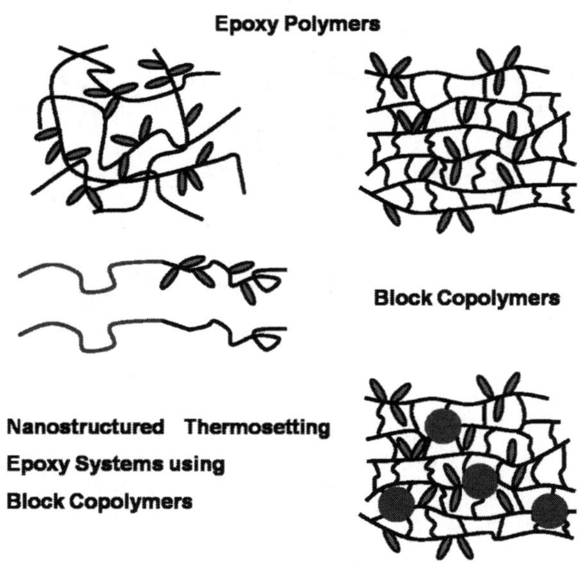

Figure 9.4 Schematic illustration of the different types of azopolymers studied in this chapter.

9.2 Azobenzene-Containing Epoxy Systems

In the case of thermoplastic epoxy systems, two strategies for incorporation of an azochromophore into polymeric matrices were investigated: (1) covalent attachment of the dye to the polymer as a comonomer; and (2) solution mixing of the dye as a guest in the polymer host. Results of induced birefringence measurements on two series of epoxy-based copolymers and two guest–host systems are presented. In addition, the benefits of covalent azodye-polymer attachment, and the influence of different main-chain segments on the molecular reorientation are discussed. The time to achieve birefringence, as well as the time to randomize the orientation, depends on a series of factors, which were also analyzed. Light intensity is one of these factors. The more photons a sample absorb, the faster the writing or erasing process [46]. The influence of film thickness is another factor to take into account, and the effects of varying azobenzene groups concentration were also studied.

In the case of thermosetting epoxy systems, the question as far as applications arises if a stable photoorientation can be induced in the polymeric form of the material and if the crosslinks have any influence on the photokinetics of the light-induced process. In addition, the azobenzene group can be attached in different polymer backbones either in the main-chain or as a side chain, allowing the mechanical and processing properties of conventional polymers to be combined with the optical properties of azobenzene groups. In an effort to establish materials with the desired optical properties, the photoorientation induced via irradiation with linearly polarized light was investigated in a series of crosslinked epoxy-based azopolymers, each with distinct differences in the molecular structure of the unit building blocks. All the epoxy-based networks analyzed have the chromophore placed in a branch, also called crosslinked side chain [47]. It is discussed to what extent the polymer structure can influence chromophores phoroorientation, and it is showed that some polymeric epoxy systems provide good results in optical storage and stability.

Figure 9.5 shows the chemical structures of the azochromophore (AZO), the epoxy resin and the amines used to prepare the different epoxy systems. The general strategy used to obtain the different azobenzene-containing epoxy systems was as follows: (1) First, an azoprepolymer (PAZO) was synthesized by reaction between an azoamine selected as photosensitive molecule and an epoxy resin, diglycidyl ether of bisphenol A (DGEBA), in an amine/epoxy stoichiometric ratio equal to 0.5 to generate reaction products with epoxy groups in the extreme of chains. (2) Afterwards, the resulting PAZO was blended with DGEBA and two different monoamines, BA and MT, in an amine/epoxy stoichiometric ratio equal to 1 to prepare azo-modified thermoplastics with different azobenzene groups concentration. Along with the monoamines, three different diamines, MXDA, D230 and D400, were used as hardeners with the aim of developing crosslinked azopolymers, following the same experimental procedure as for the thermoplastic systems. A brief account of the compositions and thermal properties of the developed materials modified with azobenzene moieties can be found in Table 9.1.

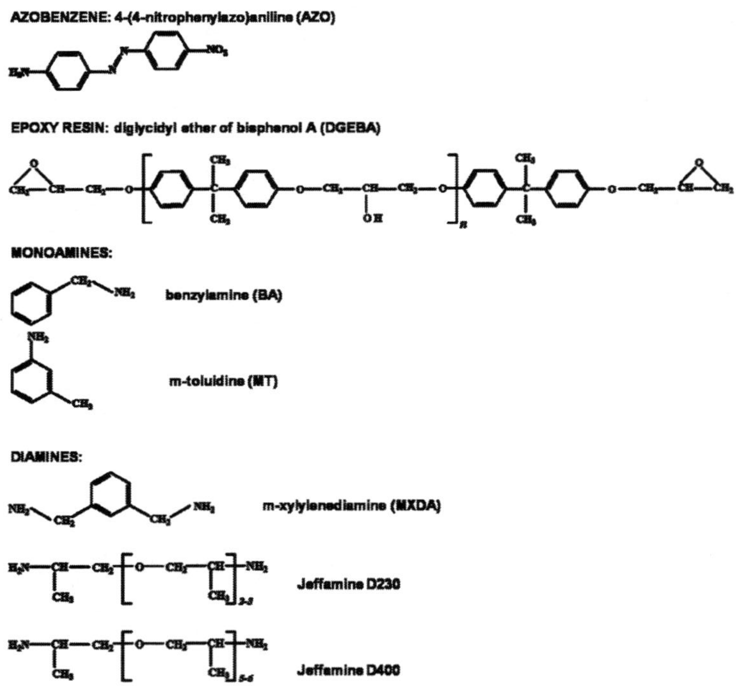

Figure 9.5 Chemical structures of the reactants used.

Table 9.1 Compositions and thermal properties of epoxy systems

Monoamines	**Thermoplastics**	**AZO (wt%)**	$T_{reaction}$ **(°C)**	T_g **(°C)**
	PAZO	26	180	28
BA	PAZO10/DGEBA/BA	2.6	60	58
	PAZO20/DGEBA/BA	5.2	60	58
	PAZO50/DGEBA/BA	13	60	59
	DGEBA/BA (AZO)*	5.2	60	58
MT	PAZO10/DGEBA/MT	2.6	100	79
	PAZO20/DGEBA/MT	5.2	100	85
	PAZO50/DGEBA/MT	13	100	76
	DGEBA/MT (AZO)*	5.2	100	85
Diamines	**Thermosets**	**AZO (wt%)**	$T_{reaction}$ **(°C)**	T_g **(°C)**
MXDA	PAZO10/DGEBA/MXDA	2.6	100	106
	PAZO20/DGEBA/MXDA	5.2	100	106
	PAZO50/DGEBA/MXDA	13	100	101
	PAZO70/DGEBA/MXDA	18	100	103
	PAZO/MXDA	24	100	103
D230-D400	PAZO50/DGEBA/D230	13	140	85
	PAZO50/DGEBA/D230-D400 (85-15)	13	140	77
	PAZO50/DGEBA/D230-D400 (70-30)	13	140	73
	PAZO50/DGEBA/D230-D400 (55-45)	13	140	68
	PAZO50/DGEBA/D230-D400 (40–60)	13	140	62

*Guest–host systems containing 5.2 wt% AZO dispersed.

Birefringence was induced in films of the developed azo-containing materials under ambient conditions using a linearly polarized argon laser operating at 488 nm (pump beam) with a polarization angle of 45° with respect to the polarization direction of a low power He-Ne laser operating at 632.8 nm (probe beam) (Fig. 9.6). The power of the pump beam used in the experiments was varied between 6 and 20 mW on a spot of 0.4 mm^2 and the change in the transmission of the probe beam, which passed through the sample between two crossed polarizers, was measured with a photodiode. The induced birefringence (Δn) was determined by measuring the reading beam transmission ($T = I/I_0$) according to:

$$\Delta n = (\lambda/\pi d) \sin^{-1} (I/I_0)^{1/2},$$

where λ is the wavelength of the reading beam, d is the film thickness, I is the intensity of the reading beam after the second polarizer and I_0 is the transmitted intensity of the reading beam between parallel polarizers in absence of birefringence.

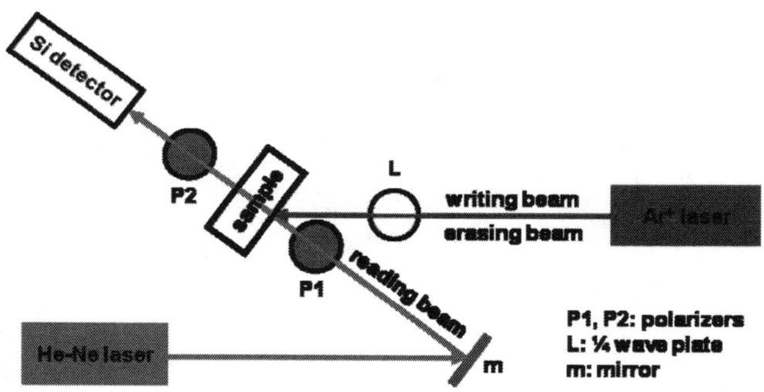

Figure 9.6 Experimental setup built for induced birefringence measurements.

9.2.1 Thermoplastic Epoxy Systems

The glass transition temperature (T_g) range of the designed epoxy thermoplastics is from 58 to 85°C (Table 9.1), as found by differential scanning calorimetry, which should favor a good orientational temporal stability of the photoactive chromophores, of great interest for applications in optoelectronics, when more stable photoinduced birefringence is required at ambient and slightly elevated temperatures. Representative writing-erasing curves of optically induced and subsequently eliminated birefringence for PAZO50/DGEBA/BA and PAZO50/DGEBA/MT are shown in Fig. 9.7.

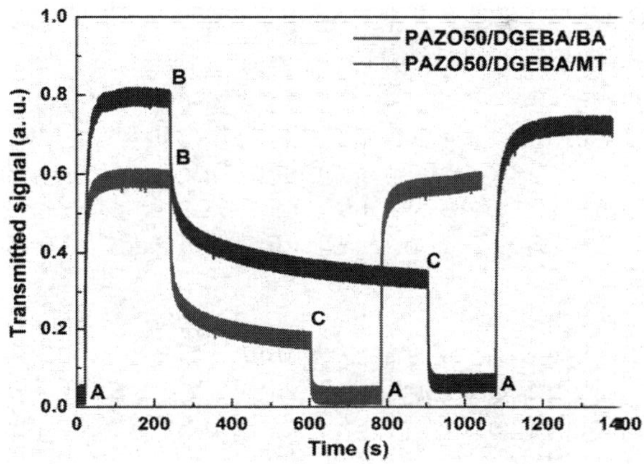

Figure 9.7 Typical writing and erasing curves for: (a) PAZO50/DGEBA/BA (970 nm thickness), and (b) PAZO50/DGEBA/MT (850 nm thickness). The pump beam (6 mW) is *on* at point A and *off* at B. Circularly polarized light was introduced at point C.

Azobenzene molecules can be oriented by irradiation with linearly polarized light of appropriate wavelength. Upon absorbing this light, azobenzene groups undergo a series of *trans-cis-trans* isomerization cycles. Those chromophores with dipole moment perpendicular to the polarization direction of the light electric field do not absorb light to undergo further isomerizations. At the end of several cycles, a net population of azobenzene molecules is oriented in this perpendicular direction, giving rise to a birefringence in the film structure. The birefringence formation can be inferred by the change in transmittance of a probe beam that passes through the sample between crossed polarizers (P1 and P2 in Fig. 9.6). To investigate the kinetics of chromophore orientation in the developed epoxy films, the creation of birefringence was followed over time through three different irradiation regimes (Fig. 9.7): when the linearly polarized orienting beam was turned *on* (A), after it was turned *off* (B) and, finally, when irradiated with circularly polarized light to randomize the induced orientation (C). The transmitted signal was normalized between 0 and 1. At the beginning of the experiment there was no transmission of the probe beam, since the chromophores were randomly distributed. At point A, the pump beam was turned *on* and the probe beam was transmitted through the polarizer-sample-polarizer system due to the birefringence induced in the film. When the pump beam was turned *off* at point B, molecular relaxation took place, but a considerable number of azobenzene molecules remained oriented, thus leading to a stable birefringence pattern. Finally, at point C, in order to remove the remaining birefringence, circularly polarized light was introduced, which completely randomized the induced chromophores orientation.

All the polymeric films investigated had comparable thickness values, on the order of 1000 (±100) nm. It was found that birefringence values increase with PAZO concentration. Since the birefringence is due to the anisotropy and photoinduced orientation of the azo units, a higher value of birefringence should appear when the azobenzene content increases. The birefringence increases with PAZO weight fraction independently of the environment in which the azobenzene moiety was located, the bonding between the dye and the matrix, and the power of the writing beam. This means that at a power of 6 mW saturation is reached for all the systems analyzed. For guest–host systems, the reached photoinduced birefringence was comparable with the values observed for the covalently bonded polymers, which confirmed that the azobenzene moieties linked to the epoxy matrix had the sufficient mobility to be aligned.

The writing process is associated with the *trans-cis-trans* isomerization rates, quantum yields, and the local mobility of the azobenzene group. Since the same chromophore is used in all the polymers analyzed, and the energy absorbed by the polymer films at the irradiation wavelength is the same, it is expected that the writing process should be affected by PAZO concentration, laser excitation intensity, and T_g of the polymer sample. Therefore, all these factors dependence on the growth of the birefringence signal was investigated. The averaged time-derivative of the optically induced birefringence was analyzed for achieving a quantitative representation of the photoorientation kinetics. The analysis was performed for the covalently bonded polymers and the guest–host systems, for the time to achieve 80% of the saturation value. To compare the rate of writing, it is essential that films of comparable thickness be used. The derivative at each point was numerically calculated by averaging the slopes of two adjacent data points; the average of all values up to 80% of saturation was calculated for each sample. Slower writing times are reflected by lower values of the curve slope. These results are summarized in Table 9.2.

Table 9.2 Matrix effects on the writing dynamics

Samples	Thickness (nm)	Writing slope (10^{-5})	
		6 mW	20 mW
DGEBA/BA (AZO)*	1000 1100	1.48	2.96
PAZO20/DGEBA/BA	950 1000	0.267	0.825
PAZO50/DGEBA/BA	970 970	2.25	5.71
DGEBA/BA (AZO)*	1030 1050	0.234	0.616
PAZO20/DGEBA/MT	980 950	0.055	0.097
PAZO50/DGEBA/MT	850 900	1.28	2.95

*Guest–host system containing 5.16 wt% AZO dispersed.

The increase of PAZO concentration provoked an increment in the writing kinetics. It is clear that the kinetics of photoorientation depends on the molecular arrangement of the environment. This was expected since the reorientation of the azochromophore around its short molecular axis induced by photoinduced isomerization cycles is dependent on the local free volume distribution. The T_g can be a guide to predict the relative size of the free volumes. The rotational motion of *trans* isomers is considered to be restricted into a small free volume. During irradiation, slope values in copolymers and guest–host systems with BA were much higher compared with those in MT-based systems. These results were an indication that the *trans* isomers rotate more easily in polymers with lower T_g (Table 9.1). The magnitude of the writing slope in BA-based copolymers is five times larger than the MT-based copolymers for samples with 20 wt% PAZO. The copolymers compared have a difference in T_g values of 27°C ($T_{g\,PAZO20/DGEBA/BA}$ = 58°C, $T_{g\,PAZO20/DGEBA/MT}$ = 85°C). However, for 50 wt% PAZO content, the difference in T_g of the copolymers is smaller ($T_{g\,PAZO50/DGEBA/BA}$ = 59°C, $T_{g\,PAZO50/DGEBA/MT}$ = 76°C), i.e., 17°C. The writing slope is only two times greater in BA-based copolymer than in MT-based copolymer, consistent with the smaller free volume differences indicated by measured T_g values. Also of interest is the fact that guest–host systems exhibit faster writing response (five times) compared with the copolymers containing the same AZO concentration (20 wt% PAZO).

Another interesting result that can be extracted from Table 9.2 is that the growth rate depends on the laser power of the writing beam. The efficiency of alignment is seen to be directly proportional to the number of photons involved in the writing process, confirming the statistical nature of the process. On turning *off* the laser beam (point B in Fig. 9.7), the birefringence of the copolymers decreases nearly 50% for MT-based system and 60% for BA-based system, independent of film thickness, PAZO concentration and intensity of the writing beam. This stable birefringence pattern corresponds to the 'store' step. In the case of the guest–host systems, the transmission signal decreased to 65% for DGEBA/MT (AZO) and 75% for DGEBA/BA (AZO) of the maximum. This low residual signal is characteristic of guest–host films because the azobenzene moiety has a high mobility since the chromophores are not functionalized into the polymer backbone. Part of the decrease is associated with the thermal reorientation of some azobenzene groups. This thermal reorientation tends to restore the thermodynamically favored disordered state and is probably made possible by some local heat associated with the laser light which is dissipated through the film when the laser is turned *off*. Depending on the mobility of azobenzene chains in matrices, different relaxations were observed. Then, according to the different T_g of the polymers analyzed, the remaining birefringence in MT-based copolymers is greater than that in BA-based copolymers. The lower the T_g value, the higher the molecular mobility and lower the remaining birefringence. No power dependence on the decay of the birefringence signal was found. This was expected as the writing laser is already *off*.

The time to remove the remaining birefringence seems to be independent of the nature of the copolymer, PAZO concentration, power of the beam, and film thickness. One could try to erase a written sample by heating. The written information is stable below the T_g of the copolymer. For amorphous polymers it is enough to reach T_g and restore complete disorder in the film [48]. The overall thermal stability of the induced reorientation is very important. It is clear that heating of a written sample destroys the induced orientation, and as one gets closer to the T_g, all reorientation is lost. For this reason, it is important to synthesize polymers with high T_g to be used for instance in reversible optical storage applications.

9.2.2 Thermosetting Epoxy Systems

The studies involving the two series of crosslinked epoxy-based azopolymers synthesized pointed to a clear relation between the optical storage properties and the structure of the network to which the azobenzene group is attached. Since the writing and relaxation times for azopolymers are usually in the range of a few seconds and are basically inherent to the mechanism of orientation, the following discussion will be focused on the maximum and the residual fraction of the induced birefringence. These two parameters are the most important for practical applications. Table 9.3 summarizes some optical storage features of films from the two series synthesized.

To understand the mobility of the chromophore in the polymer network and clarify the effect of crosslinking on the orientation behavior, in situ birefringence measurements were performed during writing cycles (linearly polarized pump *on*), relaxation cycles (pump *off*), and photoinduced erasing processes (circularly polarized pump *on*).

It is important to determine the relationship between the maximum level of anisotropy that can be induced and the concentration of the chromophores in the polymeric material. As shown in Table 9.3, there was an increase in the birefringence level as the azobenzene groups concentration increased. Since Δn is due to the anisotropy and photoinduced orientation of the azo units, a larger number of photoactive chromophores in the polymer chain generated a higher birefringence. The level of maximum birefringence that can be achieved was about 3×10^{-2} (for PAZO/MXDA, 24 wt% AZO), which is similar to previously published systems, such as the studies of the photoorientation in melamine networks reported by Stracke et al. [49]. There, it was shown that the maximum photoinduced birefringence obtained was 3.4×10^{-2} for materials containing more than 50 wt% chromophore. In a series of papers, Rochon et al. [50–52] demonstrated that for amorphous polymers containing electron-donor substituents on the azobenzene group, the maximum variation in refractive index obtained was 8×10^{-2}. Most studies were performed on a Disperse Red 1-containing polyacrylate with 85 wt% chromophore. Then, based on the reported values of birefringence levels for amorphous polymers, the birefringence level achieved in the present study is high.

Table 9.3 Optical storage characteristics of films from the series of azonetworks

Samples	M_c (g/mol)	Δn (10^{-2})	$t_{80\%}^{WRITE}$ (s)	rem (%)
PAZO	—	2.80	10.6	50
PAZO10/DGEBA/MXDA	472	0.63	37.5	66
PAZO20/DGEBA/MXDA	537	1.13	27.5	66
PAZO50/DGEBA/MXDA	798	2.07	9.80	66
PAZO70/DGEBA/MXDA	1070	2.46	7.00	64
PAZO/MXDA	1680	2.94	7.80	61
PAZO50/DGEBA/D230	875	2.21	7.20	58
PAZO50/DGEBA/D230-D400 (85-15)	897	2.23	8.10	57
PAZO50/DGEBA/D230-D400 (70-30)	918	2.34	7.30	56
PAZO50/DGEBA/D230-D400 (55-45)	938	1.92	4.60	56
PAZO50/DGEBA/D230-D400 (40–60)	958	2.00	4.30	55

Note: Films were irradiated at 6 mW.

The level of induced birefringence achieved for PAZO (26 wt% AZO) was in the order of 2.8×10^{-2}, slightly lower than the ones obtained for networks with comparable chromophore content, indicating that there is no influence of crosslinking on the maximum value attainable. It should be taken into account that the saturated level of Δn is the result of a combination of two processes: photoselection, which increases birefringence, and thermal randomization, which decreases its level. These two processes reach an equilibrium state while illumination is *on*, and the preponderance of one or the other determines the saturated level. The possibility for the azobenzene molecules to move around is great for PAZO which has a T_g of around 28°C. Consequently, its birefringence level was lower when comparing with networks.

Also of importance is to know how fast the birefringence can be induced in the film by the writing laser. The time interval necessary to achieve 80% of saturated birefringence, $t_{80\%}^{\mathrm{WRITE}}$, as a function of chromophore concentration is shown in Table 9.3. Similar writing times are observed for networks when comparing with PAZO. For an interpretation of this result we have to conclude that the photoorientation of the azobenzene groups is not influenced by the crosslinking itself but that it depends rather on the local free volume distribution around the chromophores. As the azobenzene molecule was not forming part of the bridges between the polymer chains, it had a high mobility in the free volume of the azonetwork. Similar results were obtained for crosslinkable polymethacrylates containing azobenzene and acrylic groups on separate side chains to maintain adequate azobenzene mobility in the networks [53]. In addition the order of magnitude of the time to reach the maximum birefringence is in agreement with the ones reported for cast films containing *push-pull* azodyes [54]. A decrease in the time necessary to reach the maximum birefringence was observed with the increase in AZO content. This behavior can be understood in terms of a thermal effect. The increase in the sample absorption induced a sample heating as more chromophore was incorporated into the polymer. This sample heating allowed a higher molecular mobility, which results in a faster writing process.

When the writing laser beam was turned *off*, the birefringence decayed to a stable level within 30 s for all the samples analyzed. Part of this decrease can be associated with thermal reorientation of some azobenzene groups [48]. Regarding the birefringence conserved after relaxation, it is interesting to note a slight decrease in the remaining birefringence (rem) upon increasing the azobenzene concentration in the networks. This tendency is probably related with the crosslinking density of the polymer matrix. Higher crosslinking restrains the mobility of oriented azobenzene groups for randomization. PAZO exhibited the minimum remaining birefringence (around 50%). Thus, crosslinking plays an important role in the stability of the photoinduced orientation.

Previous studies on photoorientation processes in amorphous polymers addressed the role of T_g and polymer structural effects, including the main chain rigidity, the nature of the connection of the chromophore to a rigid or semirigid or a flexible main chain, the free volume, the free volume distribution, or a combination of these [55]. The polymer structure and T_g are not the only important parameters for polymers, but the molecular weight and its distribution are also important. In fact, the glass relaxation is characterized by T_g, which is affected by several factors including the molecular weight, swelling, and crosslinking. In this investigation, all networks were synthesized using the same azoprepolymer, PAZO, in which the chromophore is tethered at both ends to the polymer chain. The connection of PAZO to network precursors is possible thanks to the presence of epoxy groups in the extreme of chains. The difference between the azonetworks is the structure of the diamine employed. The presence of aromatic rings in the backbone of MXDA-cured samples leads to a less flexible chain structure than for D230/D400-based systems. Therefore, in order to know to what extent the polymer structure can influence chromophore photoorientation in the series of epoxy-based networks containing a constant concentration of chromophore, the photoinduced birefringence for both types of thermoset systems was calculated as a function of the molecular weight between crosslinks, M_c, of azo-networks. The theoretical molecular weight between crosslinks was calculated assuming full conversion by the following equation previously employed by Lesser and Crawford [56]:

$$M_c = \frac{2\left(M_e + \sum_{f_a=2}^{\infty} \frac{M_{fa}}{f_a} \cdot \varphi_{fa}\right)}{\sum_{f_a=3}^{\infty} \varphi_{fa}},$$

where M_e is the epoxy equivalent weight of the resin, M_{fa} is the molecular weight of the f_a-th functional amine, φ_{fa} is the mol fraction of amine hydrogens provided by the f_a-th functional amine and f_a is the amine functionality.

Thermosets with the same chromophore concentration but different backbones exhibited similar levels of induced birefringence under the same irradiation conditions (Table 9.3). This is an indication that for the networks analyzed here, the maximum induced birefringence is only function of the overall chromophore concentration. These experimental results are consistent with the ones reported by Takase et al. [53] when comparing maximum birefringence of azobenzene-containing polymethacrylate copolymers before and after crosslinking. They found that the maximum value achieved during the linearly polarized laser irradiation is only a function on the azobenzene content. The writing process is associated with the trans-cis-trans isomerization rates, quantum yields, and the local mobility of the azobenzene groups. Since the same chromophore is used in all the polymers analyzed, and the energy absorbed by the polymer films at the irradiation wavelength is the same, it is expected that the writing process should be affected by the molecular structure and T_g of the polymer sample. During irradiation, $t_{80\,\%}{}^{\mathrm{WRITE}}$ values decreased as M_c increased (or T_g decreased). These results are an indication that azobenzene molecules rotate more easily in polymers with lower T_g. Concerning the polymer structural effects on the dynamics of the writing process, the network having 5 wt% AZO is 15 times slower when comparing with guest–host systems analyzed in the previous section. That is to say, $t_{80\%}{}^{\mathrm{WRITE}}$ values increase significantly because of the linkage effect. This behavior is attributed to the restriction of the dynamics of the azobenzene moiety by bonding covalently to a polymer chain and/or the presence of crosslinks between chains. Moreover, it is worth to note that the movement of azobenzene groups was facilitated in polymers with higher molecular weight between crosslinks, which was reflected in shorter relaxation times and lower remaining birefringence values (Table 9.3). This can be due to the smaller T_g values (Table 9.1) of such polymer systems. Similar results were obtained by Sekkat et al. [55] when comparing polyurethane polymers, confirming the influence of the polymer structure on photoorientation.

9.3 Azobenzene-Containing Block Copolymers

Combining the excellent properties of azopolymers with microphase separation, azobenzene-containing block copolymers might find diverse uses in advanced technology as well as newly promising nanotechnology. As discussed above, the incorporation of azobenzene molecules in a polymer matrix can be done in different ways [57]. In the simplest form of amorphous optically active polymer, the azobenzene guest molecules are dispersed in a host polymer matrix. Such guest–host systems are both flexible and cost effective as they only require mixing of the constituents to produce the desired material [58]. However, the effectiveness of conventional guest–host systems is limited by crystallization and aggregation of chromophores, which affects the photoorientation process. Moreover, in these simple systems the temporal and thermal stabilities of the induced birefringence are typically poor, as shown before, leading to a decrease in the overall response of the material. In contrast to guest–host systems, in azo-functionalized block copolymers the chromophores are covalently linked to the polymer backbone, which can prevent the aggregation of azobenzene units [59]. Nevertheless, from a practical point of view, azo-functionalized block copolymers are significantly less attractive than guest–host systems, owing to greater complexity of sample preparation and higher cost of the process.

Therefore, the development of well-defined nanostructures involves the synthesis of BCP with controlled macromolecular architecture, molecular weight distribution, and composition. Several polymerization methods, such as anionic, cationic, free radical and metal-catalyzed polymerizations, have been explored to build azobenzene-containing BCP that meet these requirements [4]. Alternatively, we present here the functionalization via epoxidation of commercial BCP, based on poly(styrene-b-butadiene-b-styrene) (SBS), modified with azobenzene moieties by one-step facile reaction between the epoxy and azoamine groups. This is a novel and simple way of preparing azo-functionalized block copolymers as well as very versatile considering that the epoxy group is receptive to a wide range of reagents. Therefore, this functional group can react with different kinds of chromophores with electron-donor groups containing hydrogen, such as amines, amides, acids, anhydrides, and phenols, giving rise to a large variety of azobenzene-containing BCP, taking into account also the great diversity of commercial available SBS-type copolymers. In addition, we investigate the influence of covalent attachment of the azobenzenes by studying the morphology and photoaddressing behavior of the resulting azobenzene-containing BCP, with respect to solution mixing of the azochromophores without chemical bonding as guest–host systems.

On the other hand, in order to combine the advantages of covalently linked systems and guest–host systems, concepts widely exploited in supramolecular chemistry become useful. In fact, non-covalent interactions offer a new way to link the active molecules to the polymer backbone. Thus, intermolecular interactions between the polymeric matrix and the azobenzene molecules can prevent the aggregation of the chromophores, enhancing the photoorientation of the active groups and the remaining induced birefringence [60]. Taking this into account, PAZO was selectively introduced into one self-assembled phase of another block copolymer, poly(styrene-b-4-vinylpyridine) (PS4VP) via intermolecular interactions. Polystyrene is a non-polar polymer, whereas poly(4-vinylpyridine) is a polar polymer containing a pyridine group that can form hydrogen bonds with the azopolymer. Consequently, novel optically active materials were developed combining the advantages of covalently functionalized polymers and the easy-to-process alternative provided by guest–host systems. Furthermore, the confinement of the azobenzene groups in one of self-assembled block copolymer domains could have effects on the photoorientation process due to the restriction imposed by a confining geometry on the motion and rearrangement of the azobenzene moieties [61]. Therefore, the influence of covalent bonding and non-covalent intermolecular interactions as well as the effect of chromophores confinement on birefringence were studied.

9.3.1 Azobenzene-Containing Block Copolymers Based on Poly(Styrene-b-Butadiene-b-Styrene)

Different epoxidized SBS triblock copolymers, SBSepX (X being the degree of epoxidation (mol%) with respect to PB double bonds obtained by high resolution ^1H NMR spectroscopy), were obtained by epoxidation of PB-blocks. This reaction was carried out using hydrogen peroxide in the presence of an in situ prepared catalyst system in a water/dichloroethane biphasic mixture, following a procedure described elsewhere [62]. Then, the resulting epoxidized copolymers were reacted with AZO in stoichiometric ratios, between epoxy groups and active hydrogens of the azoamine, at 110°C during 24 h under vacuum conditions. Figure 9.8 shows a schematic illustration of the synthesis procedure followed. In addition, guest–host systems of SBS containing AZO on the order of 15 (± 2) wt% were also prepared.

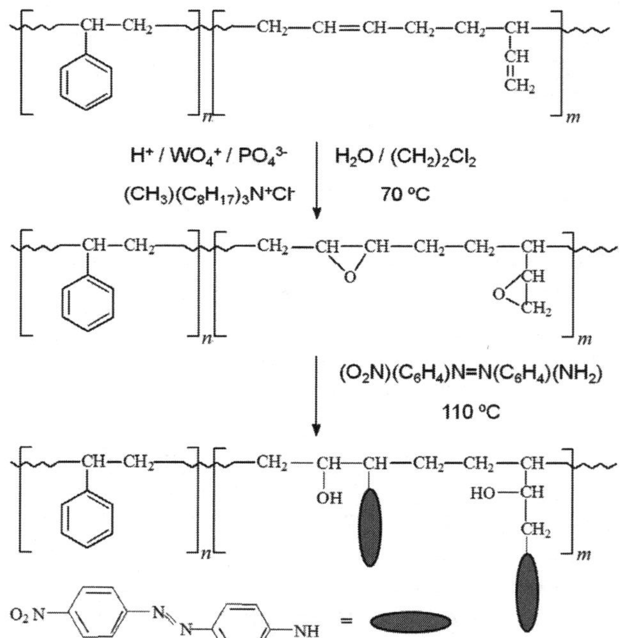

Figure 9.8 Synthesis of azo-containing block copolymers.

The characteristics of the modified block copolymers investigated can be found in Table 9.4. The samples morphologies, generated by self-assembly of the block copolymers, were investigated using tapping mode atomic force microscopy (TM-AFM). In order to obtain repeatable results, different areas of the films were scanned. Taking into account the resemblance of height and phase TM-AFM images of each sample, only phase images are shown.

Table 9.4 Characteristics of the block copolymers used

Sample	Epoxidation degree (mol%)	PS (wt%)	PB (wt%)	PBep (wt%)	AZO (wt%)	$T_{g\,PB}$ (°C)
SBS/AZO	0	25	60	0	15	-80
$SBSep_{11}$/AZO	11	25	50	10	15	-55

In Fig. 9.9 TM-AFM phase images of the guest–host system, SBS/AZO, and the functionalized system, SBSep11/AZO, after annealing at 110°C for 24 h under vacuum are shown.

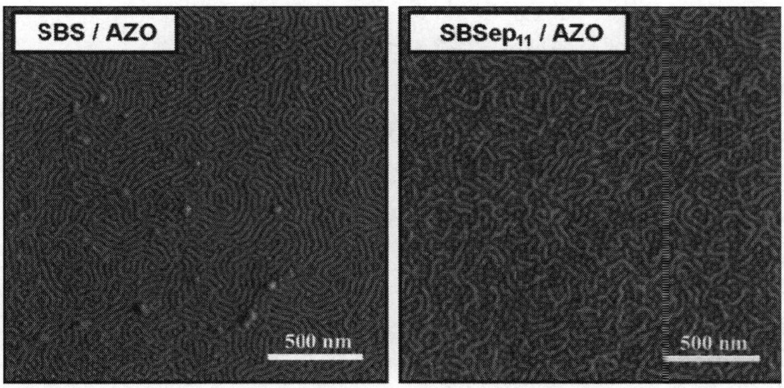

Figure 9.9 TM-AFM phase images of SBS/AZO (left) and $SBSep_{11}$/AZO (right) after annealing at 110°C for 24 h (2 μm × 2 μm).

In the guest–host system the block copolymer shows a self-assembled cylindrical structure after the annealing treatment. In particular, most of PS cylinders are positioned parallel to the continuous phase of PB-block. Nevertheless, some of them perpendicular to the surface of PB-block domains can also be observed. In addition, three different scale colors can be seen in the image: PS-block domains appear brighter than PB-block domains, since PS has a higher modulus than PB at room temperature. However, small azobenzene aggregates detected are the most rigid and, therefore, the brightest ones. Even though films preparation was made from homogeneous solutions, when solvents evaporated, migration of AZO onto the BCP surface occurred in the guest–host system. In the case of the morphology of the azobenzene-containing BCP with the chromophores covalently linked, the used epoxidation degree (11 mol%) did not modify the morphological features of the raw BCP and, what is more important, aggregation of the chromophore was not detected. Thanks to the covalent attachment of the azobenzene groups to the BCP via addition reaction of azoamine groups to epoxy groups, completely homogeneous azobenzene-containing block copolymers were achieved.

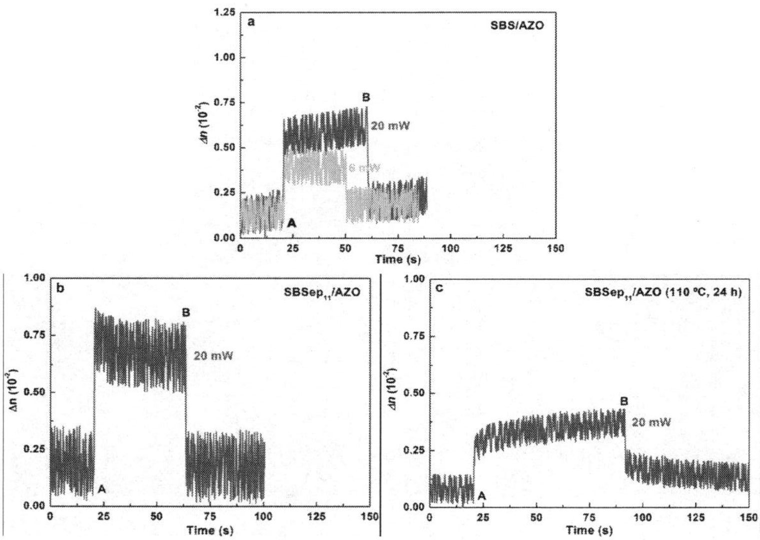

Figure 9.10 Writing-relaxing curves of: (a) SBS/AZO, (b) $SBSep_{11}$/AZO, and (c) $SBSep_{11}$/AZO after annealing at 110°C for 24 h. The writing beam is *on* at point A and *off* at B.

The birefringent properties of the block copolymers modified with azobenzene groups were also evaluated. In Fig. 9.10 writing-relaxing sequences are plotted. Different responses versus argon laser irradiation were seen. The guest–host system and the epoxidized SBS with AZO previous thermal treatment showed similar behaviors. The photoinduced birefringence was rapidly built up to the saturation level but, when the writing beam was turned off at point B, Δn quickly fell off indicating the complete randomization of the chromophores orientation. After annealing of epoxidized SBS with AZO samples, a slowdown of the azobenzenes orientation rate was noticed as well as a lower value of Δn. Quantitatively, a maximum optical birefringence of just $\Delta n = 0.43 \times 10^{-2}$ was obtained. This result is probably due to the high molecular mobility at room temperature of the epoxidized PB-block with azobenzene, since its T_g is on the order of −55°C. However, as was proven, the azobenzenes movement was somewhat restricted in the systems where they were covalently attached to the BCP from one end of the molecule, compared to the guest–host systems where the chromophores had total freedom of movement. Additionally, for SBSep11/AZO after thermal treatment, when the writing beam was turned off (point B), Δn rapidly fell off initially but, immediately after, the relaxation process of the photoinduced orientation was also slower compared to the guest–host system and SBSep11/AZO without annealing, demonstrating once more the influence of chemical bonding.

9.3.2 Azobenzene-Containing Block Copolymers Based on Poly(Styrene-b-4-Vinylpyridine)

PS4VP is an excellent BCP host due to the ability of the pyridine group to form hydrogen bonds with chromophores containing hydroxyl groups. A push-pull azoamine, AZO, with high dipole moment was selected as chromophore. In addition, the azoprepolymer, PAZO, bearing hydroxyl groups, was also used to develop materials with reversible birefringence. First, the morphologies of PS4VP/AZO and PS4VP/PAZO binary systems, generated by self-assembly of the BCP, were investigated using TM-AFM (Fig. 9.11).

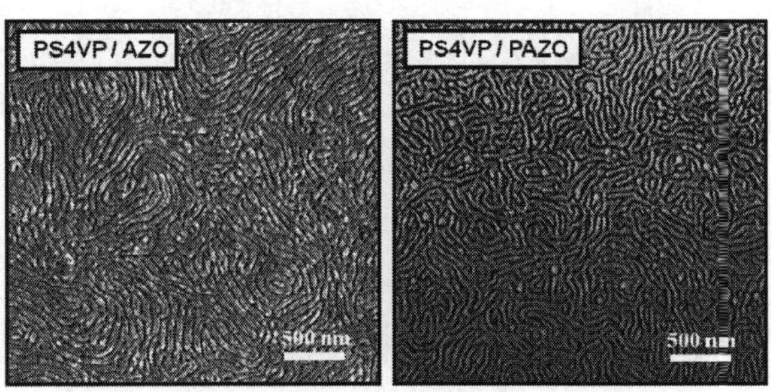

Figure 9.11 TM-AFM phase images of PSP4VP/AZO (left) and PSP4VP/PAZO (right) (3 μm × 3 μm).

Both guest–host systems display a self-assembled structure where most microdomains are parallel oriented to the film surface. Two different scale colors can be distinguished: PS blocks appear brighter than P4VP self-assembled domains [63]. Additionally, high miscibility of AZO and PAZO in the BCP was achieved without showing any segregation. Taken above into account, both the chromophore and azoprepolymer might be confined in P4VP self-assembled phase as consequence of intermolecular interactions.

To corroborate this assumption, binary systems with PS as host matrix were analyzed. Figure 9.12 shows optical microscopy images of PS4VP/AZO and PS/PAZO.

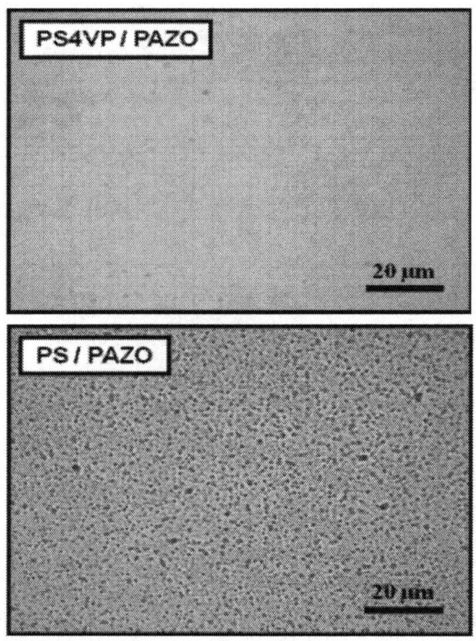

Figure 9.12 Optical microscopy images of PSP4VP/PAZO (up) and PS/PAZO (down).

Homogeneity at macroscopic level was reached only in PS4VP/AZO system, confirming the preferential affinity between PAZO and P4VP. In fact, in Section 9.2.2.1, binary systems of AZO and SBS were studied and it was shown that AZO aggregates were detected on the block copolymer surface.

Moreover, the formation of hydrogen bonds with pyridine ring was verified by FTIR spectroscopy following the symmetric ring stretching mode of the pyridine group at 993 cm^{-1} (Fig. 9.13). In the binary systems, a new band arises slightly above 1000 cm^{-1}, which can be attributed to hydrogen bonds formation with pyridine ring, as was also reported by Kuo S-W et al. [64]

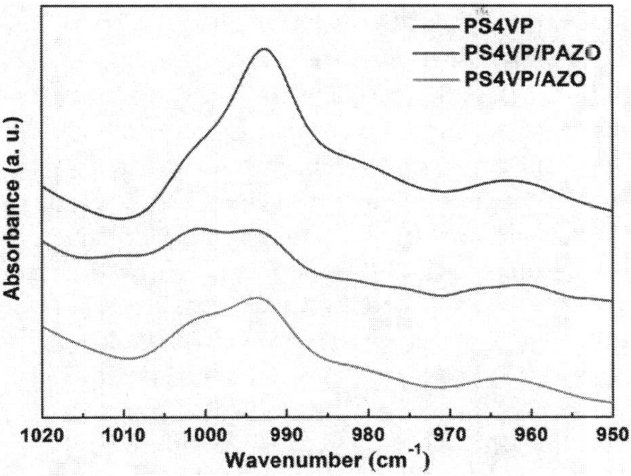

Figure 9.13 FTIR spectra of PS4VP, PS4VP/PAZO and PS4VP/AZO in the range of 1020-950 cm^{-1}.

With regard to the optical properties of the designed materials, from the point of view of applications, the maximum birefringence (corresponding to a writing mechanism) and the remaining birefringence (corresponding to a storing mechanism) are the most important parameters. Therefore, our investigation focused primarily on the enhancement of those properties. Writing-relaxing-erasing sequences were obtained for PAZO, PS4VP/PAZO and PS4VP/AZO and a summary of the optical response shown by all the systems can be found in Table 9.5.

Table 9.5 Optical features of samples as a function of AZO content

Sample	AZO (wt%)	Δn (10^{-2})	Δn_N (10^{-2})	rem (%)
PAZO	26	2.70	0.10	30
PS4VP/PAZO	3.90	0.79	0.20	76
PS4VP/AZO	15	1.10	0.07	39

As expected, the birefringence level increased as the AZO content was higher since larger number of photoactive units generated higher photoinduced orientation. Here it should be pointed out that as Δn values depend on the azobenzene content, for comparative purposes, the birefringence normalized to the chromophore content (Δn_N) was also calculated. When comparing the normalized values, a significant increase of birefringence level was achieved for PS4VP/PAZO binary system. Specifically, Δn_N of PS4VP/PAZO is twice and three times higher than that of PAZO and PS4VP/AZO, respectively. This evidence is a clear consequence of strong non-covalent interactions between the host matrix and the guest azoprepolymer. The glass transition temperature of PAZO is around 30°C. Therefore, this azoprepolymer has molecular mobility at room temperature, which contributes to thermal randomization, counteracting the orientation process and decreasing the saturated level of birefringence [65]. Hydrogen bonding between the hydroxyl groups of PAZO and pyridine moieties prevented randomization and enhanced the birefringence level.

Concerning the stability of the induced anisotropy, the remaining birefringence after turning off the writing beam (rem) was determined. As shown in Table 9.5, the remaining birefringence of PAZO increased from 30% to almost 80% when the azoprepolymer was as guest in the BCP host. Moreover, the remaining birefringence of PS4VP/AZO binary system was slightly higher than that of PAZO. This can be better understood taking into account the presence of hydrogen bonding interactions also between pyridine and AZO amine groups. These intermolecular interactions seem to be more effective than covalent ones in PAZO to retain the induced photoorientation. It should be also pointed out that AZO molecules were confined in self-organized domains of BCP. Such local confinement may restrain the azobenzene mobility due to interactions with the nanodomains wall, which might reduce azobenzene moieties randomization. Consequently, confinement effects exerted by the BCP could also contribute to the stability of the optically induced anisotropy. Finally, the optical response of binary systems with polystyrene (PS/PAZO and PS/AZO) was also analyzed. However, the lack of interaction with PS matrix resulted in the aggregation of chromophores, leading to poor optical properties: low Δn and total randomization of the induced photoorientation after turning *off* the writing laser.

9.4 Azobenzene-Containing Nanostructured Thermosetting Epoxy Systems

This section is oriented towards the study of novel nanostructured epoxy systems containing azobenzene groups for reversible optical storage. In particular, azobenzene-containing epoxy/amine formulations were modified with epoxidized SBS to promote the compatibilization of the polybutadiene block with the epoxy matrix. Thus, nanostructured epoxy thermosets containing azobenzene groups were achieved, due to reaction-induced microphase separation of polystyrene block. Optical and morphological properties of the resulting nanostructured materials were evaluated and discussed considering several variables, such as the epoxidation extent, the epoxidized SBS content and the azobenzene concentration in the epoxy matrix.

First, the azoprepolymer, PAZO, was blended in different ratios (10, 20 or 50 wt%) with DGEBA, the aromatic diamine 4,4′-methylene-bis(3-chloro-2,6-diethylaniline) (MCDEA) as hardener, and various amounts of SBSepX (prepared in the same way as described in Section 9.2.2.1) to develop different thermosetting systems. All samples were then cured at 140°C during 24 h and post-cured at 165°C for 2 h. An amino-hydrogen-to-epoxy stoichiometric ratio equal to 1 was maintained for all prepared systems. Epoxy thermosets without BCP were named according to their content in azoprepolymer as PAZOy (y being PAZO ratio (wt%)). Nanostructured systems were named according to their content in both PAZO and epoxidized SBS as SBSepXz/PAZOy (z being SBSepX ratio (phr)).

In this case, azobenzene groups were introduced via chemical bonding between PAZO and DGEBA/MCDEA system, the azoprepolymer being therefore part of the epoxy matrix. All the epoxy-based networks studied had the chromophore placed in a branch, as illustrated in Fig. 9.4 of the Introduction section. As published by Ocando et al. [66], two different mechanisms can be involved through morphology development during curing as a function of epoxidized PB content in DGEBA/MCDEA systems modified with epoxidized SB block copolymers. On the one hand, reaction-induced microphase separation of PS block leads to long-range order nanostructures and, on the other hand, combination of both self-assembly of epoxidized PB block in the initial mixture and RIMS of PS block can give rise to vesicles or long worm-like micelles with a bilayered structure [66]. Here, additionally, the presence of the azoprepolymer provoked a decrease of miscibility of initial mixtures before curing, leading to different morphologies as a function of PAZO content owing to the modification of interactions between groups with respect to the modified matrices without PAZO.

Figure 9.14 shows TM-AFM phase images of cured epoxy systems modified with different amounts of SBSep (30 and 50 phr) containing several PAZO ratios (10, 20 or 50 wt%). Prior to TM-AFM analysis reactive solutions were drop cast into polytetrafluoroethylene moulds of 4 cm × 4 cm × 1 cm. After that, the solvent was removed and the curing schedule described above was performed. Samples of cured mixtures were prepared using an ultramicrotome (Leica Ultracut R) equipped with a diamond knife.

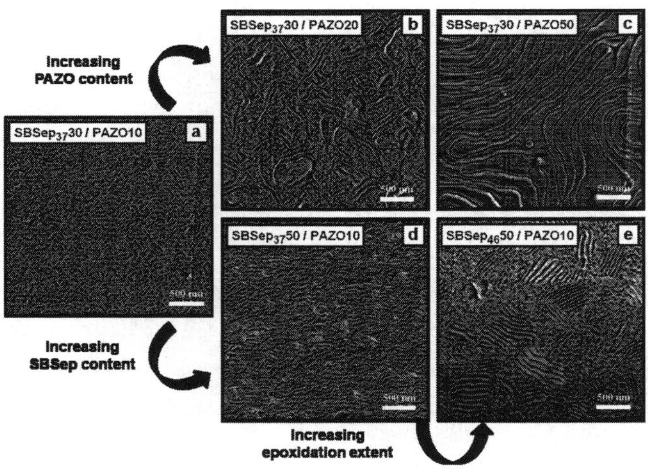

Figure 9.14 TM-AFM phase images of nanostructured epoxy thermosets as a function of PAZO content, SBSep content and epoxidation extent (3 μm × 3 μm).

The epoxidation extent of 37 mol% is above the miscibility threshold with the epoxy resin, which is set to 27 mol% using MCDEA as hardener [67]. In all cases microphase separation was achieved. SBSep3730/PAZO10 cured epoxy system showed micellar nanodomains of PS block in the epoxy matrix. The increase of the azoprepolymer content led to higher amount of segregated domains. These morphological variations were probably related to the fact that increasing PAZO content, the miscibility between the block copolymer and the epoxy system before curing decreased. Here it should be pointed out that PS homopolymer is miscible with DGEBA/MCDEA system at temperatures higher than 90°C [66].

Therefore, taking into account the initial miscibility of blocks with the epoxy system before curing, for 37 mol% of epoxidation extent one can expect the formation of segregated structures in the epoxy matrix through self-assembly of epoxidized PB occurred prior to curing, followed by fixing these structures through reaction-induced microphase separation of PS chains during network formation. Results confirmed that for investigated systems this process was favored by the increase of PAZO content. Thus, various morphologies, through combination of self-assembly of PB block and RIMS of PS block, were generated as consequence of variations in interactions between groups in the initial mixtures.

Increase of the epoxidation extent provoked higher miscibility between epoxidized PB block and epoxy matrix and, consequently, long-range order nanostructures were achieved. Indeed, when increasing the epoxidation extent from 37 up to 46 mol%, the system containing 10 wt% pazo and 50 phr BCP revealed PS cylinders inside the matrix mixture.

According to previous work done in our group [66], as well as by other authors, such as Dean et al. [40], who established that for low contents of block copolymer phase transitions from spherical micelles to worm-like micelles and, finally, to vesicles took place as the volume fraction of immiscible block increased, results presented here suggest that 46 mol% of epoxidation extent was needed to achieve long-range order nanostructuring because of the higher compatibility of the more epoxidized SBS with the epoxy-precursor mixture. Consequently, for 46 mol% of epoxidation extent morphologies were originated through reaction-induced microphase separation of PS block starting from a miscible mixture before curing. On the contrary, 37 mol% of epoxidation extent was not enough to make compatible all not randomly-epoxidized PB chains with the epoxy-precursor mixture before curing.

Thus, morphologies were developed through self-assembly before curing followed by fixing the nanostructures via phase separation of PS chains through curing reaction, leading to different segregated structures depending on the content of block copolymer and the azoprepolymer. Furthermore, miscibilization of initial mixtures was favored not only as increasing the epoxidation extent, but also as decreasing PAZO content, since a significant amount of randomly-epoxidized PB chains could not be miscibilized before curing when high quantities of PAZO were incorporated into the epoxy system.

In relation to the optical storage abilities of azobenzene-containing materials, the information is stored in the form of photoinduced anisotropy in this kind of systems [68]. Working on the principle that some of the fundamental attributes a material requires to be employed as an optical memory media are high storage capacity, fast response, stability during writing-erasing cycles and long-term stability. The parameters we wanted to investigate were the maximum level of achievable birefringence and the rate of birefringence growth, as well as the thermal relaxation rate and the long-term stability of the induced birefringence. In addition, there are a series of aspects affecting those variables, including chemical factors (such as type of azochromophore, azobenzene content, T_g, structure of the polymeric main chain, ...) and physical ones (such as irradiation wavelength and intensity, film thickness, history of the sample, operating temperature ...) [4].

Table 9.6 shows the birefringence values and glass transition temperatures of studied systems. Since birefringence depends on the azobenzene quantity, Δn_N values, the birefringence data normalized to the azodye content, are also reported. As can be seen, Δn data varied from 0.2×10^{-2} to 1.9×10^{-2} as the azochromophore increased from 1.72 to 12.9 wt%. These birefringence values can be considered adequate for optical applications taking into account the low azobenzene concentrations used and the reported values for this kind of materials [69–71]. Nevertheless, it should be noted that no significant differences in Δn_N could be seen between parent azobenzene-containing thermosets and the ones modified with epoxidized SBS. Therefore, nanostructuring did not have a significant influence on the maximum level of achievable birefringence, which means the interactions among azobenzene groups were similar in both kinds of systems, because in all cases they were covalently linked to the epoxy network via chemical bonding between PAZO and DGEBA/MCDEA system. However, a slight increase in Δn_N for all systems was perceptible when increasing PAZO content, probably due to an increase of interactions between azobenzene units, as a consequence of the increase in azodye concentration.

Table 9.6 Birefringence values and glass transition temperatures of samples

Sample	AZO (wt%)	T_g (°C)	Δn (10^{-2})	Δn_N (10^{-2})
PAZO10	2.58	155	0.33 ± 0.08	0.13
$SBSep_{37}10/PAZO10$	2.35	142	0.27 ± 0.03	0.11
$SBSep_{37}30/PAZO10$	1.98	121	0.24 ± 0.04	0.12
$SBSep_{37}50/PAZO10$	1.72	83	0.21 ± 0.01	0.12
$BSep_{46}50/PAZO10$	1.72	79	0.20 ± 0.04	0.11
PAZO20	5.16	153	0.85 ± 0.14	0.16
$SBSep_{37}10/PAZO20$	4.69	141	0.81 ± 0.12	0.17
$SBSep_{37}20/PAZO20$	4.30	134	0.64 ± 0.09	0.15
$SBSep_{37}30/PAZO20$	3.97	122	0.55 ± 0.13	0.14
$SBSep_{46}20/PAZO20$	4.30	130	0.64 ± 0.01	0.15
PAZO50	12.9	139	1.90 ± 0.12	0.15
$SBSep_{37}5/PAZO50$	12.3	138	1.70 ± 0.08	0.14
$SBSep_{37}10/PAZO50$	11.7	134	1.70 ± 0.12	0.15
$SBSep_{37}30/PAZO50$	9.92	121	1.45 ± 0.32	0.15
$SBSep_{46}5/PAZO50$	12.3	138	1.80 ± 0.11	0.15

In general, relaxation processes are easier to be studied than orientation ones since effects of heating and isomerization rates are small when the writing beam is turned off. Parameters that have a clear influence on the remaining birefringence are glass transition temperature and cooperative effects that may cause relaxation times to be longer with the increase in azochromophore content [68]. Figure 9.15 shows the percentage of remaining birefringence after turning off the writing beam at 6 mW as a function of SBSep37 ratio and azobenzene concentration in the epoxy thermosets, regardless of films thickness. At higher azochromophore contents, the epoxy systems exhibited higher remaining birefringence.

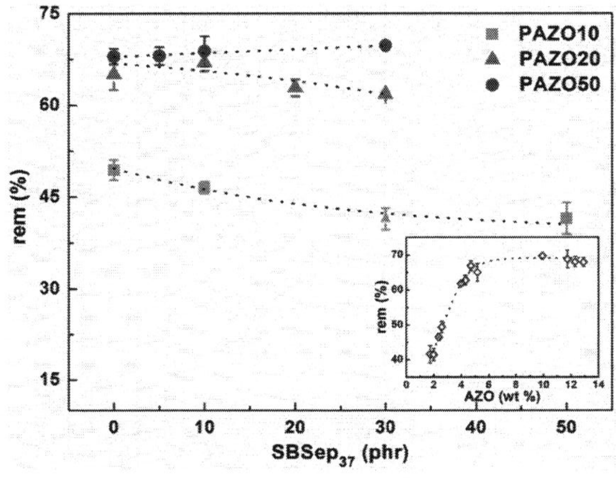

Figure 9.15 Remaining birefringence after turning *off* the writing beam as a function of $SBSep_{37}$ and PAZO content (in the inner graph is plotted the dependence on azobenzene ratio).

Furthermore, as can be seen in Table 9.7, the remaining birefringence increased with the azobenzene concentration in samples with similar T_g values, suggesting that, for the systems studied, the main factor affecting the remaining birefringence was the azodye content.

Table 9.7 Remaining birefringence of thermosets containing 30 phr $SBSep_{37}$ and different amounts of PAZO (10, 20 or 50 wt%)

Sample	AZO (wt%)	T_g (°C)	rem (%)
$SBSep_{37}$30/PAZO10	1.98	121	41.4 ± 1.7
$SBSep_{37}$30/PAZO20	3.97	122	61.8 ± 0.4
$SBSep_{37}$30/PAZO50	9.92	121	69.7 ± 0.7

However, a *plateau* for AZO contents above 4 wt% was reached, which could indicate that, for these epoxy thermosets and in the range of values evaluated, the remaining birefringence depended mainly on cooperative interactions among azobenzene groups up to an azodye concentration from which the remaining birefringence kept constant.

It should be pointed out that T_g values of the thermosets prepared varied from 79 to 155°C, which should favor a good orientation stability of the photoactive chromophores at room and slightly elevated temperatures. Hence, we also studied the long-term stability of the induced birefringence. The residual fraction of birefringence after 45 h of turning off the writing beam at 6 mW for a film of TAZ50-SBSep465 was investigated. During the first 4 min of turning off the writing beam, birefringence was relaxed around 30%. Then, after 10 h the birefringence relaxation was about 70% and, though not shown here, it kept stable during several weeks.

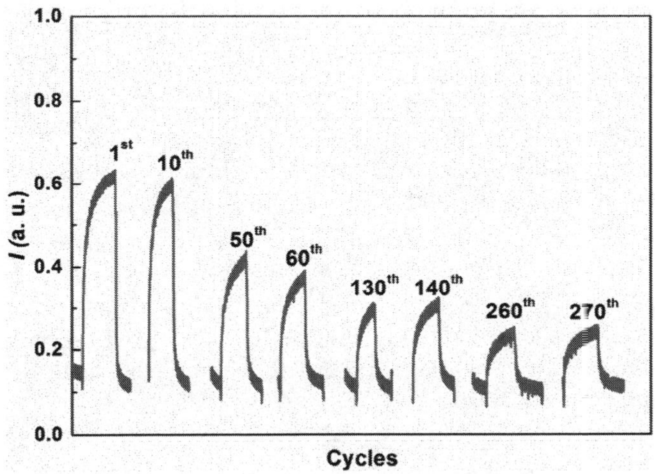

Figure 9.16 Successive writing-erasing sequences for a film of SBSep$_{37}$10/PAZO20. The power of the writing beam was 6 mW.

Our last interest was to study the rewriting stability of the samples. In particular, the maximum transmitted intensity of the reading beam after passing through a film of TAZ20-SBSep3710 during 270 writing-erasing cycles was analyzed (Fig. 9.16). During the first 10 cycles up to around 30 cycles, similar transmitted signal was achieved. After 60 writing-erasing sequences in the order of 60% of maximum intensity was still registered and after 140 cycles 50% of maximum intensity was achieved. Finally, after 270 cycles almost 40% of maximum transmitted signal was reached, which means that after 270 cycles there was still a significant birefringence that could be used to distinguish the on from the off state. Therefore, the ability of this nanostructured epoxy-based thermoset containing azobenzene groups for rewriting was found adequate after consecutive cycles.

References

1. Natansohn, A., and Rochon, P. (2002). Photoinduced motions in azo-containing polymers, *Chem. Rev.,* **102**, pp. 4139–4175.
2. Ikeda, T. (2003). Photomodulation of liquid crystal orientations for photonic applications, *J. Mater. Chem.,* **13**, pp. 2037–2057.
3. Yager, K. G., and Barrett, C. J. (2006). Novel photo-switching using azobenzene functional materials, *J. Photochem. Photobiol. A: Chem.,* **182**, pp. 250–261.
4. Zhao, Y., and Ikeda, T. (2009). *Smart Light-Responsive Materials: Azobenzene-Containing Polymers and Liquid Crystals* (Hoboken, NJ: John Wiley & Sons).
5. Yu, Y., Nakano, M., and Ikeda, T. (2003). Directed bending of a polymer film by light, *Nature,* **425**, p. 145.
6. Ichimura, K., Oh, S.-K., and Nakagawa, M. (2000). Light-driven motion of liquids on a photoresponsive surface, *Science,* **288**, pp. 1624–1626.
7. Lim, H. S., Han, J. T., Kwak, D., Jin, M., and Cho, K. (2006). Photoreversibly switchable superhydrophobic surface with erasable and rewritable pattern, *J. Am. Chem. Soc.,* **128**, pp. 14458–14459.
8. Lim, H. S., Lee, W. H., Lee, S. G., Lee, D., Jeon, S., and Cho, K. (2010). Effect of nanostructure on the surface dipole moment of photoreversibly tunable superhydrophobic surfaces, *Chem. Commun.,* **46**, pp. 4336–4338.
9. Goulet-Hanssens, A., Sun, K. L. W., Kennedy, T. E., and Barrett, C. J. (2012). Photoreversible surfaces to regulate cell adhesión, *Biomacromolecules,* **13**, pp. 2958–2963.
10. Fdez, de Nograro, F., Llano-Ponte, R., and Mondragon, I. (1996). Dynamic and mechanical properties of epoxy networks obtained with PPO based amines/mPDA mixed curing agents, *Polymer,* **37**, pp. 1589–1600.
11. Ramos, J. A., Blanco, M., Zalakain, I., and Mondragon, I. (2009). Nanoindentation study of interphases in epoxy/amine thermosetting systems modified with thermoplastics, *J. Colloid Interface Sci.,* **336**, pp. 431–437.
12. Leonardi, A. B., Fasce, L. A., Zucchi, I. A., Hoppe, C. E., Soulé, E. R., Pérez, C. J., and Williams, R. J. J. (2011). Shape memory epoxies based on networks with chemical and physical crosslinks, *Eur. Polym. J.,* **47**, pp. 362–369.

13. Moriya, K., Seki, T., Nakagawa, M., Mao, G., and Ober, C. K. (2000). Photochromism of 4-cyanophenylazobenzene in liquid crystalline-coil AB diblock copolymers: The influence of microstructure, *Macromol. Rapid Commun.*, **21**, pp. 1309–1312.
14. Cui, L., Tong, X., Yan, X., Liu, G., and Zhao, Y. (2004). Photoactive thermoplastic elastomers of azobenzene-containing triblock copolymers prepared through atom transfer radical polymerization, *Macromolecules,* **37**, pp. 7097–7104.
15. Frenz, C., Fuchs, A., Schmidt, H.-W., Theissen, U., and Haarer, D. (2004). Diblock copolymers with azobenzene side-groups and polystyrene matrix: Synthesis, characterization and photoaddressing, *Macromol. Chem. Phys.,* **205**, pp. 1246–1258.
16. Jiang, J., Tong, X., and Zhao, Y. (2005). A new design for light-breakable polymer micelles, *J. Am. Chem. Soc.,* **127**, pp. 8290–8291.
17. Sin, S. L., Gan, L. H., Hu, X., Tam, K. C., and Gan, Y. Y. (2005). Photochemical and thermal isomerizations of azobenzene-containing amphiphilic diblock copolymers in aqueous micellar aggregates and in film, *Macromolecules,* **38**, pp. 3943–3948.
18. Yoshida, T., Kanaoka, S., and Aoshima, S. (2005). Photo-responsive copolymers with azobenzene side groups synthesized by living cationic polymerization: Efficient amplification of photosensitivity in aqueous photo-switching system, *J. Polym. Sci.: Part A: Polym. Chem.,* **43**, pp. 5337–5342.
19. Kadota, S., Aoki, K., Nagano, S., and Seki, T. (2005). Photocontrolled microphase separation of block copolymers in two dimensions, *J. Am. Chem. Soc.,* **127**, pp. 8266–8267.
20. Morikawa, Y., Nagano, S., Watanabe, K., Kamata, K., Iyoda, T., and Seki, T. (2006). Optical alignment and patterning of nanoscale microdomains in a block copolymer thin film, *Adv. Mater.,* **18**, pp. 883–886.
21. Yu, H., Li, J., Ikeda, T., and Iyoda, T. (2006). Macroscopic parallel nanocylinder array fabrication using a simple rubbing technique, *Adv. Mater.,* **18**, pp. 2213–2215.
22. Jin, M., Lu, R., Yang, Q. X., Bao, C. Y., Sheng, R., Xu, T. H., and Zhao, Y. (2007). Preparation of side-on bisazobenzene-containing homopolymers and block copolymers via ATRP and studies on their photoisomerization and photoalignment behaviors, *J. Polym. Sci.: Part A: Polym. Chem.,* **45**, pp. 3460–3472.
23. Wang, D., Ye, G., and Wang, X. (2007). Synthesis of aminoazobenzene-containing diblock copolymer and photoinduced deformation

behavior of its micelle-like aggregates, *Macromol. Rapid Commun.*, **28**, pp. 2237–2243.

24. Yu, H., Shishido, A., Iyoda, T., and Ikeda, T. (2007). Novel wormlike nanostructures self-assembled in a well defined liquid crystalline diblock copolymer with azobenzene moieties, *Macromol. Rapid Commun.*, **28**, pp. 927–931.

25. Su, W., Zhao, H., Wang, Z., Li, Y., and Zhang, Q. (2007). Sphere to disk transformation of micro-particle composed of azobenzene-containing amphiphilic diblock copolymers under irradiation at 436 nm, *Eur. Polym. J.*, **43**, pp. 657–662.

26. Breiner, T., Kreger, K., Hagen, R., Häckel, M., Kador, L., Müller, A. H. E., Kramer, E. J., and Schmidt, H.-W. (2007). Blends of poly(methacrylate) block copolymers with photoaddressable segments, *Macromolecules,* **40**, pp. 2100–2108.

27. He, X., Sun, W., Yan, D., and Liang, L. (2008). Novel ABC2-type liquid-crystalline block copolymers with azobenzene moieties prepared by atom transfer radical polymerization, *Eur. Polym. J.*, **44**, pp. 42–49.

28. Komura, M., Watanabe, K., Iyoda, T., Yamada, T., Yoshida, H., and Iwasaki, Y. (2009). Laboratory-GISAXS measurements of block copolymer films with highly ordered and normally oriented nanocylinders, *Chem. Lett.*, **38**, pp. 408–409.

29. Wang, Y., Zhang, M., Moers, C., Chen, S., Xu, H., Wang, Z., Zhang, X., and Zhibo, L. (2009). Block copolymer aggregates with photo-responsive switches: Towards a controllable supramolecular container, *Polymer,* **50**, pp. 4821–4828.

30. Khandpur, A. K., Förster, S., Bates, F. S., Hamley, I. W., Ryan, A. J., Bras, W., Almdal, K., and Mortensen, K. (1995). Polyisoprene-polystyrene diblock copolymer phase diagram near the order-disorder transition, *Macromolecules,* **28**, pp. 8796–8806.

31. Han, C. D., Baek, D. M., Kim, J. K., Ogawa, T., Sakamoto, N., and Hashimoto, T. (1995). Effect of volume fraction on the order-disorder transition in low-molecular-weight polystyrene-block-polyisoprene copolymers. 1. Order-disorder transition-temperature determined by rheological measurements, *Macromolecules,* **28**, pp. 5043–5062.

32. Matsen, M. W., and Bates, F. S. (1996). Unifying weak- and strong-segregation block copolymer theories, *Macromolecules,* **29**, pp. 1091–1098.

33. Muthukumar, M., Ober, C. K., and Thomas, E. L. (1997). Competing interactions and levels of ordering in self-organizing polymeric materials, *Science,* **277**, pp. 1225–1232.

34. Modi, M. A., Krishnamoorti, R., Tse, M. F., and Wang, H. C. (1999). Viscoelastic characterization of an order-order transition in a mixture of di- and triblock copolymers, *Macromolecules,* **32**, pp. 4088–4097.
35. Abetz, V., and Goldacker, T. (2000). Formation of superlattices via blending of block copolymers, *Macromol. Rapid Commun.,* **21**, pp. 16–34.
36. Lodge, T. P. (2003). Block copolymers: Past successes and future challenges, *Macromol. Chem. Phys.,* **204**, pp. 265–273.
37. Yu, H., Asaoka, S., Shishido, A., Iyoda, T., and Ikeda, T. (2007). Photoinduced nanoscale cooperative motion in a well-defined triblock copolymer, *Small,* **3**, pp. 768–771.
38. Yu, H., Iyoda, T., and Ikeda, T. (2006). Photoinduced alignment of nanocylinders by supramolecular cooperative motions, *J. Am. Chem. Soc.,* **128**, pp. 11010–11011.
39. Ritzenthaler, S., Court, F., David, L., Girard-Reydet, E., Leibler, L., and Pascault, J. P. (2002). ABC Triblock copolymers/epoxy-diamine blends. 1. Keys to achieve nanostructured thermosets, *Macromolecules,* **35**, pp. 6245–6254.
40. Dean, J. M., Grubbs, R. B., Saad, W., Cook, R. F., and Bates, F. S. (2003). Mechanical properties of block copolymer vesicle and micelle modified epoxies, *J. Polym. Sci. Part B: Polym. Phys.,* **41**, pp. 2444–2456.
41. Hermel-Davidock, T. J., Tang, H. S., Murray, D. J., and Hahn, S. F. (2007). Control of the block copolymer morphology in templated epoxy thermosets, *J. Polym. Sci. Part B: Polym. Phys.,* **45**, pp. 3338–3348.
42. Fan, W., Wang, L., and Zheng, S. (2010). Double reaction-induced microphase separation in epoxy resin containing polystyrene-block-poly(-caprolactone)-block-poly(n-butyl acrylate) ABC triblock copolymer, *Macromolecules,* **43**, pp. 10600–10611.
43. Kosonen, H., Ruokolainen, J., Nyholm, P., and Ikkala, O. (2001). Self-organized cross-linked phenolic thermosets: Thermal and dynamic mechanical properties of novolac/block copolymer blends, *Polymer,* **42**, pp. 9481–9486.
44. Leibler, L. (2005). Nanostructured plastics: Joys of self-assembling, *Prog. Polym. Sci.,* **30**, pp. 898–914.
45. Jaffrennou, B., Portal, J., Méchin, F., and Pascault, J P. (2008). Nanostructured poly(urethane)s and poly(urethane-urea)s from reactive solutions of poly[Styrene-b-Butadiene-b-(Methyl Methacrylate)]-triblock copolymers, *Eur. Polym. J.,* **44**, pp. 3439–3455.

46. Natanshon, A., Xie, S., and Rochon, P. (1992). Azo polymers for reversible optical storage. 2. Poly[4′-[[2-(acryloyloxy)ethyl]ethylamino]-2-chloro-4-nitroazobenzene], *Macromolecules,* **25**, pp. 5531–5532.
47. Oliveira Jr., O. N., dos Santos Jr., D. S., Balogh, D. T., Zucolotto, V., and Mendonça, C. R. (2005). Optical storage and surface relief gratings in azobenzene containing nanostructured films, *Adv. Colloid Interface Sci.,* **116**, pp. 179–192.
48. Natansohn, A., Rochon, P., Pezolet, M., Audet, P., Brown, D., and To, S. (1994). Azo polymers for reversible optical storage. 4. Cooperative motion of rigid groups in semicrystalline polymers, *Macromolecules,* **27**, pp. 2580–2585.
49. Stracke, A., Wendorff, J. H., Maher, J., and Rafler, G. (2000). Dielectric and photoorientation studies in semicrystalline melamine monomers and amorphous melamine networks containing azobenzene chromophores, *Macromolecules,* **33**, pp. 2605–2609.
50. Rochon, P., Gosselin, J., Natansohn, A and Xie, S. (1992). Optically induced and erased birefringence and dichroism in azo-aromatic polymers, *Appl. Phys. Lett.,* **60**, pp. 4–5.
51. Natansohn, A., Rochon, P., Gosselin, J., and Xie, S. (1992). Azo polymers for reversible optical storage. 1. Poly[4′-[[2-(acryloyloxy)ethyl]ethylamino]-4-nitroazobenzene], *Macromolecules,* **25**, pp. 2268–2273.
52. Meng, X., Natansohn, A., and Rochon, P. (1996). Azo polymers for reversible optical storage. 11 poly{4,4′-(1-methylethylidene)bisphenylene 3-[4-(4-nitrophenylazo)phenyl]-3-aza-pentanedioate, *J. Polym. Sci. Part B: Polym. Phys.,* **34**, pp. 1461–1466.
53. Takase, H., Natansohn, A., and Rochon, P. (2001). Effect of crosslinking on the photoinduced orientation of azo groups in thin polymer films, *J. Polym. Sci. Part B: Polym. Phys.,* **39**, pp. 1686–1696.
54. Tawa, K., Kamada, K., and Ohta, K. (2000). Azo-dye-structure dependence of photoinduced anisotropy observed in PMMA films, *J. Photochem. Photobiol. A: Chem.,* **134**, pp. 185–191.
55. Sekkat, Z., Yasumatsu, D., and Kawata, S. (2002). Pure photoorientation of azo dye in polyurethanes and quantification of orientation of spectrally overlapping isomers, *J. Phys. Chem. B,* **106**, pp. 12407–12417.
56. Crawford, E. D., and Lesser, A. J. (1998). The effect of network architecture on the thermal and mechanical behavior of epoxy resins, *J. Polym. Sci. Part B: Polym. Phys.,* **36**, pp. 1371–1382.
57. Oliveira Jr, O. N., dos Santos Jr, D. S., Balogh, D.T., Zucolotto, V., and Mendonça, C. R. (2005). Optical storage and surface-relief gratings

in azobenzene-containing nanostructured films, *Adv. Colloid Interface Sci.*, **116**, pp. 179–192.

58. Priimagi, A., Shevchenko, A., Kaivola, M., Rodriguez, F. J., Kauranen, M., and Rochon, P. (2010). High and stable photoinduced anisotropy in guest-host polymer mediated by chromophore aggregation, *Opt. Lett.*, **35**, pp. 1813–1815.

59. Sáiz, L. M., Orofino, A. B., Ruzzo, M. M., Arenas, G. F., Oyanguren, P. A., and Galante, M. J. (2011). Controlling mobility and birefringence of azo chromophores in epoxy polymers, *Polym. Int.*, **60**, pp. 1053–1059.

60. Priimagi, A., Cattaneo, S., Ras, R. H. A., Valkama, S., Ikkala, O., and Kauranen, M. (2005). Polymer-dye complexes: A facile method for high doping level and aggregation control of dye molecules, *Chem. Mater.*, **17**, pp. 5798–802.

61. Tong, X., Cui, L., and Zhao, Y. (2004). Confinement effects on photoalignment, photochemical phase transition, and thermochromic behavior of liquid crystalline azobenzene-containing diblock copolymers, *Macromolecules*, **37**, pp. 3101–3112.

62. Serrano, E., Larrañaga, M., Remiro, P. M., Mondragon, I., Carrasco, P. M., Pomposo, J. A., and Mecerreyes, D. (2004). Synthesis and characterization of epoxidized styrene-butadiene block copolymers as templates for nanostructured thermosets, *Macromol. Chem. Phys.*, **205**, pp. 987–996.

63. Potemkin, I. I., Kramarenko, E. Yu, Khokhlov, A. R., Winkler, R. G., Reineker, P., Eibeck, P., Spatz, J. P., and Möller, M. (1999). Nanopattern of diblock copolymers selectively adsorbed on a plane surface, *Langmuir*, **15**, pp. 7290–7298.

64. Kuo, S.-W., Lin, C.-L., and Chang, F.-C. (2002). The study of hydrogen bonding and miscibility in poly(vinylpyridines) with phenolic resin, *Polymer*, **43**, pp. 3943–3949.

65. Fernández, R., Mondragon, I., Galante, M. J., and Oyanguren P. A. (2009). Bonding and molecular environment effects on photoorientation in epoxy-based polymers having azobenzene units, *Eur. Polym. J.*, **45**, pp. 788–794.

66. Ocando, C., Tercjak, A., Martín, M. D., Ramos, J. A., Campo, M., and Mondragon, I. (2009). Morphology development in thermosetting mixtures through the variation on chemical functionalization degree of poly(styrene-b-butadiene) diblock copolymer modifiers. Thermomechanical properties, *Macromolecules*, **42**, pp. 6215–6224.

67. Ocando, C., Tercjak, A., Serrano, E., Ramos, J. A., Corona-Galván, S., Parellada, M. D., Fernández-Berridi, M. J., and Mondragon, I. (2008).

Micro-and macrophase separation of thermosetting systems modified with epoxidized styrene-block-butadiene-block-styrene linear triblock copolymers and their influence on final mechanical properties, *Polym. Int.*, **57**, pp. 1333–1342.

68. Mendoça, C. R., dos Santos Jr, D. S., Balogh, D. T., Dhanabalan, A., Giacometti, J. A., Zilio, S. C., and Oliveira Jr, O. N. (2001). Optical storage in mixed Langmuir-Blodgett (LB) films of azopolymers and cadmium stearate, *Polymer,* **42**, pp. 6539–6544.
69. Fernández, R., Mondragon, I., Galante, M., and Oyanguren, P. (2009). Influence of chromophore concentration and network structure on the photo-orientation properties of crosslinked epoxy-based azopolymers, *J. Polym. Sci. Part B: Polym. Phys.*, **47**, pp. 1004–1014.
70. Gimeno, S., Forcén, P., Oriol, L., Piñol, M., Sánchez, C., Rodríguez, F. J., Alcalá, R., Jankova, K., and Hvilsted, S. (2009). Photoinduced optical anisotropy in azobenzene methacrylate block copolymers: Influence of molecular weight and irradiation conditions, *Eur. Polym. J.*, **45**, pp. 262–271.
71. Sáiz, L. M., Orofino, A. B., Ruzzo, M. M., Arenas, G. F., Oyanguren, P. A., and Galante, M. J. (2011). Controlling mobility and birefringence of azo chromophores in epoxy polymers, *Polym. Int.*, **60**, pp. 1053–1059.

Index